THE NATURE OF BIOLOGICAL SYSTEMS AS REVEALED BY THERMAL METHODS

Hot Topics in Thermal Analysis and Calorimetry

Volume 5

Series Editor:
Judit Simon, *Budapest University of Technology and Economics, Hungary*

The titles published in this series are listed at the end of this volume.

The Nature of Biological Systems as Revealed by Thermal Methods

Edited by

Dénes Lörinczy
University of Pécs, Biophysical Department,
Faculty of Medicine, Hungary

KLUWER ACADEMIC PUBLISHERS
DORDRECHT / BOSTON / LONDON

A C.I.P. Catalogue record for this book is available from the Library of Congress.

ISBN 1-4020-2218-2 (HB)
ISBN 1-4020-2219-0 (e-book)

Published by Kluwer Academic Publishers,
P.O. Box 17, 3300 AA Dordrecht, The Netherlands.

Sold and distributed in North, Central and South America
by Kluwer Academic Publishers,
101 Philip Drive, Norwell, MA 02061, U.S.A.

In all other countries, sold and distributed
by Kluwer Academic Publishers,
P.O. Box 322, 3300 AH Dordrecht, The Netherlands.

Printed on acid-free paper

Printed in the Netherlands.

Contents

Preface

After a kind motivation by Judit Simon (Editor-in-Chief of the *Journal of Thermal Analysis and Calorimetry*, Kluwer Academic Publisher) and negotiations with possible contributors – lasting for more than one year – it was decided to write a book about the application of thermal methods in biology. Its aim was to be a guide how to perform experiments and what kind of information might be gained by them. We tried to collect information that could be achieved only during a long personal practice. In this way scientists from biology and medicine , e.g., who are not so skilled in physics and mathematics may realize very soon the beauty and power of this tool at one hand. On the other hand, those scientists with better background in natural sciences can be more sensitive to find out exciting biological problems.

The recent situation in the literature of thermal methods (as techniques) and their application to biological problems is such that there are plenty of monographs discussing the working principles of different types of thermal analysis and calorimetry. Such books mainly deal with the general principles and present applications typical for inorganic materials. Moreover, there are some good, but relatively old reviews from the field of food physics and from different sections of biology. But it is known that the 'devil is hidden in the details': therefore, a beginner in the field of biological thermal analysis or calorimetry should 'find out' everything by his own when the principles of thermodynamics are tried to be applied to biological systems. These are highly organized and very complex objects where water and the different types of weak interactions among the macromolecules (dipoles, H-bonds, van der Waals forces etc.) make the interpretation of thermal events rather difficult.

After many discussions with colleagues at various international conferences during the last one and half year I do hope that our book will find an interested acceptance in the bio-community due to the choice of topics and authors. Moreover, the following reasons support my expectation:

- Biological calorimetry and of course thermal analysis find an increasing interest in the natural sciences community also, but both are still stepmother like treated in textbooks, monographs and journals.
- The spectrum of the book is rather broad, expanding from polymers and food over tissues to whole organisms in their active state.
- It presents macroscopic methods for rather inhomogeneous material where micromethods are often impossible or senseless.
- Thermal analysis as well as calorimetry are non-invasive and impose only limited or even no restrictions at all on the systems under research.

- The book may stimulate corresponding research and perhaps establish better contacts between very distant fields like Food Industry and Medicine, e.g.

We do not know of any book with such an orientation in the field of thermal analysis applied to life sciences.
The scientific problems discussed in this monograph are organized in four parts.

Part I. renders an insight into the properties of biotechnological polysaccharides combining the information from experimental data of thermoanalytical origin with that from statistical-thermodynamic models. Foods are discussed as multi-component and multi-phase systems where the heat treatment can produce transitions of compounds from one phase to another. Ingredients, starch-based biodegradable polymers, have an influence on the texture of the product, and thus the glass transition phenomena should be taken into consideration in the processing techniques. Proteins and fats are involved in the formulation of many food emulsions. Their structure and concentration have effects on the physical stability and organoleptic quality of emulsions while heating and cooling steps during the processing influence the storage quality.

Part II. presents examples how to use Differential Scanning Calorimetry (DSC) for structural and functional studies of muscle proteins. An exciting field of muscle research is discussed from different motor or regulator proteins up to highly organised muscle fibres. The cyclic interaction of myosin heads with actin filaments fuelled by ATP hydrolysis is basis of molecular mechanism of a number of events in biological motility. One may find studies on nucleotide-induced structural changes in the myosin head and in actin, simulating the different intermediate states of ATP hydrolysis. Interaction of F-actin with myosin heads, tropomyosin and other actin binding proteins serves as an example of studies on protein–protein interactions. Combination of DSC with other methods (e.g. electron paramagnetic resonance spectroscopy (EPR)) renders the molecular dynamic interpretation of global structural changes.

Part III. contains a review from the field of plant and plant tissues, thermobiochemical studies of animal cells in vitro, thermal investigations of social insects and an introduction into the world of human cartilage from the point of view of arthritis and degenerated lumbar intervertebral discs. We will see that wood as one of the most important plant products opens a new field for application of thermal analysis. Insects themselves represent more than half of the animal biomass on Earth so that their energetic impact can not be underestimated. Their energy saving e.g. by insulation of wasp nests or by the bee cluster strategy for surviving at low temperatures are also exciting thermoanalytical problems.

Part IV. demonstrates some efforts to make thermal analysis more quantitative by application of technical and theoretical improvements. The experimental heat capacity of carbohydrate–water systems is explained in terms of their molecular motion. Such an approach should also be valid for a more realistic description of heat capacities of other biological materials, including cellulose–water or protein–water systems.

A new result in connection with thermal stability of proteins is that in the statistical mechanical analysis a simple transformation following the Gibbs-Helmholtz equation $G = H - TS$ is no good approximation around the transition temperature. This suggests that the thermal transition of protein molecules is actually a phase transition. Therefore, in a correct statistical mechanical analysis the system should be deconvoluted into several thermodynamic states that satisfy the necessary condition for the Legendre transformation.

This short introduction to the content of this monograph shall just bring the reader to his favourite topic on a short way. Authors and Editor will be happy to receive comments, criticism and remarks in connection with this book to improve its quality for a possible next edition in future.

As the Scientific Editor of this volume and author of some chapters, I would like to thank all the staff of the *Journal of Thermal Analysis and Calorimetry* for the help, which was given to me during the technical editing of this book.

Dénes Lőrinczy, Editor
December of 2003, Pécs (Hungary)

Chapter 1

Order-disorder conformational transitions of carbohydrate polymers

The calorimetry contribution to understand polysaccharide solution properties

A. Cesàro*, F. Sussich L. and L. Navarini**

Laboratory of Physical and Macromolecular Chemistry, Dept. BBCM and UdR-INSTM, University of Trieste Via Giorgieri 1, I-34127 Trieste, Italy

Introduction

Polysaccharides have often been treated as *'the poor relations'* in comparison with other highly important biopolymers, nucleic acids and proteins. It is not yet clear whether this axiom was, at least in the past, generated by the conviction/belief that the application of quantitative methods of structural, functional and biological investigations could only seldom be used. As a consequence, the studies of physico-chemical properties and their interpretation seem to have been limited, much more than would have been expected in view of the intrinsic peculiarity of the complex chemical structure of many polysaccharides.

Among these limitations, we would like to focus here on the use of thermodynamic approaches that are very well established for the characterisation of the 'molecular domains' of biomacromolecules, and which are relevant for the energetics and the structural organisation, let us say, of globular proteins (for example, Privalov, 1980a, b). It is never adequately appreciated that thermodynamics, while not providing any information on the detailed structural organisation of molecules, does infact give a body of mathematical correlations between all the properties of the system and is therefore able to identify, from among several models, the one(s) compatible with the observed experimental data. Modelling of macroscopic rheological behaviour (e.g. for physical gels) is one of the most appealing aims. It is therefore rather surprising to notice that literature tends to gloss over the correct use of thermodynamic tools or even provides a misinterpretation of the calorimetric determination of the enthalpy of the helix-coil transition in polysaccharides. This fact prompts us to clarify how much can be gleaned from

* cesaro@units.it

** Permanent address: ILLYCAFFE S.p.A. Via Flavia 110, 34100 Trieste, Italy

D. Lörinczy (ed.), The Nature of Biological Systems as Revealed by Thermal Methods, 1–30.

the data collected on this transition. For this to be done, some original data produced in the authors' laboratory are presented together with a review of data obtained from literature. The aim is to instigate a full analysis of calorimetric data on the conformational transition of polysaccharides in order to provide relevant information on structural parameters that are not easily determined otherwise.

Theory

Many biopolymers undergo conformational transitions as a function of composition and/or temperature. Conformational transitions are *'in themselves'* conceptually analysed as phase transitions, since the polymer state is characterised by a difference in the structural and thermodynamic properties. We shall briefly summarise the experimental results which can be obtained by differential scanning calorimetry on the helix ↔ coil (in some cases including gel ↔ sol) conformational transition in linear biopolymer chains.

Before analysing some of these conformational transitions which occur in ordered polysaccharides, let us briefly recall some underlying concepts which have dealt mainly with the helix ↔ coil 'phase' transition of polypeptides and proteins (Poland, 1978; Cantor and Schimmel, 1980). In the case of globular proteins, it is worth mentioning the original observation that the contribution of the individual aminoacids to the Gibbs free energy of the native species is often about 200–600 J/mol of aminoacid (2–6 J/g) lower than for the denatured random coil form. Due to this low free energy difference of the monomers, the temperature of transition between the two species is higher than the ambient temperature (i.e., $T_m > 25°C$) only if the whole macromolecule can be thermodynamically considered a single domain, a fact that brings the total free energy difference to the order of 40–60 kJ per mole of protein. The hypothesis was therefore made that denaturation is a 'cooperative' process between two distinct states, which are thermodynamically defined and in equilibrium with each other at the transition temperature. The confirmation of the validity of this hypothesis, by means of DSC, represented one of the most significant milestones in the thermodynamics of biopolymer systems. It is also useful to make a reference to the current understanding of the peculiar thermodynamic behaviour of nanostructured systems which have nowadays an increasing relevance. The dependence of the melting temperature on the dimension of polymeric crystals at nanometers size is theoretically predicted by the Thompson-Gibbs equation and experimentally known since long time (Keller *et al.* 1993); crystalline lamellae show melting temperatures linearly decreasing with the inverse of the lamella thickness in the nanoscale. Similarly, for isolated polypeptide chains the helical stability (in helicogenic solvents) is predicted to asymptotically increase with chain length. Figure 1 shows the chain length dependence of the polypeptide helix-coil transition temperature (adapted from Cantor and Schimmel) and in the same plot the melting temperature (mirror-scale) of polyethylene lamellae as a function of the number of methylene units in the chain thickness.

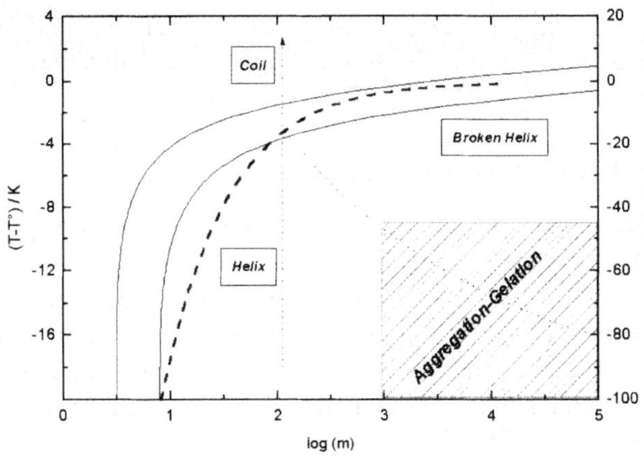

Fig. 1 Generalised phase diagram of transition temperature (arbitrary scale) for the helix-coil transition as a function of logarithm of degree of polymerisation *m* (adapted from Cantor and Schimmel). Regions for coil, helix, broken helix and aggregation state are shown. Dotted curve show the dependence of melting temperature (arbitrary scale) of paraffin-polyethylene system

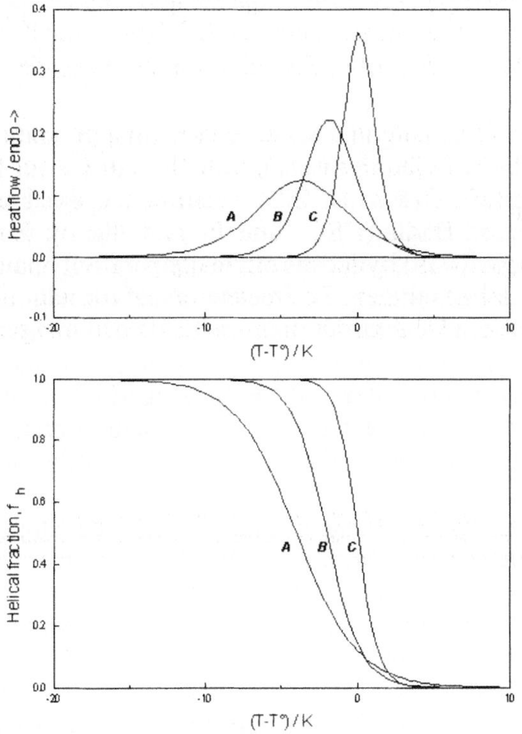

Fig. 2 Schematic DSC curves and dependence of helical fraction as a function of temperature for the three cases with increasing cooperativity (from A to C)

Order-disorder (melting) transition of helical conformations can be therefore traced by heating scans in a calorimeter (Fig. 2). The complete statistical thermodynamic description of the heat capacity curve provided by a DSC experiment showed that not only can all the information on the transition be obtained directly from the analysis of the shape of these curves without any additional data being used (Freire and Biltonen, 1978), but also that the resolution of the intrinsic structural energetics of the biopolymer and ligand binding interaction is possible by a global linkage analysis of two-dimensional DSC (Straume and Freire, 1992). This is due to the fact that the thermodynamic value of the enthalpy of the process can be written as

$$\langle \Delta H \rangle = \sum_{i-1}^{n} \Delta H_i \left(\frac{1}{Q} \right) \exp\left(-\Delta G_i \,/\, RT \right) \qquad (1)$$

by means of the elementary enthalpy ΔH_i and the probability of each step (given by Gibbs energy difference ΔG_i and the partition function Q). While the reader is referred to the ample literature, the following paragraphs outline a summary of the fundamental concepts and equations for the practical, simple, use of the experimental DSC data on the helix-coil transition.

THERMODYNAMIC STABILITY AND HELIX-COIL TRANSITION IN BIOPOLYMERS

Whenever biopolymers have a regular sequence of units, the stability of ordered helical structures is also a function of chain-length with a critical value above which the helix is interrupted (Poland, 1978). This concept was introduced, before the above mentioned findings for globular proteins, by the Zimm-Bragg theory (Zimm and Bragg, 1959) by means of the cooperativity parameter σ. This parameter essentially defines the excess free energy of formation of an isolated helical conformation with respect to the same process occurring as a neighbour of a helical sequence, for which the free energy change associated is described by the parameter s. Terms like 'initiation' and 'propagation' of a cooperative helical conformation were then suggested. The σ parameter is related to the sharpness of the change in any property measured as a function of a variable inducing helix-coil transition.

The original statistical-mechanical matrix model developed for the helix-coil transition in linear polypeptides has already been generalised to include other parallel phenomena such as, for example, the zippering of ordered chains in double or triple helices (Poland, 1978; Cantor and Schimmel, 1980). It has been also used to treat the binding of small iodine molecules into the amylose core that effectively induce the ordering conformational transition (Cesàro *et al.* 1986). In these theoretical approaches, the partition function is written as $Q = \underline{P}\ \underline{U}^{m}\ \underline{Q}$, where the statistical weight matrix $\underline{U}$ is properly indexed for every nearest-neighbour interaction on the polymer of chain-length m, each element in the matrix giving the relative probability (statistical weight) for finding site i ($1 < i < m$) in a particular state, helical (h) or coil (c). Proper differentiation of the partition function with respect to the statistical weights give the thermodynamically averaged quantities which characterise the helical features of the chain in terms of the average number of helical segments, $<N_h>$, and the average number of monomers in a helical segment, $<N^o>$, defined by:

$$<N_h> = \mathrm{dln}Q / \mathrm{dln}s \quad \text{and} \quad <N^o> = \mathrm{dln}Q / \mathrm{dln}\sigma \tag{2}$$

These two quantities are sufficient to model the long polymer chain into few or several helical segments of defined length according to the value of σ (smaller σ fewer the number of broken helix, see Fig. 3).

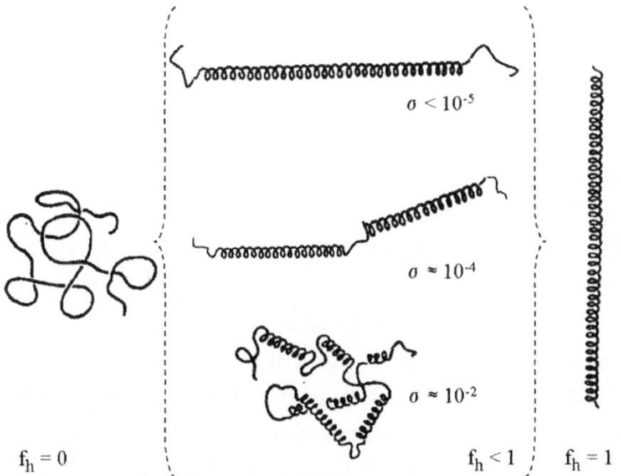

Fig. 3 Representation of dimensional (conformational) properties of chains undergoing coil-to-helix transition with different cooperativity

Without going into further detail of these theoretical approaches (Poland, 1978; Cantor and Schimmel, 1980), the prediction is that the cooperativity of the transition depends on the parameter σ, but also on the chain-length m, whilst the average transition temperature depends on m and mainly on the value of s. To underline the role of the chain length on the stability, let us remind that the phase diagram reported in Fig. 1 shows not only the stability of ordered helical conformations, but also the breadth of the transition, as a function of the variables m and T (for fixed values of σ and s).

It is also important to recognise the consequences that changes in the value of the parameter σ have in the dimensional properties (and in all other properties related to chain topology, e.g. rheology). This correlation has been theoretically clarified by Flory and co-workers (Flory, 1969), by calculating the chain dimensions (radius of gyration) of polypeptides with different cooperativity as a function of the helical fraction (related to s). Figure 4 shows the relative dimensional changes of an idealised polymer chain as a function of the helix fraction f_h for

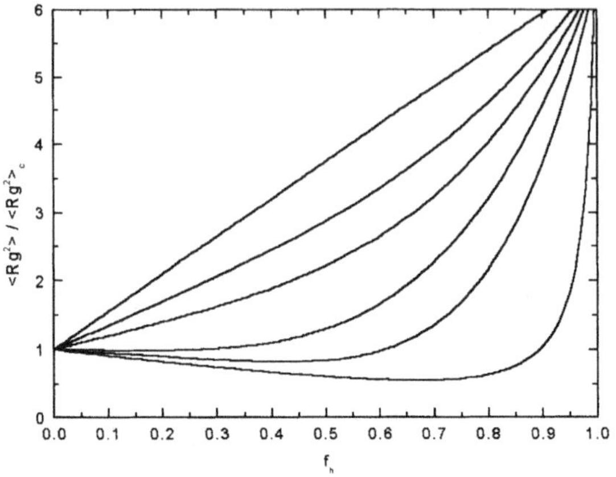

Fig. 4 Relative changes of dimensional properties (given by the square radius of gyration) as a function of the helical fraction for different values of cooperativity. The cooperativity parameter σ changes from 1 to 10^{-5} from top to bottom

different cooperativity values. Once more, the non-linear change (very abrupt, for $\sigma \approx 10^{-3}$ or smaller) emphasises the influence of the cooperativity on other physical properties, a matter of great importance for both the scientific implications and the technological applications. It is surely intriguing to note that the radius of gyration of a polymer is intrinsically related to its dynamical properties and that there is a linear log-log dependence of the average chain correlation time with the chain dimension (Cesàro *et al.* 2002). As a conceptual speculation, the increasing rapidity of helical chain collapse as a function of cooperativity closely reminds the phenomenon of fragility of supercooled liquids in a scaled Arrhenius plot of temperature dependence of segmental relaxation times (Angell, 1997). Long range correlation in fragile liquids and in cooperative helical chains are the key-parameters that will have to be further analysed in order to explore the usefulness of this conceptual correlation.

THEORETICAL ANALYSIS OF MICROCALORIMETRIC DATA

Since the earliest experiments (for example, the 'denaturation' of poly-γ-benzyl-L-glutamate, Ackermann; 1969), DSC experiments have always been more frequently used to characterise the helix-coil transition process in biopolymers. It was immediately noted that the heat of transition evaluated from DSC experiments differs from that evaluated by using the van't Hoff isochore for the apparent equilibrium constant. This discrepancy is a direct consequence of, and theoretically related to, the existence of 'molecular blocks of monomer units' which undergo a phase transition, with a change in enthalpy which is greater than the unitary change (i.e., per residue) by a factor of $\sigma^{-1/2} \approx N°$, which has been defined as the number of monomer units in a cooperative segment. Calorimetric measurements directly provide the value of $N°$ as the ratio of the apparent van't Hoff heat of transition and the calorimetric one. Simplistically speaking, this is also the reason why, although the specific heat of fusion of ice is 1.436 kcal/mol, a van't Hoff analysis of the temperature dependence of, let us say, the density in the melting region would provide an 'apparent' heat of fusion that is higher by very many orders of magnitude, given the size of the thermodynamic domains (crystals) undergoing the transition.

The most simplified approach gives the length of the cooperative unit in terms of the specific *excess* heat capacity of the system at the transition mid-point T_m and of the specific enthalpy change for the transition Δh:

$$\frac{\Delta H^{\mathrm{vH}}}{\Delta H^{\mathrm{cal}}} = \frac{4RT_{\mathrm{Tm}}^{2}\,\Delta c_{\mathrm{p}}^{\mathrm{Tm}}}{\Delta h^{2}} = N^{\circ} \tag{3}$$

where ΔH^{vH} is the van't Hoff enthalpy of the 'equilibrium process', defined in terms of the partition function Q:

$$\Delta H^{\mathrm{vH}} = RT^{2}\frac{\partial \ln Q}{\partial T} \approx RT^{2}\frac{\mathrm{d}\ln K}{\mathrm{d}T} \tag{4}$$

where K is the 'a-dimensional' equilibrium constant of the process which, according to the measurement method, can only be defined by the fraction of the species in the state h or in the state c, f_{h} and f_{c}. That is, the equilibrium constant, $K = f_{\mathrm{i}} / (1-f_{\mathrm{i}})$, is expressed through any experimental value sensitive to the molecular state of the system, such as the intensity of the absorption, the dichroic or fluorescence band, as well as structural and thermodynamic properties.

In all cases, K is defined as a fractional ratio of the final state to the initial state. The definition of the molecular size of the species undergoing the transition is a consequence of the statistical mechanical analysis of the 'cooperativity' of the process. In other words, because of the definition of the equilibrium constant, the molecular weight enters the van't Hoff equation only for the determination of the enthalpy change involved in the process.

From the structural point of view the regularity of primary structure involves the possibility that the chains with ordered helical conformations may form supramolecular structures, either of single or multiple strand type. However, it must be noted that the above mentioned analysis does not give the number of chains involved in the helical domain, but only the average number of monomers in the domain. Nonetheless, it has been shown that, in some cases, it is possible to use the concentration variable as an additional parameter to reveal such a further stage of helix dimerisation or multiple aggregation. Theoretical work on some of these processes has been published (Poland, 1978; Kidokoro and Wada, 1987; Robert *et al.* 1989). Complex transitions can be analysed within the framework of the polysteric model for conformational transitions, as has already been done for the polysaccharide succinoglycan (Burova *et al.* 1996). Although this is very important for many hydrocolloids, we are not considering here the statistical mechanical analysis that can be carried out on biopolymers which, in addition to the helix-coil transition, exhibit these further associations of helical segments in larger aggregates and/or supramolecular structures. Therefore, at this level of interpretation, we wish to underline that, while the stability and size of the thermodynamic domains are clearly defined through the DSC experiments, the actual molecularity of the process may still need to be supported in the model by other evidence. Past investigators resort to interpretation of experimental data, such as those given by light-scattering determination of the 'mass per unit length', or poly-electrolytic assessment of the 'charge per unit length' in the case of charged biopolymers. Recent exploitation of non-contact atomic force microscopy (AFM) in the tapping mode to solvated biopolymers opens a new avenue for direct access to molecular conformational data.

SOME RECENT DISPUTE ON THE VAN'T HOFF ENTHALPY

Only the fundamental aspects have been reported of the theoretical background which accompanied the development of calorimetric analysis of the cooperative conformational transitions of biopolymers and of discrepancy between van't Hoff and calorimetric enthalpies. However, at the end of this brief outline it seems more than appropriate to comment on some recent dispute about this question. Argumentation and rumours schematically concern two problems: *i*) the presence of small heat capacity changes that, even if not clearly discerned, induce discrepancies between ΔH_{vH} and ΔH_{cal}: *ii*) the possible intrinsic discrepancy of the two-state model. To be clear since the very beginning, none of the major criticisms and argumentation refers to biopolymer cooperativity to a first instance.

The original thermodynamic revisitation of the van't Hoff assumptions was made about ten years ago (Weber, 1996). Literature rejections was almost immediate and his argumentation was thereafter shown to have been unproperly developed as 'basic premise of his argument was incorrect, generating results fatally flawed' (Holtzer, 1997; Ragone, 1995). However, the dispute returned the question of the presence of hidden contribution to the data cast in the form of van't Hoff plot due to small values of ΔC_p. The discrepancies between the two calculated values of enthalpy, ΔH_{vH} and Δh_{cal}, are originated by different factors illustrated by several authors.

First of all, a non-vanishing heat capacity change ΔC_p introduces a curvature of the van't Hoff plot. For some binding reactions Sturtevant and coworkers (Liu *et al.* 1995, 1997; Naghibi, 1997) calculated temperature dependent ΔH_{vH} values which differ from those calorimetrically obtained; the ratio of $\Delta h_{cal}/\Delta H_{vH}$ varied from ca 0.5 to 4.3. Although a clear explanation was not provided, the indication was given that ΔH_{cal} includes all contributions from any processes underlying the reaction (including buffer or solution components) while ΔH_{vH} refers to the given 'simple' equilibrium. In a successive paper by the same authors, 'the discouraging conclusion' was reached that chemical reactions, at least in solution, are quite generally more complex than indicated by the simple chemical equations. It has been also analysed the question of whether or not the differences arise from real underlying physical reasons, or from 'more mundane' difficulties in the proper analysis of the van't Hoff data. The effect of hidden contribution, arising from small ΔC_p values into the van't Hoff analysis, may bias the slope even if apparent curvature is not produced. On the other hand, good calorimetric and van't Hoff data might be used to infer the existence of a ΔC_p small in magnitude. Let us also explicitly mention that the differences are often more illusory than real.

The main reason to have reported the above comments is dictated by the necessity of completely differentiating between doubts and argumentation about possible discrepancies and the 'real' large differences that are found when 'nanosize-organised' systems are disrupted by temperature and their decomposition is studied by calorimetry or followed by measuring the change in any composition-dependent properties. In the latter case, the very large values of ΔH_{vH} are uniquely, although not precisely, interpreted in terms of collapse of the macrostructure involving a large number of molecular units, while calorimetric output can be normalised by any arbitrary unit amount (generally weight or mole of substance). The term cooperativity unambiguously defines the melting of finite nano-ordered species as well as the disordering of linear Ising chains (Zimm and Bragg, 1959).

Differential scanning microcalorimetry

Several high sensitivity instruments are available from different producers. In a typical run, calorimetric cells (sample and reference) are heated up with scan rates ranging between 0.5–1 down to 0.01 K/min. Several scan rates (temperature-time profiles) are usually investigated to optimise the proper equilibration time with the best signal-to-noise ratio, as low scanning rates produce small heat flow signals (energy per unit time). Distortion of the shape of the heat capacity function can be effectively corrected by the approximate Tian equation (Calvet and Prat, 1963). However, in view of the low scanning rates commonly used and the instrumental characteristic times (of the order of 100 s), this correction is taken as negligible under most experimental conditions.

Scanning microcalorimetry studies of helix ↔ coil transition of polysaccharides

Microbial polysaccharides are biotechnologically produced and have a paramount relevance in industrial food and non-food applications (Sutherland 1998). Their 'quality' resides in their reproducible chemical structure (contrary to many plant gums) and their ecological properties (contrary to many synthetic polymers). In addition to the valuable physical properties (they act as emulsion stabilisers, gelling agents, inhibitors of crystal formation, viscosity controllers), many of them exhibit biological properties which have been positively explored in biomedicine.

```
xanthan                                →4)-β-D-Glc-(1→4)-β-D-Glc-(1→
Mo = 937.45 (824.8)                                          3
                                                             ↑
                                                             1
              β-D-Man-4,6pyr-(1→4)-β-D-GlcA-(1→2)-α-D-Man-6-OAc

schizophyllan       →3)-β-D-Glc-(1→3)-β-D-Glc-(1→3)-β-D-Glc-(1→
Mo = 648.64                                         6
                                                    ↑
                                                    1
                                                 β-D-Glc

gellan              →3)-β-D-Glc-(1→4)-β-D-GlcA-(1→4)-β-D-Glc-(1→4)-α-L-Ram-(1→
Mo = 742.52 (632.64)          ↑
                     (6-Oacetyl)0.5 and (2-L-glycerate)

succinoglycan       →4)-β-D-Glc-(1→4)-β-D-Glc-(1→3)-β-D-Gal-(1→4)-β-D-Glc-(1→
Mo = 1367.88 (1297.28)                                                  6
                                                                        ↑
                                                                        1
              β-D-Glc-4,6pyr-(1→3)-β-D-Glc-(1→3)-β-D-Glc-(1→6)-β-D-Glc

CPS Rhizobium TA-1                                   β-D-Gal
Mo = 972.96                                             1
                                                        ↓
                                                        6
                    →3)-β-D-Man-(1→3)-β-D-Gal-(1→4)-β-D-Glc-(1→
                                                        6
                                                        ↑
                                                        1
                                          β-D-Gal-(1→4)-β-D-Gal

Agarose             →3)-β-D-Gal-(1→4)-3,6-anhydro-α-L-Glc-(1→
Mo = 306.32

Amylose[1]          →4)-α-D-Glc-(1-[→4)-α-D-Glc-(1]-→4)-α-D-Glc-(1→
Mo = 162.16
```

[1] a trisaccharide has been reported for the structural unit to represent an essentially linear polymer, while M_o refers to the single monosccharide repeat unit

Fig. 5 Scheme of the structural architecture of polysaccharide repeat units. Molecular mass of the repeat unit is also indicated (numbers in parentheses refer to the units without non-sugar substituents)

The following sections report on the results of some DSC analyses of the cooperativity of the helix ↔ coil transition (Table 1), together with some relevant structural information of the polysaccharides studied. The list (see Fig. 4 for names and formulas) is not intended to be exhaustive, but only to cover a range of polysaccharides on which investigations have been accumulated, with some preference to microbial polysaccharides as biased by the authors' experience and of polymers importance. Figure 5 reports also the molecular masses of the repeating unit of the polymer in the 'idealised' native form, with a stoichiometric amount of non-sugar substituents as indicated. The molecular mass of the repeat unit, free from non-sugar substituents, is also shown since sample preparation may often include hydrolytic removal of these substituents.

Table 1 Thermodynamic data for the helix-coil transition of some polysaccharides

Polysaccharide	T_m/°C	ΔH/J g^{-1}	σ	N^o
Xanthan (0.01 M NaCl)	45	12	$\approx 10^{-5}$	$\approx$300
Schizophillan (in water)	$\approx$135	$\approx$27	$2.5 \cdot 10^{-5}$	200
Gellan	75	25	$\approx 10^{-2}$	$\approx$8
CPS *Rhizobium* TA-1	47	22	$7 \cdot 10^{-5}$	120
Succinoglycan (0.1 M NaCl)	71	17	$4 \cdot 10^{-5}$	1 50
Agarose	40	18	$1.5 \cdot 10^{-4}$	80
Amylose (-iodine-triiodide)	50	57	$4 \cdot 10^{-2}$	$\approx$5

Although the ordered structure in solution cannot be precisely determined, the assumption is usually made that it is essentially preserved from the helical form in the solid state. Therefore, a brief account of the helical parameters is given for each polymer (Rao *et al.* 1998).

Xanthan: X-ray data are not conclusive, although indicating that has a 5-fold helix symmetry and pitch of 4.7 nm (*c*-axis); none of the several models (single, double, parallel, antiparallel, left-, right-) provide acceptable X-ray fit.

Schizophyllan: A structure similar to that of hydrated curdlan is usually assumed, with $c = 1.878$ nm ($h = 0.314$) given by a 6-fold, parallel, right-hand, triple helix.

Gellan: Both native gellan and de-esterified (acetyl and glycerate) gellan (K^+ form) have been studied by X-ray, giving essentially a three-fold helix with c=2.815 nm ($h = 0.913$, 3-fold, left-handed, half staggered, parallel, double helix)

Succinoglycan: No data have been clearly published on X-ray fiber diffraction of succinoglycan. It has been quoted (Borsali *et al.* 1995) that it is a 'single helix' with a repeat length $h = 1.92$ nm, while most recent data (in solution) substantiate the existence of a double helix with a pitch of about 2 nm per repeat unit (Nakanishi and Norisuye, 2003).

CPS *Rhizobium* TA-1: The polysaccharide forms a 2-fold single helix of pitch 2.02 nm; since it is stabilised by a series of hydrogen bonds that involve the side chains, it has the appearance of a pseudo-double-helix.

Agarose: The original proposal is of a 3-fold, left-handed, half-staggered, parallel, double helix) with a pitch of c=0.95 nm ($h = 0.633$). Another set of data on dried films was interpreted as extended single helices with h ranging from 0.89 to 0.97 nm.

Amylose: Based on the energy contours several helical polymorphs are possible in view of the external conditions. The so-called hydrated 'V' amylose is characterised by a left-handed helical conformation with a pitch of 0.8 nm involving six residues per turn ($h = 1.33$ nm).

XANTHAN

Xanthan is a microbial polysaccharide produced by *Xanthomonas campestris*, the first bacterial polysaccharide to be food-approved by FDA in 1969 (and by EC in 1980). Its primary structure is constituted by a cellulose-like backbone of $(1 \rightarrow 4)$–β-D-glucose residues with a trisaccharidic side chain composed by mannose, glucuronic acid and mannose, attached at C(3) and linked on alternate glucosyl residues. The proximal α-D-mannose residue is usually acetylated on C(6) while the distal β-D-mannose may present a pyruvic acid residue in ketal linkage at C(4) and C(6). The proportion of these substituents can be easily modified by mild chemical treatments (acidic or alkaline hydrolysis) or by changing strain and culture conditions (Sutherland 1998).

Although its peculiar thermally stable viscosity behaviour was immediately appreciated in many technological applications, its conformation in the native state was a matter of debate for a long time. Nowadays the most credited stable conformation in solution is that of a double stranded chain (Berth *et al.* 1996 and reference therein).

The thermally induced order-disorder transition of xanthan in aqueous salt solution has been detected by a number of physical methods, such as viscosity, optical rotation, differential scanning calorimetry. In particular, given the ionic character of the polysaccharide, the influence of the ionic strength on the transition temperature has been largely investigated in order to analyse its polyelectrolytic behaviour in the frame of polyelectrolytic theories. The reader is addressed to the basic theoretical background here not reported (Anderson and Record, 1990, 1995; Paoletti *et al.* 1985) and to its application to succinoglycan (Burova *et al.* 1996) and to other polysaccharides (Benegas *et al.* 1998). This type of analysis, corroborated by many independent measurements from different authors, has univocally assigned the conformational transition largely as a double-helix to coil.

A detailed statistical mechanical analysis was offered by Brant and coworkers to elucidate the thermodynamic aspects of the conformation and of helix stability of xanthan fractions subjected to thermal treatments (Hacche *et al.* 1987) up to a throughful exploration of its rheological properties (Lee and Brant 2002a, 2002b, 2002c). The transition has been seen as a partial melting of the double strand: from light scattering a decline in M_w as a function of temperature, on passing through T_m, was not seen even if expected and evidences showed that a significative amount of the dimers dissociate in water only at 95°C (Kawakami *et al.* 1991).

The possibility of analysing several results on xanthan homologous samples with different acetyl groups and/or different pyruvyl substituents opens an interesting opportunity to verify, within the accuracy of the experimental data, some basic axioms of the helix-coil conformational transition of charged biopolymers. Among all these data, the enthalpy of melting ranges from ca 9 to 12 J/g (Christensen *et al.* 1993, Paoletti *et al.* 1983).

The contents of acetyl and pyruvyl residues strongly affect the polysaccharidic solution properties (Holzwarth 1979, Shatwell *et al.* 1990a, 1990b). These authors showed that acetyl groups have a stabilising effect on the ordered conformation and therefore increase the transition temperature while opposite effect can be attributed to pyruvate substituents; acetyl groups have the major effect on the shift in temperature. As for charged polysaccharides the transition temperature increases with increasing salt concentration and at constant salt concentration (below 1M) the transition temperature decreases with increasing pyruvate content (Kitamura *et al.* 1991).

Another series of xanthan derivatives has been prepared (Christensen, 1993) by depleting the terminal β-mannose residue in the side chains to a variable extent (f_M from 1.0 to 0), whereas the rest of the molecule remains essentially unchanged (however, also the acetyl group was always removed). The conformational transition of these samples, studied by optical rotation and calorimetry, has been analysed both in terms of the Zimm-Bragg theory and of the polyelectrolytic theory of conformational transition of charged polymers (Anderson and Record, 1990, 1995).

Values of the cooperativity parameter σ were evaluated from these data. Although the results are quite reproducible over different sample preparations, the scattering of the data as a function of f_M does not allow to extract a clear dependence (if any) on the content of the terminal mannose units. Taking for granted the self-consistency of calorimetric data alone, then σ should range between 10^{-4} and 10^{-5} with an upward parabolic curvature. The higher cooperativity for the unmodified sample and for the fully modified sample with respect to those partially modified is amply justified in terms of the perturbation of the ordered state, given by a statistical distribution of structural modifications.

The ionic strength dependence of the transition temperature would therefore return the non-ionic contribution to the transition enthalpy and the changes in the charge density of the polymer due to the conformational transition. Given the structural parameters that enter into equations, it is mandatory that the actual conformational states are known and that the transition occurs between two structurally defined states. The possibility of a time-dependent mixed population of double- and single-stranded makes difficult to properly analyse these data species, as previously suggested by Brant and co-workers. Under these circumstances, the unusual temperature dependence of light scattering data from fractionated xanthan samples was interpreted with the formation of both linear and cyclic structures, later confirmed by AFM investigations (McIntire and Brant, 1997).

SCHIZOPHYLLAN

Non-ionic glucans with a ß-1,3 sequence of glucose are produced by many microorganisms and include the curdlan family and the scleroglucan-schizophyllan family. Schizophyllan primary structure consists of linearly linked ß-1,3-D-glucose residues with one ß-1,6-D-glucose side chain every third main-chain residues. In water and at room temperature the polysaccharide exists as triple helices (Norisuye, 1980; McIntyre and Brant, 1998) made up by three chains interacting by intermolecular hydrogen bonds with the side arms outward of the helix. This ordered conformation is still preserved even in presence of DMSO up to 70% (Kitamura and Kuge, 1989) while a triple-helix to single-coil transition occurs at higher DMSO contents (Sato *et al.* 1983). Conformational transitions can be also thermally induced, in various solvent compositions; the polymer presents a highly cooperative order-disorder transition in the side-chain conformation at low temperature and a dissociation-disordering transition at high temperature.

Regarding the low temperature transition, in water schizophyllan exhibits considerable changes in optical rotation (OR) and heat capacity at about 6°C (Itou *et al.* 1987). The small endothermic peak appears also in the experiments made by Bot *et al.* (2001), as well as in those reported by Yoshiba *et al.* (2002). Itou *et al.* (1986) proposed a long distance organisation of the side chain in which water molecules plays an important role, organisation that evolves toward a disordered form by increasing temperature. The fact that this conformational transition at low temperature is affected by the substitution of water molecule with D_2O, supports this hypothesis (Itou *et al.*, 1987).

In a detailed calorimetric study (Kitamura and Kuge, 1989), high-sensitivity DSC was shown to be a very useful tool for investigating thermally induced conformational transitions of this polysaccharide. The original paper reports a set of DSC curves depicting the phase diagram for the conformational transition of a low molecular weight ($Mw = 1.34\ 10^5$) schizophyllan in water-DMSO mixtures. Values of ΔH_{cal} are reported for the two transitions at several solvent compositions. For the triple-helix to coil transition a value of 27 J/g (in water at $T \approx 135$°C) can be extrapolated from experimental data as a function of T and solvent composition. The ratio of the van't Hoff to calorimetric enthalpy, related to the size of the cooperative unit, raises from ca 70 up to 200 with increasing water concentration. In the context of this study, Kitamura and Kuge (1989) neatly showed how to reconcile the previous literature results, which seemed inconsistent only because they were incomplete. Not only does the complete phase diagram of schizophyllan in water-DMSO clarify such a discrepancy, but moreover, the direct calorimetric determination of transition enthalpy has provided further insight to the energetics and cooperativity of the two conformational processes.

In a more recent work (Kitamura *et al.* 1996) it is proved that ΔH_{cal} is independent on the pH of the solution and, at a constant pH, is also independent on added salt. The ratio $\Delta H_{vh} / \Delta H_{cal}$ leads to a cooperative units size of about 300 for pH below 10 that decrease to a value of approximately 30 with increasing pH. Although an analogous diminishing in the cooperative unit size was observed due to addition of DMSO, the cooperative length is less sensitive to the addition of DMSO and the given explanation concerns the preferential solvation of the polymer by

DMSO. The solvent effect on the cooperativity has also been shown by Hirao *et al.* (1990); as the DMSO content increases in a schizophyllan/D_2O solution, the transition shifts towards higher temperature but the cooperativity in terms of σ is almost the same for the different solvent compositions under study. The transition at higher temperatures is characterised by an asymmetry of the DSC curves which, however, could be accounted for on the basis of a simultaneous conformational transition and dissociation process.

GELLAN

Gellan is a bacterial polysaccharide produced by the micro-organism *Sphingomonas elodea* with a primary structure consisting of a regular sequence of tetrasaccharide repeat units in the backbone composed of glucose, glucuronic acid and rhamnose at a molar ratio of 2:1:1. The native polysaccharide contains an acetyl and an L-glyceryl as substituent on one of the glucose unit and forms a soft and elastic gel. Deacylation by alkaline treatment gives gellan in its commercial form. The commercial polymer is able to form rigid and clear gels which are in some respects comparable with those formed by agarose. X-ray diffraction studies have shown that in the solid state gellan exists as an extended intertwined, three-fold left handed double helix (Chandrasekaran *et al.* 1995). The glyceryl group enhances the stability of the double helix, whereas the acetyl group does not interfere with packing arrangement and hence has no structural influence.

Identification of ordered conformations of gellan in solution is complicated by its ability to aggregate and form intermolecular ordered structure. However, several facts argue in favour of the double helix conformation in dilute solution at low temperatures; from small-angle X-ray scattering of commercial gellan in aqueous solution, the relative linear mass density (polymer concentration 1.0–1.5%) at 10°C was reported twice that at 60°C (Yuguchi *et al.* 1996). The increase of polymer concentration (2.9 and 5.7%) led to larger values confirming the further association of double helices. This double helix association, essential for gel formation, is controlled by the type of counter ions. The cations role (for deacylated gellan in sodium salt form in the presence of calcium and potassium ions) has been recently studied by transmission electron microscopy (Atkin *et al.* 2000). The cation type and the cation:carboxylate concentration ratio (below, above or at the stoichiometric equivalence) have a profound influence on polymer morphology with evidence of lateral aggregation of the thermodynamically stable conformation of gellan in salt-free aqueous solution (double helix and double-helical duplexes).

Differential scanning calorimetry studies of 10 mg/mL gellan in the absence of added salt showed single thermal transition on heating and cooling that has been attributed to coil-helix transitions. At polymer concentrations higher than 32 mg/mL DSC heating curves show two endothermic peaks; the lower temperature transition was attributed to aggregate-helix melting and the high temperature transition to helix-coil melting (Miyoshi, *et al.* 1995a, b). Similar studies (Mazen *et al.* 1999) showed that at fixed polymer concentration (10 mg/mL) and low ionic strength (0.01 M NaCl) thermograms of deacylated gellan are characterised by a single peak on heating and cooling at relatively low temperature (ca. 34°C) with a very small hysteresis. The enthalpy of the process, attributed to the helix-coil transition, was reported to be 9.5 J/g. On increasing the salt concentration to 0.05 M a second peak at higher temperature appeared on heating (42°C) and at 0.1 M two well separated peaks were present in the thermograms. The first peak (for 0.1 M NaCl at ca 53°C), attributed to the helix coil transition appeared nearly located at the same temperature as the peak on cooling runs. The second DSC peak (for 0.1 M NaCl at ca. 75°C) was related to the formation of large aggregates of double helices. The enthalpy of the conformational transition increases progressively with the salt concentration (up to 18 J/g) as expected for polyelectrolytes.

In order to avoid contributions related to secondary aggregation process, deacylated gellan in tetramethylammonium (TMA) salt form (salt concentration range: 0.0025–0.5 M TMACl) in dilute solution (polymer concentration: 0.5–0.7 mg/mL) has been recently studied by means of high sensitivity DSC (Grinberg *et al.* 2003). Some common features of the transition have been observed at every salt concentration like, for instance, the presence of a single heat absorption (with position, size and shape depending on ionic strength) and the λ-like profile (long tail to the left of the maximum and an apparent break point to the right of it) with a very sharp maximum of the thermograms. Similar profiles were observed for a highly purified sample of sodium salt gellan (10–15 mg/mL) in salt free solution (Miyoshi, *et al.* 1999). Moreover, no distortion of the transition λ-like profile was observed in the whole range of ionic strength confirming that the double helices of the TMA gellan are not capable of aggregation. Both transition temperature and enthalpy increased with increasing salt concentration as expected for charged linear biopolymers. In particular, the enthalpy of the process has been reported to be confined in the range 5–10 kJ/mol within the investigated ionic strength range. By analysing the profile of the conformational transition with a model which considers two sources of cooperativity of the double-helix transition (stacking and loop factors) the authors led to the conclusion that the cooperative unit of gellan involves about eight repeating units (close to the persistence length of the disordered gellan chain) with a cooperativity parameter (0.62±0.01) indicating that the partial unfolding of the double helix in its middle section (loop effect) dominates the cooperativity of gellan transition. Moreover, in order to fit the Poisson-Boltzman model to the experimental free energy of transition against the concentration of the salt at $T = 273$ K it is necessary to suggest that the effective linear charge density of gellan in the coil conformation is larger that that estimated for the fully extended chain.

Native gellan (0.8 acetyl and 0.8 glyceryl substituents per repeat unit) conformational transition has been also investigated by DSC (Mazen *et al.*

1999). Calorimetric data confirm that the native gellan is a double helix with a much higher thermal stability than the deacylated one due to the role of the glycerate groups (deacetylation have nearly no influence on the stability of the double helix). In facts both transition temperature and enthalpy have been reported significantly higher than those of commercial sample. The enthalpy ranges from 18 to 22 J/g (polymer concentration 10 mg/mL), passing from 0.01 to 0.1 M NaCl. Contrary to deacylated polymer, the ionic strength dependence of transition temperature of native gellan is relatively small and on increasing salt concentration, only one peak on thermograms has been observed. The role played by glyceryl groups, by perturbing the conformational transition, has a correspondence in the modification of rheological properties and of the packing density in the solid state.

CAPSULAR POLYSACCHARIDE FROM RHIZOBIUM TRIFOLII TA-1

The chemical structure of *Rhizobium trifolii* capsular polysaccharide (TA-1-CPS) is characterized by a trisaccharide in the chain backbone which possesses two branches on the same glucosidic residue. The most dramatic solution property exhibited by TA-1-CPS is, by far, its ability to form aqueous thermo-reversible gels in a wide range of polymer concentration (down to ≈ 0.1 g/L). In particular, due to the non-ionic character of the polysaccharidic chain, gels can be formed in the absence of ionic co-solutes and show a remarkable gel strength. A number of experimental observations (Cesàro *et al.*, 1987; Gidley *et al.*, 1987), in particular on the thermal and rheological behaviour of TA-1-CPS in the presence of co-solutes (urea, salt, or sucrose), suggest that at least three different levels of structure may be involved in the process of aqueous gel formation. While the first level was referred to as local chain conformational ordering, it was thought that the second one involved 'intermolecular ordering between conformationally ordered segments'. This structure has been shown to resist shear and such denaturants as urea. The third level of structure provides for the three-dimensional gel network and is labile under moderate shear and in concentrated urea solution: it involves supramolecular aggregation. Evidence for a complex aggregation in the development of the gel structure has also been accumulated from independent experimental work (Cesàro *et al.*, 1987 and Gidley *et al.*, 1987). In particular, both the hysteresis and the temperature dependence of the rigidity (storage) modulus in water and in aqueous urea solution support the presence of an intermediate step for the formation of an aggregate structure. From the structural point of view, although the quality of the diffraction pattern of the TA-1-CPS did not at first permit a good resolution of its structure, the layer line spacings show that the chain has a 2-fold helical symmetry with a chain repeat axis of 0.98 nm per repeating unit (Lee *et al.* 1992).

For this polysaccharide, calorimetric experiments were carried out under similar conditions and with three different high-sensitivity DSC instruments. The polysaccharide was repeatedly heated and cooled, and the thermal curves were almost completely reproducible. The DSC results showed a very sharp transition at around 47°C. The transition, indeed, was reversible, sharp but asymmetrical (related to the aggregation). From the direct calorimetric heat of transition of 22.2±0.2 J/g and the van't Hoff enthalpy of about 1255 kJ/mol, which was estimated according to the procedure outlined above, the molecular weight of the cooperative unit $N°$ resulted as 57000. This value brings the length of the cooperative segments to about 57 nm which stabilises the ordered helical conformation in the gel structure.

Therefore, gelling properties of CPS arise from a stabilised array of energetically favourable overlaps between the side chains, while the stereoregular non-ionic main-chain maintains a helical conformation in water, which, most probably is the same as that found for CPS fibres by means of the new X-ray diffraction data (Chandrasekaran *et al.* 1992). It is interesting to note that the presence of side chains on a polysaccharide backbone is normally considered a perturbing factor with respect to gelation. Examples can be taken from literature and include the welan-rhamsan family as well as the curdlan-schizophyllan case. The case of TA-1-CPS, however, points to the opposite. In fact, either the cleavage or modification of the side chain destroys the ability of this polysaccharide to form a gel. A scrutiny of the results concerning the conformational transition induced by temperature on derivatives, obtained by sidechain partial modification, led to the conclusion that the gel stability decreases linearly with the side chain modification, which must destroy the functionality of the arms in the intermolecular cross-linking process (Cesàro *et al.*, 1989).

Let us quote here that the same strain of *Rhizobium trifolii* produces an abundant quantity of exocellular ionic polysaccharides (TA-1-EPS) which also exhibits a ionic strength dependent conformational transition (Crescenzi *et al.* 1987a, 1987b). In addition, other microrganisms offers polysaccharides with the same primary structure but naturally differing in the amount of non-sugar substituents (Faleschini 1988; Cosani *et al.* 1989; Cesàro *et al.* 1992). Although this would have offered an interesting case of homologous sample, detailed analysis of calorimetric data has not been published.

SUCCINOGLYCAN

Succinoglycan is a microbial exopolysaccharide produced by several strains of soil bacteria belonging to the genera *Alcaligenes*, *Pseudomonas*, *Agrobacterium* and *Rhizobium*. The polymer chains are made up of octasaccharide repeat units. Four monosaccharides of every repeat unit (three *D*-glucose and one *D*-galactose residue) make up the backbone, where the galactosylated glucose residue serves as a branching point bearing the tetrasaccharide side chain composed of *D*-glucose residues at position 6. Two charged non-carbohydrate substituents (succinate half-ester and 1'-carboxyethylidene acetal) are located in this side chain, whereas O-acetyl groups, when present, may be found in the backbone. Succinoglycan does not form gels, but give rise to extremely viscous solutions or weak gels (Cesàro *et al.* 1992), however at sufficiently high polymer concentration and in dependence of sample origin and thermal history of its aqueous solutions substantial aggregation can occur leading eventually to the formation of thermoreversible gel (Boutebba *et al.* 1999).

According to most recent light scattering and viscometric data (Kaneda *et al.* 2002; Nakanishi and Norisuye 2003), succinoglycan in salt solution (0.01 and 0.1 M NaCl at 25°C) behaves as a rod-like polymer with a persistence length of 50–180 nm and a molar mass per contour length of 1510 nm^{-1} (corresponding to that of a double-helix). At 75°C, the polymer behaves as a worm-like chain with a persistence length and a molar mass per contour lenght of 10 nm and 750 nm^{-1} respectively. From these data, the polysaccharide is considered to be a dimer that has ordered structure of double helical nature. However, in salt-free solution, it was suggested that succinoglycan behaves as a single chain in relatively low polymer concentration (Borsali *et al.* 1995).

Up to now, the thermally induced order-disorder conformational transition of succinoglycan has been studied by high sensitivity calorimetry only in one detailed study, even if scarce calorimetric data have frequently been included in studies aimed at characterizing the polymer solution properties. The complex nature of the succinoglycan order-disorder conformational transition has been studied (Burova *et al.* 1996) by examining the concentration dependence of the transition temperatures and the shape of the excess heat capacity curves obtained by high-sensitivity adiabatic DSC (5–100°C, heating rate 1 K min^{-1}). Thermograms of succinoglycan in salt-free solution at polysaccharide concentrations of less than ca. 2 mg/mL, have been satisfactorily described by the two state model suggesting the transition mechanism to be of the single helix-coil type. At higher polymer concentration, the transition curves become characterised by a marked asymmetry and are described by a polysteric model which includes two stages: the cooperative dissociation of the helix dimer and subsequent melting of helix monomer. At NaCl concentrations 0.01 and 0.1 M thermograms have been well fitted by the polysteric model within the whole studied range of polymer concentration (0.1–3.5 mg/mL).

The theoretical profile of the transition was therefore calculated by using an application of the general allosteric approach developed by Gill's group (Robert *et al.*, 1989) According to this model, based on the theory of the helix ↔ coil transition, the fitting approach gives the number N° of octasaccharides in the cooperative unit of succinoglycan. Furthermore, the results suggest that the average length of a succinoglycan helix increases with salt concentration but does not depend on polysaccharide concentration. The average value of N° changes from 85±25 in water to 150±20 in NaCl 0.1 M, in agreement with the indications of the scattering data. As an example, a chain of succinoglycan with molecular weight 5.4 10^6 in aqueous 0.1 M NaCl solution includes more than 10 «independent» helix segments each made up of an average of 150 repeat units. Therefore, the stability of such a system, consisting of a sufficiently large number of these extended segments, should be fairly insensitive to moderate variations in the chain-length. For this reason, the effect of the molecular weight heterogeneity of the sample on the profile and parameters of the conformation transition for succinoglycan appeared to be negligible according to the authors (Burova *et al.*, 1996).

The total value for the enthalpy of transition of succinoglycan double helix to single coil chain (average upper limit of 17.34 J/g) is given by the sum of the double helix dissociation enthalpy and the melting enthalpy. The contribution of the dissociation enthalpy of 6.67±0.6 J/g has been calculated. The contribution of melting enthalpy is strongly influenced by polymer concentration in salt-free solution (from 5.34 J/g at 0.1 mg/mL to 10.67 J/g at about 2.0 mg/mL with a S-shape profile). In 0.01 M NaCl the melting enthalpy pass from 8.00 J/g at 0.1 mg/mL to 10.67 J/g at 2.5 mg/mL. A constant value of 10.67 J/g in the whole range of polymer concentration is observed in 0.1 M NaCl. The peculiar behaviour emerging from this calorimetric study, not only permits to interpret some previously unclear issues but also to reconcile some discrepancies of previous studies. For instance, the transition enthalpy reported by Ridout *et al.* for native succinoglycan, 9.47 J/g, the value reported by Fidanza *et al.* 14.40 J/g and the values reported by Boutebba *et al.* 17.1–18.8 J/g are presumably obtained under different experimental conditions.

The cooperativity parameter has been reported to range from 14±6 10^{-5} (salt-free solution) to 4.4±1.2 10^{-5} (0.1 M NaCl) indicating that the melting process of succinoglycan helices is a highly cooperative transition in comparison with other biopolymers (Burova *et al.* 1996). The presence of non-carbohydrate substituents has stimulated investigations on the role played by O-acyl and pyruvyl residues on the stability of the succinoglycan ordered conformation. Preliminary DSC studies on acetyl-containing succinoglycan sample in salt-free aqueous solution (Ridout *et al.* 1997) revealed that removal of the acetyl substituents does not improve the cooperativity of the transition and reduce the stability of the helix whereas removal of succinyl groups raises the thermal stability of the helix, increases the transition enthalpy and improves the cooperativity of the transition. The latter behaviour has been suggested explainable in terms of the reduction in charge density on the polysaccharide chain.

ALGAL GALACTANS: AGAROSE

Agarose is a neutral algal polysaccharide, ideally constituted by alternating residues of 1,4-linked 3,6-anhydro-α-L-galactopyranose and 1,3-linked β-D-galactopyranose. The polysaccharide is the neutral member of the agar family extracted from red algae and is usually associated with the sulphated galactans (carrageenans), which however, in addition to the charged sulphate groups, present a different sugar stereochemistry. The widespread empirical use of agarose in the preparation of neutral gelled substrate for the electrophoretic separation of valuable biological material has often obscured the efforts made to elucidate the gel microstructure and its molecular architecture (Maaloum *et al.*, 1998). The gel is stable at room temperature and is characterised by a high rigidity. The hysteresis between the melting temperature (in the range of 85–95°C) and the gelling temperature (25–35°C) is rather pronounced, although recovery of the gel properties usually occurs after heating and cooling cycles. Furthermore, while the specific enthalpy of melting does not change with polysaccharide concentration, the gelling temperature increases asymptotically with concentration, with a limiting upper temperature of about 40°C. Unfortunately, despite the different temperature range for the melting and gelling phenomena, no clear cut analysis has been made for the two distinct processes, and the tacit underlying assumption seems to be that the system is effectively in a sort of 'delayed' equilibrium. Comparative data for the agarose gelling process can be taken from the temperature dependence of the dichroic absorption as reported by Fujii *et al.*, (2000) and from the calorimetric DSC data reported by Rochas (1987). When the calorimetric enthalpy is set at 18.28 J/g (5.6 kJ/mol of repeat units), the van't Hoff enthalpy obtained from the CD data is 450 kJ/mol, giving a value of about 80 units for the cooperative length. It is important to stress that this analysis cannot provide any information as to whether the helix-coil transition involves single or double helices (either with intertwined or side-by side geometry); this information can only be obtained from direct structural analysis. The value of 80 units relates only to the meaning of the size of the cooperative block which melts simultaneously, no matter whether it is a single linear chain of 80 units or 10 associated chains each of 8 units. It is, therefore, rather surprising that the divergence between the calorimetric enthalpy and the van't Hoff enthalpy has been taken, at different times by both these authors, as evidence for the contribution of the helix-helix interaction among agarose fibres, which is claimed to be considerably larger than the conformational contribution in the coil-helix transition. Needless to say, also the size of the molecular unit in the equilibrium constant has been misinterpreted, taking into account a hypothetical macromolecular weight of 120 000 for the agarose polysaccharide. Given the context of a scaling analysis of rheological properties of agarose gel, the molecular dimension of the ordered blocks may also play a role in the 'chicken wire' network responsible for the elastic properties. It has also to be taken into account that most recent calorimetric data give a 'moisture' dependent heat of transition, providing a limiting value at high water content of 57.6 J/g (Cooke *et al.* 1996). Whether the data above reported could be re-evaluated in the frame of a suitable model of helix-coil transition, is matter of future debate.

STARCH AMYLOSE

The essentially linear α(1–4) glucose polymer is named amylose; together with the highly branched amylopectin constitutes the polysaccharidic component of starch. By specifically referring to amylose molecule (i.e., in the absence of amylopectin) we wish to avoid the confusion of attributing to starch as a whole the properties of individual components, a problem that seems quite common in the literature when 'starch properties' are intended to refer to amylopectin or amylose alone.

No proofs have been presented for an amylose regular helix conformation in solution, but rather for a propensity to form loose and irregular single-stranded pseudo-helices (for an extensive review, refer to Bank sand Greenwood, 1975; for update conformational data Cesàro *et al.* 2002). This view has been supported in the late seventies by Monte Carlo simulation (Jordan *et al.* 1978) in agreement with several experimental evidences which ruled out the presence of even short regular helical sequences (Cesàro *et al.* 2002). The most recent direct evidence of amylose coonformation has been offered by AFM imaging of a monodisperse samples (enzymatic synthesis) for which the distribution of contour lengths has been evaluated. From these data a number-average contour length of 231 ± 101 nm indicates that the amylose chain is contracted in a worm-like pseudo-helical conformation (McIntyre and Brant, 1999).

As a consequence of this loosely regular chain topology, the observation of a single helix-to-coil cooperative conformational transition is practically unrealistic. Nonetheless, given the relatively high statistical weight of the helical conformational state, the possibility that helical sequences are formed under given circumstances has to be taken into consideration. Such a condition is specifically verified in the peculiar case of self-assembled nanostructured array of iodine atoms in rods, reported below.

Let us also mention that, contrary to the first belief, amylose component in native starches is largely in a conformationally disordered state interspersed within the amylopectin crystallites in a still unknown way. Therefore, 'natural' stability toward a helical conformation of amylose is not founded on experimental evidence. However, literature offers few papers in which the calorimetric evaluation of the melting enthalpy of starches with different amylose content (low and high amylose starches) is elaborated in terms of 'Zimm-Bragg' model (Waigh *et al.* 2000; Matveev *et al.* 2001). Having clarified that the melting refers to the amylopectin fraction and that the amylopectin lamellar crystallites are composed by intertwined double-helix 'physically limited in the chain length', it may be useful to report the cooperativity parameters for this system. We resort therefore to the comparison made in Section 2 between single helix model and nanoscrystalline species (see Fig. 1).

The blue iodine-amylose complex (a tri-iodide induced poly-iodine complex) has been the first helical structure proposed among biopolymers on the basis of X-rays. Proofs for the structure of the complex have been accumulated in favour of a linear array of iodine species inside the annular cavity formed by the helical conformation of amylose chain (Banks and Greenwood, 1975). An induced-fit model (Cesàro *et al.* 1986) has been proposed which makes the complex properties dependent on iodine/tri-iodide ratio, degree of polymerisation of amylose, temperature, ionic strength and other factors, in addition to the total iodine/amylose ratio (Rendleman 2003; Cesàro *et al.* 1980, 1986). This model relies on the known dependence of the wavelength λ_{max} of the strong absorption and circular dichroic bands on the mean degree of polymerisation m of the amylose chains, reaching an asymptotic upper limit near 640 nm for m >100. The value of λ_{max} also depends on the degree of saturation of the complex and moreover on the concentration of iodide ion. These changes have been related to the changes in chain length of the polyiodine arrays in analogy with the Kuhn model for the polyenes. Theoretical fit of binding isotherms suggests that N° should reach the limit of about 30 glucose units ($\sigma = 4\ 10^{-2}$) for long chains, which means about 5 iodine units accommodated in a helical cavity of 4 nm.

Direct calorimetric measurements on the complex formation have been performed and literature data reviewed (Cesàro *et al.* 1980). The enthalpy of complexation is found to be constant in the range of reaction conditions which leave λ_{max} unchanged, and it varies with chain length m in a way that mimics the dependence of λ_{max} on the degree of polymerisation. Therefore the rather large enthalpy change (–71 kJ/mol of bound I_2) must sustain its largest contribution from the cooperative interactions between the atoms of the linear bound iodine chains and a much smaller contribution from interactions of the bound species with the polymer chain. Given the cooperativity of the complex formation, the isosteric heat of binding depends on the degree of complexation, θ. Direct microcalorimetric determinations show that the integral heat of reaction from $\theta = 0$ to 0.2 ($\Delta H_{\theta=0.2}$) is ca. 13 kJ/mol more negative than the value of $\Delta H_{\theta=1}$ under the same experimental conditions. Other enthalpic data on complex formation were derived from the van't Hoff plot of the apparent equilibrium constant as a function of the temperature, AH_{vH}, and range from –34 to –71 J/g of complex (–42 to –87 kJ/mol of bound molecular iodine). It has been pointed out that the difference in the enthalpic data may reflect the rearrangement of tri-iodide equilibrium (Cesàro *et al.* 1980), according to the warning expressed on page 8.

Correlation between structure, function and technological application

In very dilute solution the viscosity behaviour of polymers is mainly determined by their conformation (size and shape). Experimental evidence supports the earl hypothesis that chain conformation and rigidity can be qualitatively predicted on the basis of viscosity data. Although a wide range of rheological experiments, including small deformation oscillatory, steady and transient shear, have been performed at finite concentrations and under well defined experimental conditions (for a reference, Lapasin and Pricl, 1995), still little effort has been done in attempt to correlate the rheological behaviour (slow dynamics) of polysaccharide solutions in terms of chain conformational dynamics. Solution dynamical properties of disordered polysaccharides can be modelled within a reasonable degree of accuracy and their rheological behaviour at finite concentrations is that of a typical network of flexible chains. The lack of ordered secondary structure lead to entangled networks in which the topological constrains govern the rheological behaviour. Ordered regions along the polysaccharide chains seem to be the prerequisite to establish intermolecular interactions which lead to non-transient polymer networks. Whenever ordered conformations are stable in solution, both solution properties and rheological behaviour display features that arise not only from the stiffening of the chains in the ordered conformation, but from the likely interchain interactions between the ordered 'array(s)' of chains. The strength of these interactions reflects in the behaviour of the gel (weak to true gel).

The influence of polymer conformation and rigidity on the rheological properties is especially evident in the semi-dilute concentration regime. The rheological behaviour of flexible linear polysaccharides in good solvent is that of an entangled network of physically interacting chains, being the Corx-Merz rule obeyed except at very low polymer concentration (Lapasin and Pricl, 1995). In the same concentration regime and under conditions of an ordered conformation, worm-like polysaccharides with low flexibility (i.e. with persistence length > 60 nm) show a remarkable deviation from the behaviour described for flexible molecules. In particular, the viscoelastic spectrum resembles that of a typical gel system with both moduli G'' and G' nearly frequency independent ($G' > G''$ in the whole range of frequency accessible) and the Corx-Merz superposition rule fails.

From a technological point of view, it is relevant to mention the importance played by polysaccharides in the food industry. Polysaccharide thickeners are widely used to modulate and to control texture and mouthfeel of viscous food systems, ranging from sauces and beverages to dairy products. Several in-mouth perceived textural characteristics, like thickness and sliminess of polysaccharide food models, have been successfully correlated to rheological parameters; the correlation, however, resulted to be dependent on polysaccharides conformation. In facts, disorder random coil polysaccharides or rigid ordered polymer like xanthan gave different responses.

In addition to texture perception polysaccharide conformation and related rheology is of crucial importance as far as flavour release is concerned (Morris, 1995). It is well known in the food industry that the quantity of flavouring required to produce the same subjective flavour intensity is often much higher in thickened or structured products than in fluid systems. A similar suppression of perceived intensity is also well established for taste attributes (sweet, sour, bitter, and salty). Random coil polysaccharide solutions, ordered polysaccharide weak gels and true gels show relevant differences in the flavour/taste release. In particular, according to Morris, the increased viscosity in ordered/gelling polysaccharides hinders the mixing process by which flavour/taste molecules diffuse (Morris, 1995).

Conclusions

Direct calorimetric determinations of both the heat of transition and of its partial derivative have been proved to be irreplaceable methods to quantify the cooperative character of the polymeric helical chain. Evidence is in favour of a true temperature-driven phase transition occurring in polysaccharides under given circumstances. Besides the formal presentation of the thermodynamic analysis of the well-known process of helix-coil transition in linear biopolymers, accurate DSC data (even alone) can give the structural information of the cooperativity parameter (and therefore of N°) from the evaluation of calorimetric ΔH and ΔH_{vH}. Reference to the theoretical approaches shows that $N^{\circ} \approx \sigma^{-1/2}$, at least for chains with a degree of polymerisation m larger that N°, which becomes an effective number of monomer units energetically (and topologically) correlated. Therefore, the larger N°, the larger is the topological constraint in an ordered biopolymer chain.

Several investigations on the rheological properties of concentrated solutions of polysaccharides have produced data which can be interpreted on the basis of the polymer conformation and chain rigidity. In fact, worm-like polysaccharides with low flexibility show a remarkable deviation from the behaviour described for flexible statistically disordered chains. The viscoelastic spectrum of the former resembles that of a typical gel system, indicating that non-transient supramolecular structures also occur, although these weak-gel systems flow upon increasing shear. This behaviour, typical of so-called weak gels, has been claimed to reflect the occurrence in the polymer network of weak non-covalent intermolecular forces. Persistence of these weak interactions over an extended length of the chain is conceivable within a fraction of regular structures. It is the authors' intention to claim that the energetics and the cooperativity of the phase transition are just another aspect of the macroscopic solution behaviour of the gelling ordered polysaccharides. Molecular description of these structures is not yet achieved and other experimental data are necessary to substantiate these hypotheses.

Acknowledgements

The paper has been prepared with financial support of M.U.R.S.T. and of University of Trieste. F. S. is grateful to INSTM (Florence) for research grant.

References

Ackermann, T. (1969) Physical States of Biomolecules: Calorimetric Study of Helix-Random Coil Transitions in solution in H. D. Brown (ed), *Biochemical Calorimetry*, Academic Press, New York, 121–148.

Anderson, C. F. and Record, M. T. (1990) Ion Distributions around DNA and Other Cylindrical Polyions - Theoretical Descriptions and Physical Implications *Annu. Rev. Biophys. Biophys. Chem.*, 19, 423–465.

Anderson, C. F. and Record, M. T. (1995) Salt Nucleic-acid Interactions *Annu. Rev. Phys. Chem.*, 46, 657–700.

Angell, C. A. (1997) Entropy and Fragility in Supercooling Liquids, *J. Res. Natl. Inst. Stan.*, 102, 171–185.

Atkin, N., Abeysekera, R. M., Kronestedt-Robards, E. C. and Robards, A.W. (2000) Direct visualization of changes in deacylated Na^{+}gellan polymer morphology during the sol-gel transition, *Biopoymers*, 54, 195–210.

Banks, W. And Greenwood, C. T. (1975) *Starch and its components*, Edinburgh University Press, Edinburgh.

Benegas, J. C., Cesàro, A., Rizzo, R. and Paoletti, S. (1998) Conformational stability of biological polyelectrolytes: evaluation of enthalpy and entropy changes of conformational transition, *Biopolymers*, 45, 203–216.

Berth, G., Dautzenberg, H., Christensen, B. E., Harding, S. E., Rother, G. and Smidsrød, O. (1996) Static light scattering studies on xanthan in aqueous solution, *Macromolecules*, 29, 3491–3498.

Biltonen, R. L. and Freire, E. (1978) Differential scanning calorimetry, *Crit. Rev. Biochem.*, 5, 85–110.

Borsali, R., Rinaudo, M. and Noirez, L. (1995) Light-scattering and small-angle neutron scattering from polyelectrolyte solutions-the succynoglycan, *Macromolecules*, 28, 1085–1088.

Bot A., Smorenburg, H. E., Vreeker, R., Pâques, M. and Clarke, A. H. (2001) Melting behaviour of schizophyllan extracellular polysaccharide gels in the temperature range between 5 and 20°C, *Carbohydr. Polymers*, 45, 363–372.

Boutebba, A., Milas, M. and Rinaudo, M. (1997) Order-disorder conformational transition in succinoglycan: Calorimetric measurements, *Biopolymers*, 42, 811–819.

Boutebba, A., Milas, M. and Rinaudo, M. (1999) On the interchain associations in aqueous solutions of a succinoglycan polysaccharide, *Int. J. Biol. Macromol.*, 24, 319–327.

Burova, T. V., Golubeva, I. A., Grinberg, N. V., Mashkevich, A. Ya., Grinberg, V. Ya., Usov, A. I., Navarini, L. and Cesàro, A. (1996) Calorimetric Study of the Order-Disorder Conformational Transition in Succinoglycan, *Biopolymers*, 39, 517–529.

Calvet, E. & Prat, H. (1963) *Recent Progress in Microcalorimetry*, Pergamon Press. Oxford.

Cantor, C. R. and Schimmel, P.R. (1980) *Biophysical Chemistry*, W.H Freeman and Co. San Francisco, Vol 3, Chapt. 23.

Cesàro, A., Jerian, E. and Saule, S. (1980). Physicochemical studies of amylose and its derivatives in aqueous solution: thermodynamics of the iodine–triiodide complex, *Biopolymers*, 19, 1491–1506.

Cesàro, A., Benegas, J. C. and Ripoll, D. (1986) Molecular Model for the Cooperative Amylose-iodine-triiodide complex, *J. Phys. Chem.*, 90, 2787–2791.

Cesàro, A., Paoletti, S., Delben, F., Cavallo, S., Crescenzi, V. and Zevenhuizen, L. P. T. M. (1987) Thermoreversible gels of the capsular polysaccharide from Rhizobium trifolii strain TA-1, in V. Crescenzi, I.C.M. Dea, S.S. Stivala (eds), *Industrial Polysaccharides*, Gordon & Breach, N. Y., pp. 99–109.

Cesàro, A., Esposito, P., Bertocchi, C. and Crescenzi, V. (1989) The influence of side-chain modifications on the solution behavior of the capsular polysaccharide from *Rhizobium trifolii* strain TA-1, *Carbohydr. Res.*, 186, 141–155.

Cesàro, A., Gamini, A. and Navarini, L. (1992) Supramolecular Structure of Microbial Polysaccharides in Solution: from Chain Conformation to Rheological Properties, *Polymer,* 19 4001–4008.

Cesàro, A., Angioletti, C. and Ruggiero, J. R. (2002) Amylosics Chain Conformation and Dynamics in V. P. Yuryev, A. Cesàro, W. J. Bergthaller (eds.) *Starch and Starch Containing Origins.*, Nova Publishers, New York 2002, pp. 3–22.

Chandrasekaran, R., Lee, E. J., Radha, A. and Thailambal, V. G. (1992) Correlation of Molecular Architecture with Physical Properties of Gellan Related Polymers in R. Chandrasekaran (ed.), *Frontiers in Carbohydrate Research*, vol. 2, Elsevier Science Publ., London, pp. 65–84.

Christensen, B. E., Knudsen, K. D., Smidsrod, O., Kitamura, S. and Takeo, K. (1993) Temperature-Induced Conformational Transition in Xanthans with Partially Hydrolyzed Side Chains, *Biopolymers,* 33 151–161.

Cooke, D., Gidley, M. J.and Hedges, N. D. (1996) Thermal properties of polysaccharides at low moisture. II. Molecular order and control of dissolution temperature in agar, *J. Thermal Anal.,* 47 1485–1498.

Cosani A., Terbojevich, M. and Bertocchi, C. (1989) Exocellular Polysaccharides from *Rhizobium Leguminosarum* strain 1044-2 and strain Ng1a: Solution properties, in V. Crescenzi, I. C. M. Dea, S. Paoletti, S. S. Stivala and I. W. Sutherland (eds.) *Biomedical and Biotechnological Advances in Industrial Polysaccharides*, Gordon & Breach, Amsterdam, pp. 175–183.

Crescenzi, V., Dentini M. and Coviello, T. (1987) Solution Properties of Typical Microbial Polysaccharide Polyelectrolyttes in V. Crescenzi, I. C. M. Dea and S. S. Stivala (eds.) *Industrial Polysaccharides,* (Gordon & Breach, N Y) 1987, 69–97.

Crescenzi, V., Dentini, M., Coviello, T., Paoletti, S., Cesàro, A. and Delben, F. (1987) On the Solution and Gelling Behaviour of Typical Bacterial Polysaccharides, *Gazz. Chim. Ital.,* 117, 611–616.

Faleschini, P. (1988) Studio chimico-fisico delle proprietŕ in soluzione di alcuni polisaccaridi esocellulari da Rhizobium, D. Thesis, Trieste University.

Fidanza, M., Dentini, M., Crescenzi, V. and Del Vecchio, P. (1989) Influence of charged groups on the conformational stability of succinoglycan in dilute aqueos-solution, *Int. J. Biol. Macromol.,* 11, 372–376.

Freire, E. and Biltonen, R. L. (1978) Statistical Mechanical Deconvolution of Thermal Transitions in Macromolecules. I. Theory and Application to Homogeneous Systems; and II. General Treatment of Cooperative Phenomena, *Biopolymers,* 17, 463–479 and 481–496.

Flory, P. J. (1969) *Statistical Mechanics of Chain Molecules*, Interscience, New York, Chapt. 7.

Fujii, T., Yano, T., Kumagai, H. and Miyawaki, O. (2000) Scaling analysis on elasticity of agarose gel near the sol-gel transition temperature, *Food Hydrocolloids,* 14, 359–363.

Gidley, M. J., Dea, I. C. M., Eggleston, G. and Morris, E. R. (1987) Structure and Gelation of *Rhizobium* Capsular Polysaccharide, *Carbohydr. Res.,* 160, 381–396.

Grinberg, V. Y., Burova, T. V., Grinberg, N. V., Mashkevich, A. Y., Plashchina, I. G., Usov, A. I., Shusharina, N. P., Khokhlov, A. R., Navarini, L. and Cesàro, A. (2003) Thermodynamics of the double helix-coil equilibrium in tetramethylammonium gellan: High-sensitivity differential scanning calorimetry data, *Macromol. Biosci,.* 3, 169–178.

Hacche, L. S., Washington, G. E. and Brant, D. A. (1987) Light-scattering investigation of the temperature-driven conformation change in xanthan, *Macromolecules,* 20, 2179–2187.

Hirao, T., Sato, T., Teramoto, A., Matsuo, T. and Suga. H. (1990) Solvent effect on the cooperative order-disorder transition of aqueous solution of schizophyllan, a triple-helical polysaccharides, *Biopolymers,* 29, 1867–1876.

Holtzer, A. (1997) Persistent Confusion on the van't Hoff Equation, *Biopolymers,* 42, 499–503.

Holzwarth, G. and Ogletree, J. (1979) Pyruvate-free xanthan, *Carbohydr. Res.,* 76, 277–281.

Itou, T., Teramoto, A., Matsuo, T. and Suga, H. (1986) Ordered structure in aqueous polysaccharide. 5. Cooperative order-disorder transition in aqueous schizophyllan, *Macromolecules,* 19, 1234–1240.

Itou, T., Teramoto, A., Matsuo, T. and Suga, H. (1987) Isotope effect on the order-disorder transition in aqueous schizophyllan, *Carbohydr. Res.,* 160, 243–257.

Kaneda, I., Kobayashi, A., Miyazawa, K. and Yanaki, T. (2002) Double helix of Agrobacterium tumefaciens succinoglycan in dilute solution, *Polymer,* 43, 1301–1305.

Kawakami, K., Okabe, Y. and Norisuye, T. (1991) Dissociation of dimerized xanthan in aqueous solution, *Carbohydr. Polymers,* 14, 189–203.

Keller, A., Goldbeck-Wood, G. and Hikosaka, M., (1993) Polymer Crystallization: Survey and New Trends with Wider Implications for Phase Transformations, *Farady Discuss.,* 95, 109–128.

Kidokoro, S. I. and Wada, A. (1987) Determination of Thermodynamic Functions from Scanning Calorimetry Data, *Biopolymers,* 26, 213–229.

Kitamura, S. and Kuge, T. (1989) A differential scanning calorimetric study of the conformational transitions of schizophyllan in mixture of water and dimethylsulfoxide, *Biopolymers,* 28, 639–654.

Kitamura, S., Takeo, K., Kuge, T. and Stokke, B. T. (1991) Thermally induced conformational transition of double-stranded xanthan in aqueous salt solution, *Biopolymers,* 31, 1243–1255.

Kitamura S., Hirano T., Takeo K., Fukada, H., Takahashi, K., Falch, B. H. and Stokke, B. (1996) Conformational transitions of schizophyllan in aqueous alkaline solution, Biopolymers, 39, 407–416.

Lapasin R. and Pricl S. (1995) Rheology of industrial polysaccharides, Chapman & Hall, London Chapt. 4.

Lee, E. J. and Chandrasekaran, R. (1992) The pseudo double-helical structure of the gel-forming capsular polysaccharide from Rhizobium trifolii, Carbohydr. Res., 231, 171–183.

Lee, H. C. and Brant, D.A. (2002) Rheology of concentrated isotropic and anisotropic xanthan solutions: 1. A rodlike low molecular weight sample, *Macromolecules,* 35, 2212–2222.

Lee, H. C. and Brant, D.A. (2002) Rheology of concentrated isotropic and anisotropic xanthan solutions: 2. A semiflexible wormlike intermediate molecular weight sample, *Macromolecules,* 35, 2223–2234.

Lee, H. C. and Brant, D. A., (2002) Rheology of concentrated isotropic and anisotropic xanthan solutions: 3. Temperature dependence, *Biomacromolecules,* 3, 742–753.

Liu, Y. F. and Sturtevant, J. M. (1995) Significant discrepancies between van't Hoff and calorimetric enthalpies. 2, *Protein Sci.* 4, 2559–2661.

Liu, Y. F. and Sturtevant, J. M. (1997) Significant discrepancies between van't Hoff and calorimetric enthalpies. 3, *Biophys. Chem.* 64, 121–126.

Maaloum, M., Pernodet, N. and Tinland, B. (1998) Agarose gel structure using atomic force microscopy: Gel concentration and ionic strength effects, *Electrophoresis,* 19, 1606–1610.
Matveev, Y. I., van Soest, J. J. G., Niemand, C., Wasserman, L. A., Protserov, V. A., Ezernitskaj, M. and Yuryev, V. P. (2001) The relationship between thermodynamic and structural properties of low and high amylose maize starches, *Carbohydr Polymers,* 44, 151–160.
Mazen, F., Milas, M. and Rinaudo, M. (1999) Conformational transition of native and modified gellan, *Int. J. Biol. Macromol.,* 26, 109–118.
Matveev, Y. I., van Soest, J. J. G., Nieman, C., Wasserman, L. A., Protserov, V., Ezernitskaja, M. and Yuryev, V. P. (2001) The relationship between thermodynamic and structural properties of low and high amylose maize starches, *Carbohydr. Polymers,* 44, 151–160.
McIntire, T. M. and Brant, D. A. (1997) Imaging of individual biopolymers and supramolecular assemblies using noncontact atomic force microscopy, *Biopolymers,* 42, 133–146.
McIntire, T. M. and Brant, D. A. (1998) Observations of the (13)—D-glucan linear triple helix to macrocycle interconversion using noncontact atomic force microscopy, *J. Am. Chem. Soc.,* 120, 6909–6919.
Miyoshi, E., Takaya, T. and Nishinari, K. (1995) Effects of salts on the gel-sol transition of gellan gum by differential scanning calorimetry and thermal scanning rheology, *Thermochim. Acta,* 267, 269–287.
Miyoshi, E., Takaya, T. and Nishinari, K. (1995) Gel-sol transition in gellan aqueous solutions, *Macromol. Symp.,* 99, 83–91.
Miyoshi, E., Takaya, T. and Nishinari, K. (1996) Rheological and thermal studies of gel-sol transition in gellan gum aqueous solutions, *Carbohydr. Polymers,* 30, 109–119.
Morris, E.R., (1995) Polysaccharide rheology and in-mouth perception in Alistair M Stephen ed. *Food polysaccharides and their applications* M. Dekker, New York chapt., 16, 517–546.
Naghibi, H., Tamura, A., Sturtevant, J. M. (1995) Significant discrepancies between van't Hoff and calorimetric enthalpies. *Proc. Natl. Acad. Sci. USA* 92, 5597–5599.
Nakanishi, T. and Norisuye, T. (2003) Thermally induced conformation change of succinoglycan in aqueous sodium chloride, *Biomacromolecules,* 4, 736–742.
Norisuye, T., Yanaki, T. and Fujita, H. (1980) Triple Helix of of *Schizophyllum commune* polysaccharide in aqueous solution, *J. Polym. Sci. Polym. Phys. Ed.,* 18, 547–558.
Paoletti, S., Cesàro, A. and Delben F. (1983) Thermally Induced Conformational Transition of Xanthan Polyelectrolyte, *Carbohydr. Res.,* 123, 173–178.
Paoletti, S., Cesàro, A., Delben, F., Crescenzi, V. and Rizzo R. (1985) Polyelectrolytic Aspects of Conformational Transitions and Interchains Interactions in Ionic Polysaccharide Solutions: Comparison of Theory and Microcalorimetric Data. in P. Dubin (ed) *Microdomains in Polymer Solutions*, Plenum Press, New York, pp.159–189.
Poland, D. (1978) Cooperative equilibria in physical biochemistry, Oxford University Press, Oxford.
Privalov, P. L. (1980) Heat Capacity Studies in Biology, in A.E. Beezer Ed. *Biological Microcalorimetry*, AcademicPress, London, p.413–451.
Privalov, P. L. (1980) Scanning microcalorimeters for studying macromolecules, *Pure & Appl. Chem.,* 52, 479-497.
Privalov, P. L. and Potekhin, S. A. (1986) Scanning Microcalorimetry in Studying Temperature-Induced Changes in Proteins, *Methods in Enzymology,* 131 4–51.

Ragone, R. and Colonna, G. (1995) Reliability of the van't Hoff Plots, *J. Phys. Chem.*, 99, 13050–13050.

Rao, V. S. R.; Qasba, P. K.; Balaji, P. V.; Chandrasekaran, R. *Conformation of Carbohydrates*; Harwood Academic Publ.: Amsterdam, 1998; Chapter 8, pp 258–263 and references therein.

Rendleman, J. A. Jr (2003) The reaction of starch with iodine vapor. Determination of iodide-ion content of starch-iodine complexes, *Carbohydr. Polym.*, 51, 191–202.

Ridout, M. J., Brownsey, G. J., York, G. M., Walker, G. C. and Morris, V. J. (1997) Effect of o-acyl substituents on the functional behaviour of Rhizobium meliloti succinoglycan, *International Journal of Biological Macromolecules,* 20, 1–7.

Robert, C. H., Colosimo, A. and Gill, S. J. (1989) Allosteric Formulation of Thermal Transitions in Macromolecules, Including Effects of Ligand Binding and Oligomerization, *Biopolymers,* 28, 1705–1729.

Rochas, C. (1987) Calorimetric study of galactans, *Food Hydrocolloids,* 1, 215–225.

Sato, T., Norisuye, T. and Fujita, H. (1983) Triple helix of *Schizophyllum comune* polysaccharide in dilute solution. 5. Light scattering and refractometry in mixtures of water and dimethyl sulfoxide, *Macromolecules,* 16, 185–189.

Shatwell, K. P., Shuterland, I. W., Dea, I. C. M. and Ross-Murphy, S.B. (1990) The influence of acetyl and pyruvate substituents on the helix-coil transition behaviour of xanthan, *Carbohydr. Res.,* 206, 87–103.

Shatwell, K. P., Shuterland, I. W. and Ross-Murphy, S. B. (1990) Influence of acetyl and pyruvate substituents on the solution properties of xanthan polysaccharide, , *Int. J. Biol. Macromol.,* 12, 71–78

Stokke, B. T., Elgsaeter, A. Kitamura, S., (1993) Macrocyclization of Polysaccharides Visualized by Electron Microscopy *Int. J. Biol. Macromol.*, 15, 63–68.

Straume, M. and Freire, E. (1992) Two-dimensional differential scanning calorimetry, *Anal.Biochem.,* 203, 259–268.

Yoshiba K., Ishino T., Teramoto A., Nakamura, N., Miyazaki, Y., Sorai, M., Wang, Q., Hayashi, Y., Shinyashiki, N. and Yagihara, S. (2002) Ordering in aqueous polysaccharide solutions. II. Optical rotation and heat capacity of aqueous solutions of a triple-helical polysaccharide schizophyllan, Biopolymers, 63, 370–381.

Zimm, B. H. and Bragg, J. K. (1959) Theory of the Phase Transition between Helix and Random Coil in Polypeptide Chains, *J. Chem. Phys.,* 31, 526–535.

Waigh, T. A., Gidley, M. J., Komanshek, B. U. and Donald, A. M.(2000) The phase transformations in starch during gelatinisation: a liquid crystalline approach, *Carbohydr. Res.,* 328, 165–176

Weber, G. (1996) Persistent Confusion of Total Entropy and Chemical System Entropy in Chemical Thermodynamics, *Proc. Natl. Acad. Sci. USA,* 93, 7452–7453.

Chapter 2

Thermal analyses and combined techniques in food physical chemistry

A. Schiraldi*

DISTAM, University of Milan, Via Celoria 2, 20133 Milano, Italy

Introduction

Today's Thermal Analyses (TA) represent a very wide panoply of methods that allow accurate monitoring of several physical and chemical properties of a given system, which are directly affected by temperature changes. Some of these methods have been adapted to isothermal investigations and therefore are employed to check changes that take place in a given lapse of time.

The physical principles underlying thermal analyses apply to every physical state, namely gaseous, liquid and solid, provided that suitable sample holders are used. However, depending on the kind of specific physical and/or chemical property, or the physical state of the sample, different sensitivities are required. These can be achieved by improving the performances of the detectors, like Peltier elements in a calorimeter or the lever drift in a thermo-balance, but improved sensitivity is often attained through a suitable combination with another instrument, like a mass spectrometer, a gas chromatographer, an IR spectrophotometer, etc..

These improvements therefore rely on the match between different apparatuses which allow detection of different physical properties of the system investigated. Typical examples are the combinations between a thermo-balance and one instrument, which can be referred to as a specific sensor that allows the chemical analysis of the gaseous out-stream of the thermo-balance. The 'sensor' is kept at constant temperature and therefore remains unaffected by the temperature changes experienced by the sample. Other combinations allow the simultaneous evaluation of two different properties, which are both affected by temperature. The best example is the DSC-TG coupling that allows the check of mass loss and related thermal effect from the same sample.

The combination of traditional thermal analyses with other techniques is recommendable when studying food systems that always demand a careful evaluation of the instrumental outputs which usually are not easy to interpret. For example, the DSC trace of a fresh cheese (Fig. 1) shows a number of humps that

* alberto.schiraldi@unimi.it

D. Lörinczy (ed.), The Nature of Biological Systems as Revealed by Thermal Methods, 31–48.

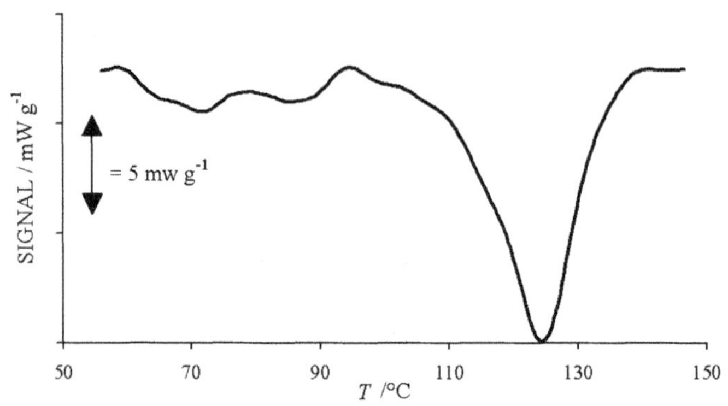

Fig. 1 DSC trace from a fresh cheese sample. The series of endotherms is related to the unfolding of fragments of partially proteolysed casein

cannot be referred to as 'noise' since each of them has (or can have) a precise physical meaning.

Foods are indeed a challenge for a researcher who aims at their characterization because they are multi-component and multi-phase systems. This means that any thermal treatment can produce displacements of compounds from one phase to another and, at the same time, trigger phase transitions. As a consequence, the study of a food system should be approached the same way as that of a whole phase diagram. A further complication comes from the fact that real foods usually host micro-organisms which can substantially contribute to the overall variability of the system when temperature is changing, or when the rate of an experimental run is comparable with the rate of the microbial growth (or death).

This chapter is a short review of the TA and relevant combined techniques, that can be of help in the study of food and related systems. Some combinations of instruments are promising arrangements that still require further adjustments to become of practical use: they are included in this review to stimulate readers to suggest possible enhancements. Each section of this presentation deals with a specific TA and its combination with other techniques.

DSC and isothermal calorimetry

It is well known that the interpretation of any DSC trace requires a preliminary assessment of the base line. The base line of the DSC trace of many food samples can be rather 'irregular', when compared to that underlying the DSC peak of the fusion of Indium (the usual standard compound). This simply means that food samples undergo changes of the heat capacity with no parallel changes of enthalpy. This kind of processes can take place several times in the course of the temperature scan, since they are relevant to different phases of the food system.

It is therefore expedient to split the trace into regions, each relevant to a given main 'signal' (like an endo- or exo-thermic peak), which are to be ana-

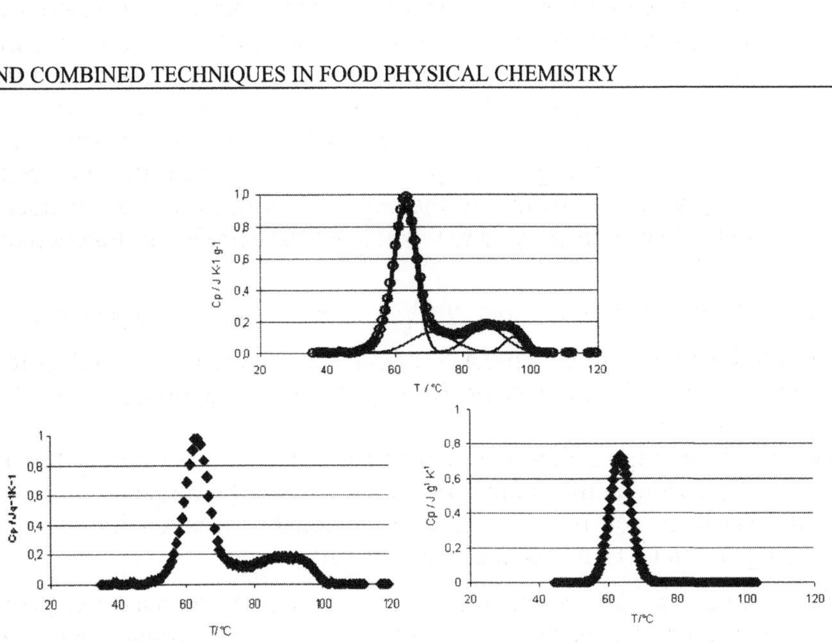

Fig. 2 Starch gelatinization in a rice kernel suspended in excess water. The relevant endothermic signal has been singled out from the overall trace by splitting it into four gaussian functions

lysed separately from one another. Within each region the base line trend can be tentatively defined with a SP-line (or even a straight line) across peak shaped signals, or with a sigmoid function when a simple transition is supposed to take place. Once the trace is accordingly scaled, it can be split into the minimum number of gaussian functions to attain an acceptable fit ($P < 0.05$). This treatment allows a tentative estimation of the enthalpy associated to each peak and the progress of the relevant transformation across the respective temperature span (Fig. 2).

The single peak can be finally interpreted according to either a thermodynamic or a kinetic approach. Comparison with DSC signals obtained from pure compounds can be of help to improve the analysis of each gaussian peak.

The glass transition temperature, T_g, is another quantity that can be determined with a DSC investigation. This transition is actually spread over a relatively wide temperature range and corresponds to the relaxation of the translation degrees of freedom within the sample investigated. The process is therefore accompanied by an increase of the heat capacity which produces the endothermic shift of the base line of the DSC trace. As a typical finding, the glass transition is followed by other phenomena (Fig. 3) that are sustained by the increased molecular mobility, like crystallization of ice (exothermic peak) or enthalpy relaxation (endo- or exothermic peak).

Food systems that undergo depletion of liquid solvent (water) on freezing host a residual liquid phase with increased viscosity where nucleation and growth of crystal phase (ice) is hindered: this liquid forms a glassy phase on further cooling [2]. The same physical interpretation applies to solvent poor systems where the large viscosity hinders the formation of crystals of the 'solute' (e.g., sugars)

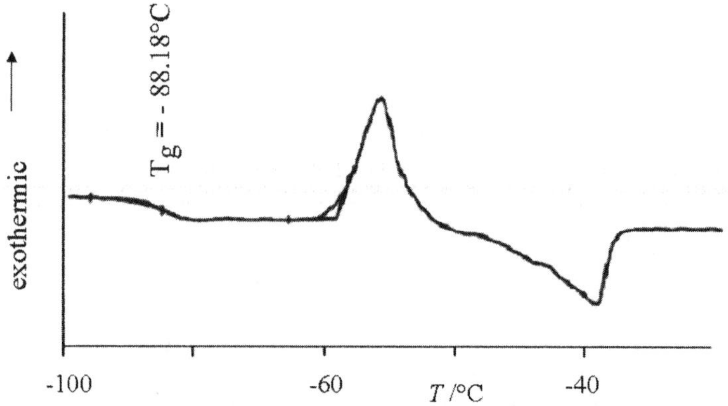

Fig. 3 DSC record from a vegetal. Upper trace: the base-line shift at Tg is followed by an exothermic peak related to a partial ice formation; the endothermic peak, at higher temperature, is related to ice melting (modified from ref. [1])

[3]. But glass transition is also observed in systems where no solvent is present: the glass transition indeed occurs in every polymer material that is brittle for $T < T_g$ and rubbery for $T > T_g$. Many foods contain biopolymers like amylose, amylopectin, gluten, etc. Therefore typical T_g shifts of the base line are found in the DSC traces of relatively dry food samples.

It is well known from polymer science that compounds with a small molecular mass can directly affect the molecular mobility of larger molecules and therefore modify the overall viscosity of the system [4], as revealed by changes of the glass transition temperature. The ubiquitous compound responsible for such effects in foods is water. It therefore is of interest the study of the aqueous binaries of a number of compounds by defining the relevant state diagram in the T–*vs.*–c(w) plane, where c(w) is the water content. The curve that fits the T_g–*vs.*–c(w) trend separates the underlying glassy region, where because of the low molecular mobility and high viscosity no transition or reaction can take place, from the upper region of the diagram (Fig. 4) where these changes can occur [5, 6].

A scheme of the various regions of the diagram can be summarized as follows. A *liquidus* curve fits the freezing points of water-rich binaries: it starts from $T = 273.15$ K for pure water and bends down with increasing solute content until it intercepts the T_g–*vs.*–c(w) curve in the point [T_g', c'(w)], that is the lowest temperature at which a liquid phase can be observed in the presence of ice crystals. At higher solute contents the viscosity of the solution would be too high for any further ice nucleation and the expected eutectic point cannot be attained. T_g' accordingly is the T_g of maximally freeze-concentrated solutions.

For c(w) < c'(w) and $T \leq T_g'$ the system is an amorphous glass. For c(w) > c'(w) and $T < T_g'$ ice can still nucleate and grow although at a much lower rate. When a sample with c(w) > c'(w) is thawed (at a given rate) from $T < T_g$, the DSC trace shows a first endothermic shift of the base-line at $T = T_g$, which can be immediately followed by an exothermic wave that corresponds to ice crystal-

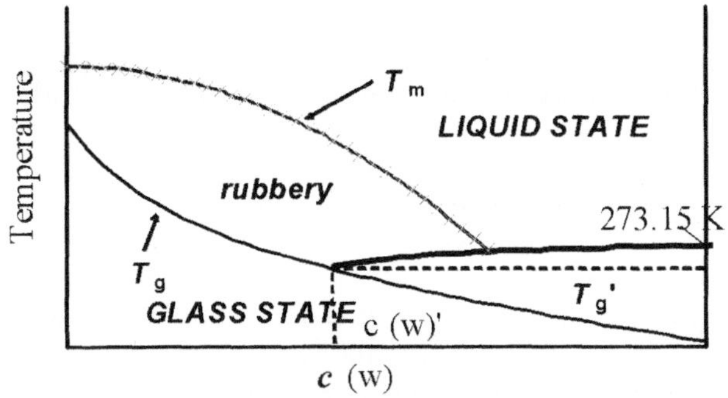

Fig. 4 Schematic view of a state diagram of an aqueous binary system that undergoes glass transition because of the high viscosity that hinder nucleation and growth of crystal phases. See text for lettering

lization. When T_g' is attained some ice melting takes place as revealed by an endothermic peak: the larger the *c*(w) the broader the endotherm. An example of such a behaviour is given in Fig. 3. The DSC trace for samples with *c*(w) < *c*'(w) shows an endothermic shift of the base-line at $T = T_g$ which can be followed by other signals according to the nature of the solute. In the case of simple compounds, like sugars and pure polysaccharides, a broad endothermic peak is observed which corresponds to the solubilization into the liquid phase. The solubility curve (*T*m-*vs.*-*c*(w) in Fig. 4). bends down from the melting point of the pure solute (when it actually exists) and crosses the curve of primary ice separation at the point [T_g', *c*'(w)]: DSC investigations indicate that this intersection may often occur at $T > T_g$' and *c*(w) > *c*'(w), as shown in the diagram in Fig. 4.

A number of applications and/or phenomena of technological interest, like freeze-drying, caking of powders, cryo-preservation, etc., have been described on the basis of the relevant phase diagrams [7], as well as an application to extrusion processing of flour [8]. A number of papers [9–16] therefore appeared where various experimental approaches to T_g, like DSC, TMA DTMA, NMR, ESR, fluorescence and phosphorescence decay, etc., were reported and sometimes compared to each other. It however should be emphasized that some spectroscopic techniques, like NMR and ESR, reveal changes related to short range molecular mobility within a 10^{-3}–10^{-6} s time scale, being practically blind for the macroscopic modifications of viscosity and specific heat at the operator's time scale which can be detected through other approaches, like thermal analyses. The comparison of the results of different techniques should therefore be considered taking into account the relevant time scale involved.

A particular emphasis has been recently given to the application of the Modulated Temperature DSC (MTDSC) to separate reversing from non-reversing heat-flow signals obtained from food systems. In MTDSC a sinusoidal temperature fluctuation is superimposed on a main increase of the temperature at a constant heating rate, β_o, so that the overall change of *T* is described by the expression

$$T = T_o + \beta_o t + A \sin(\omega t)$$

Where T_o, t, A, stand for starting temperature, time and amplitude of the temperature fluctuation, respectively, while $\omega = 2\pi\nu$ accounts for the frequency, ν, of the fluctuation. The instantaneous heating rate is therefore

$$\beta = \beta_o + A\ \omega \cos(\omega t)$$

When $A \leq (\beta_o\ \nu)/2\pi$, β is always positive. The modulated temperature program produces a heat flow trace, HF, that fluctuates with the same frequency, ν, and is the sum of two components, dubbed *reversible* and *non reversible*, respectively:

$$HF_{tot} = HF_{rev} + HF_{non\ rev}.$$

where HF_{tot} corresponds to the trace that would be obtained in a traditional DSC run performed at β_o heating rate. A Fourier analysis allows these components to be separated in the form of two orthogonal heat capacities,

$$C_p' = |C^*_p| \cos \alpha$$

$$C_p'' = |C^*_p| \sin \alpha$$

where the $|C^*_p|$ is the modulus of the heat capacity corresponding to the ratio between the amplitude of HF oscillation, A_{HF}, and the amplitude of heating rate, $(A \times \omega)$. The reversible heat flow is $HF_{rev} = \beta_o\ C_p'$.

A major information drawn from MTDSC is relevant to the heat capacity drop observed at T_g from the stress-relaxation endotherm (non-reversing signal) that is often observed on heating samples previously cooled at subzero temperatures, like frozen doughs [17]. The changes of the relaxation enthalpy are worth determining since they are related to the residual molecular mobility in quenched products and therefore with their stability and shelf-life. It has to be noticed that starch gelatinisation is seen as a totally *irreversible* process.

When dealing with an aqueous solution of a biopolymer, several conformational changes can take place above the T_g threshold, like formation of entangled chain gel, gel-sol transition, thermosetting, etc., according to the chemical nature of the compound [8]. This often implies large changes of mechanical properties: it can be clearly demonstrated by coupling the DSC record with that of volume dilation (Fig. 5).

An interesting combination of DSC concerns X-ray diffraction. Synchrotron radiation is employed in these investigations. The small (SAXS) and wide (WAXS) angle X-ray scattering are of interest in food systems, like starch gels [18, 19] and fats [20]. The X-ray beam of given wavelength is conveyed toward the calorimetric cell that is a glass capillary (Ø ~ 1.5 mm) of about 20 mL volume. Fig. 6 reports a sketched view of the very complex instrumental apparatus [21].

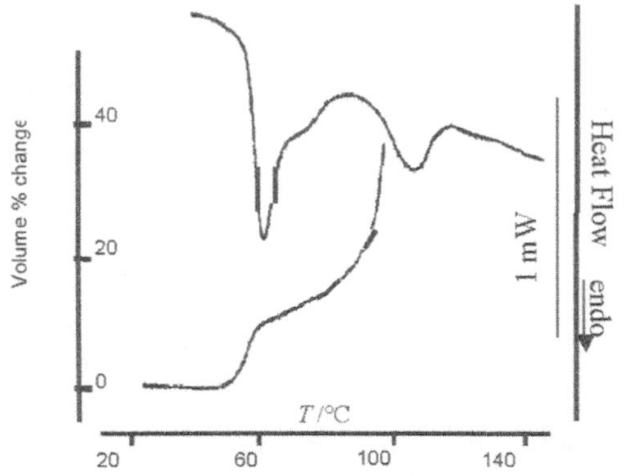

Fig. 5 DSC and TMA (Thermo-Mechanical Analysis) traces obtained at 10 and 2°C min-1 heating rate, respectively, from rice starch with 50% w/w moisture (modified from [9])

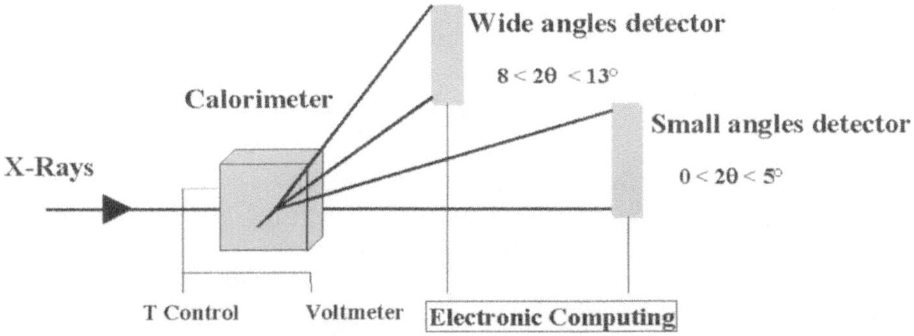

Fig. 6 Synchrotron X-ray / DSC combination. For details see ref [21]

Amylose crystallisation can be monitored during the isothermal annealing at adequate temperatures. The exothermic effect is indeed hard to detect, but the growth of B- and V-amylose crystal structure [22] can be neatly recognised in the relevant X-ray diffraction pattern. The same kind of information comes from the investigation of cocoa butter and milk fat, that contain large amounts of triglycerides showing a monotropic polymorphism related to the time and temperature of annealing [20].

In the case of amylose the DSC signal mainly concerns the fusion of amylose lipid complexes (Fig. 7), while in the case of triglicerides DSC traces are the resultant of the fusion of many coexisting crystal forms that have different melting points but very close fusion enthalpies (Fig. 8).

The combination of X-ray diffraction with DSC allows in both cases a much better view of the transformations of interest, that are directly related to the quality of the hosting food, like bread, chocolate, milk fats, etc.

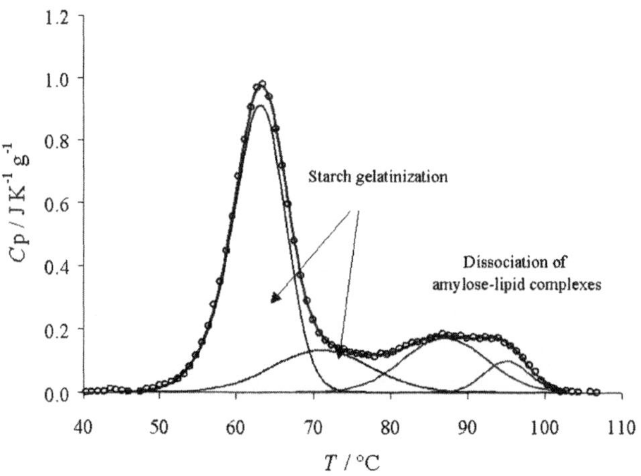

Fig. 7 Starch gelatinisation followed by the fusion of amylose-lipid complexes in a wheat flour dough sample

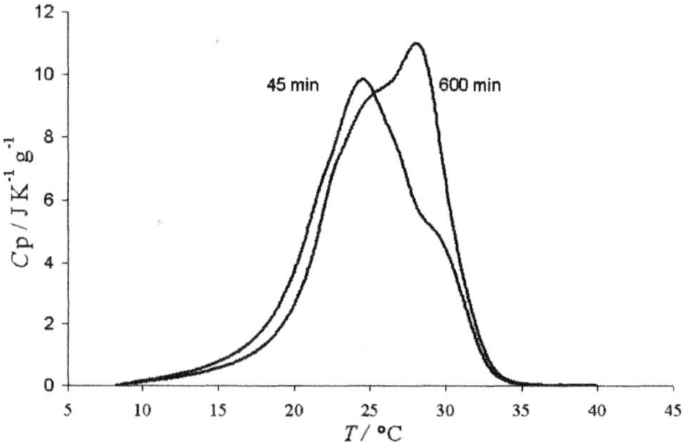

Fig. 8 Fusion endotherms of cocoa butter samples annealed at 5°C for 45 and 600 min after quenching from the liquid state

Another combination of DSC with gas analysis allows the characterization of processes with emission of volatile compounds. The instrument, that allows the simultaneous evaluation of the thermal effect and the amount of gas released, requires a fitting with a gas-chromatographer or a mass spectrometer. For this reason Calvet calorimeters are the most suitable. The gas coming out from the open sample cell is conveyed to the gas chromatographer. An independent circulation of inert gas flows through the reference cell. The combination with a mass spectrometer allows a more rapid identification of the released volatiles. The sample and reference materials are under continuous flow of gas during the experiment. The gas collection is performed at normal pressure by means of a capillary, which is kept hot to prevent condensation of the volatiles before the injection into the mass spectrometer inlet.

Simulation of heat treatment is among the aims of investigations on foods. Since most of these treatments, like cooking and baking, are carried out, or practically occur, in isothermal conditions, the investigations aimed to simulate the process should be carried out at constant temperature. Isothermal Calorimetry (IC) can be of help. The best approach requires the use of a calorimeter that can host sufficiently large cells (e.g., about 10 mL); as a standard procedure, one should first thermally equilibrate at the desired temperature the sample to be investigated and the an empty calorimetric vessel where it has to be dropped, in order to avoid initial misbalance of the instrumental output (Fig. 9). The time lag of the instrument is to be accounted before the treatment of the results (see appendix in [23]). These procedures were employed to characterize several food systems and processes, like starch gelatinization and retrogradation in cereal products, milk pasteurization, egg white denaturation, pasta and rice isothermal cooking, microbial growth, etc. [24–27].

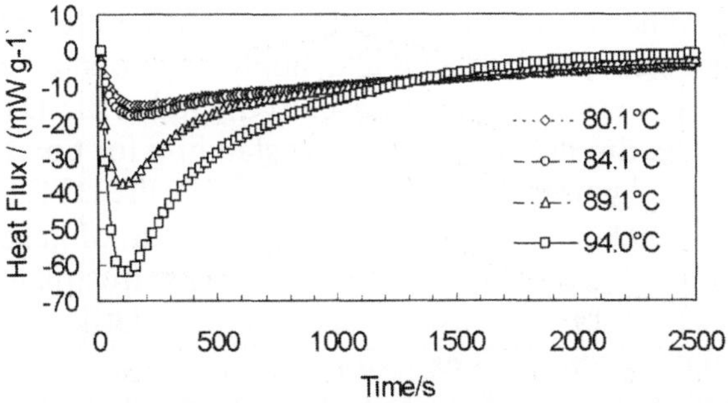

Fig. 9 Isothermal 'cooking' of rice at various temperature. The traces have to be scaled taking into account the time lag of the instrument at each single temperature [25]

Food spoilage and preparation of particular dairy foods, like cheeses, yoghurt, kyr, etc., are sustained by specific microbial activity. In these cases too IC can be of help when coupled with the traditional microbiology techniques (Fig. 10). Cultures of living organisms, like yeast and fermenting bacteria, are poured into the calorimetric cell under strictly defined conditions, to obtain a calorimetric trace that may be directly related [27] to growth and metabolic rate.

Thermogravimetry of food samples

Thermo-gravimetric analysis (TG) is the thermal analysis related to the mass loss during a temperature scan at a given heating rate. It therefore reveals the vaporization of volatiles that can be either original components of the food examined, or the products of the thermal decomposition of some of them. The released compounds stripped toward the exterior by a gentle nitrogen stream can be identified by conveying the out-flow into a gas chromatographer or an IR or

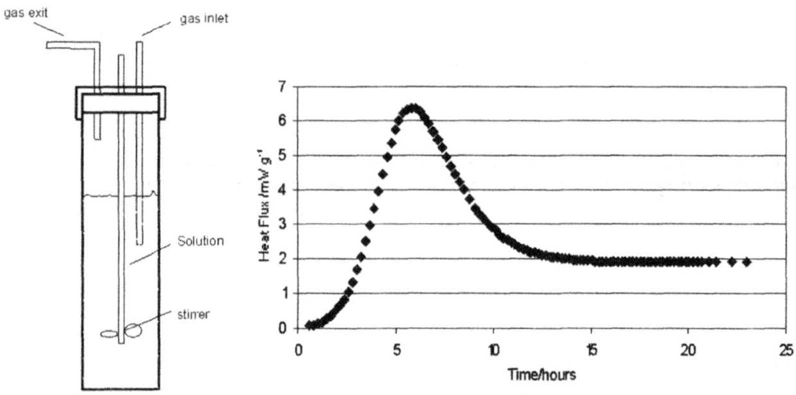

Fig. 10 Isothermal (28°C) calorimetric trace of a mixed culture *S. thermophilus* and *bulgaricus*. The 10 mL calorimetric cell is sketched to show the possibility to perform experiments under controlled atmosphere and various stirring speeds

mass spectrometer. When several volatiles leave the heated sample, the TG trace shows several down-steps, each related to a single compound.

The output of modern instruments usually includes the DTG, namely the trace of mass loss rate, dm/dt, *vs.* T, or *vs.* t (t standing for time). When several compounds are released within the same T (or t) range, the DTG record shows as a multi-peak profile, that can be deconvolved into a sum of gaussian or gaussian-like peaks, each related to a single compound. This treatment allows the assessment of the kinetics of the release for each single volatile. Food dehydration can be easily simulated with a TG experiment.

Significant improvements of classical TG can be achieved by matching the mass loss rate with the simultaneous record of the related DSC trace, that allows evaluation of the associated enthalpy (Fig. 11). In the case of food samples, most of the mass loss concerns released water, as confirmed by the finding that the associated enthalpy is always close to 2.2 kJ/g [28], namely the vaporization enthalpy of pure water.

The best application of TG in foods indeed concerns the water content. The presence of different water states can be detected with NMR and TG [29]. These techniques clearly indicate that mechanical stresses (e.g., mixing, kneading, etc.) are able to trigger water displacements within a given food. The relaxation back to the original partition can be much longer than the investigation time or the preparation process of a given final product.

Relaxation of nuclear spins is related either to spin-lattice, or spin-spin interactions, with relaxation times T1 and T2, respectively: The larger the relaxation time the higher the mobility of the molecules that host the relaxing nuclear spin.

The ^{1}H spin-spin relaxation signal recorded from dough samples with T2 in the millisecond range is mainly related to mobile water molecules that interact with flour components [30]. Protons of water molecules tightly bound to the substrates or protons of polymers have instead T2 in the range of microseconds.

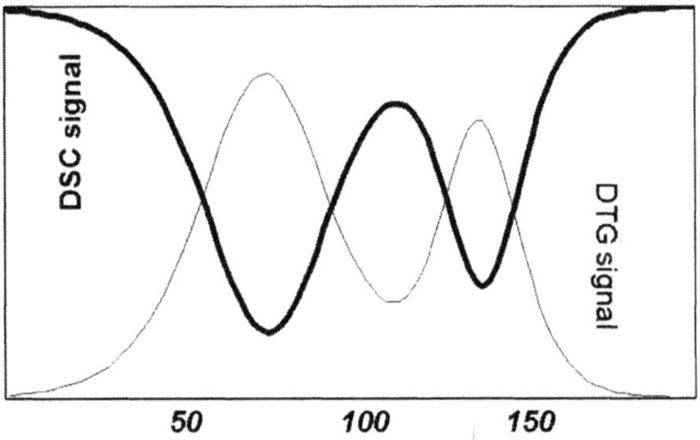

Fig. 11 TG-DSC trace of wheat dough at 2 K min^{-1} heating rate (modified from [28])

A CPMG (Carr-Purcell-Meiboom-Gill) pulse sequence therefore allows signals from mobile protons to be singled out.

The Free Induction Decay (FID) of the T2 relaxation of a given kind of nucleus can be reliably described with an exponential law. The overall FID detected from a dough sample can be split in three exponential components [31], each with its own T2 in the (2–5), (9–18), and (50–200) ms range, respectively. This means that at least three main kinds of water molecules can be distinguished because of differences of their local molecular mobility (Fig. 12).

The DTG trace of a wheat flour dough significantly changes when the starting moisture content is increased. The high temperature peak shifts toward lower temperature (Fig. 13) with increasing the overall dough moisture [28].

DTG traces of a freshly mixed dough and from a dough let at rest for a couple of hours are different. The high temperature peak shifts toward lower temperature (Fig. 14) when the dough is over mixed: an 'intermediate' shoulder appears between the main components of the signal, but it comes back to the starting position after a two hour rest [28].

The overall result of such a behaviour is that the DTG trace can be indeed used as a record of the water partition attained before the experimental run. The release of water that takes place in the course of the TG run strongly depends on the starting conditions of the system. The water partition is not substantially modified during the TG run in spite of important changes that take place, like starch gelatinization and gluten reticulation. This is because, during a TG run, starch and gluten transitions mainly involve the next neighbouring water molecules.

This why such a good agreement is found between the conclusions drawn from NMR and TG experiments, that should more often compared and combined to each other whenever the molecular mobility is poor and transitions imply short range displacements of the solvent.

Water displacements take place within the bread crumb during the shelf life in sealed bags [32]. TG reveals dramatic changes of the trace (Fig. 15).

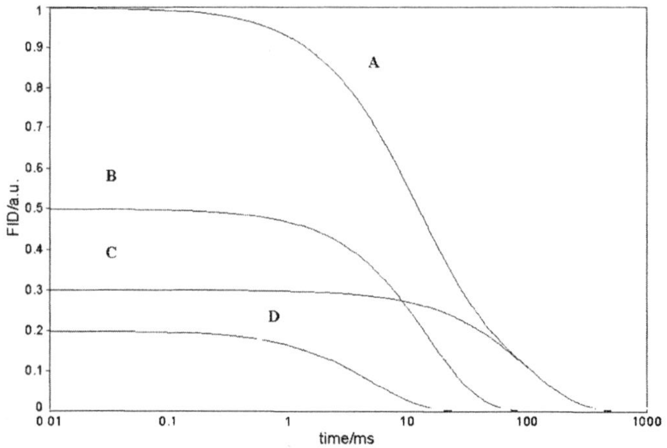

Fig. 12 FID Decay: (A) overall FID decay; (B) T2=15 ms component; (C) T2=100 ms component; (D) T2=5 ms component

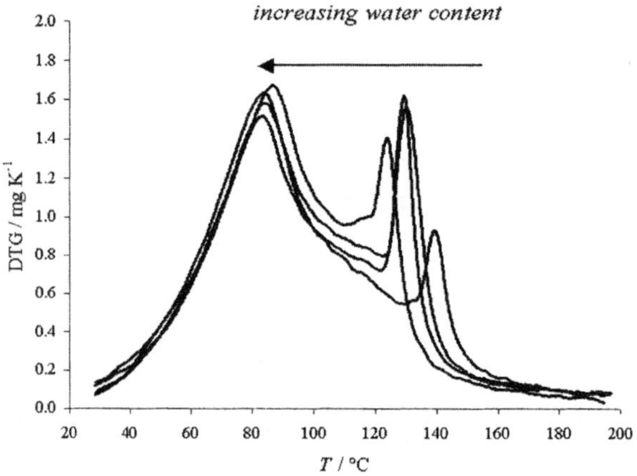

Fig. 13 DTG traces from wheat flour dough of different water content (from 40 to 50% *w/w*)

When the standard pans of the thermobalance are replaced with Knudsen cell and a dehydration experiment is carried out at constant temperature (25°C) under dynamic vacuum, the DTG trace obtained is directly related to the thermodynamic water activity [32]. Since almost every food can be referred to as a system far from the true thermodynamic equilibrium, the term 'relative humidity', RH, is indeed more appropriate than 'water activity'.

The traditional approaches to RH are usually based on the direct or indirect determination of the water vapour pressure in the relevant head space. As shown above, water in foods can be found in various states; because of the lack of a true thermodynamic equilibrium, such states imply different RH levels, that can remain unbalanced for hours or days. The highest RH state is often the only one

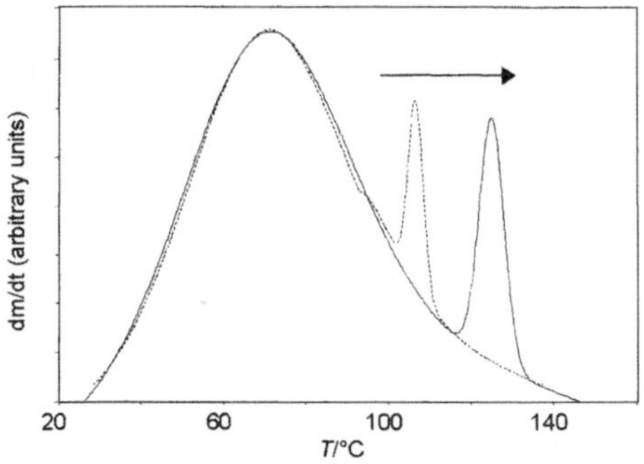

Fig. 14 DTG traces from an overmixed wheat flour dough. The high *T* peak occurs at a lower temperature (dashed curve), but it comes back in the traces from dough samples let at rest for a couple of hours

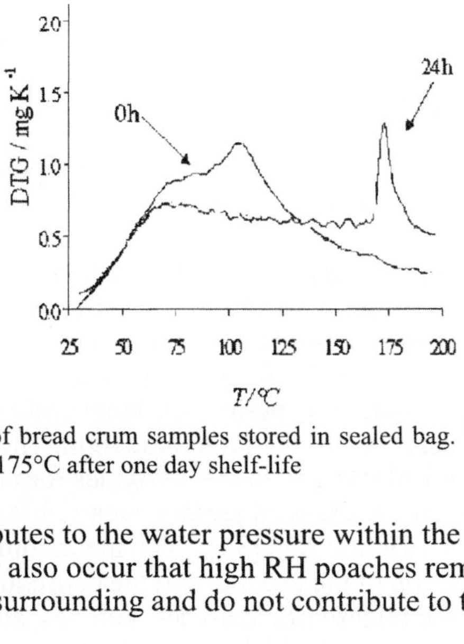

Fig. 15 DTG traces of bread crum samples stored in sealed bag. The high *T* peak is shifted up to 175°C after one day shelf-life

that actually contributes to the water pressure within the head space of the sample; however, it can also occur that high RH poaches remain imbedded in an almost impermeable surrounding and do not contribute to the water content of the head space.

As a result the RH detection is related only to the water fraction that has an easy access to the exterior atmosphere. Because of the water displacements, the situation can change in few hours or days, revealing that water properties in foods are time dependent.

Knudsen TG allows a quick detection of the apparent RH of a given food sample and, as expected, allows one to understand that only one fraction of the overall mixture actually contributes to the pressure in the head space. In the case of a wheat flour dough, dehydration during a Knudsen TG run (Fig. 16) involves only the moisture fraction that is responsible for the low-*T* DTG peak in the standard TG experiment (Fig. 13, 14).

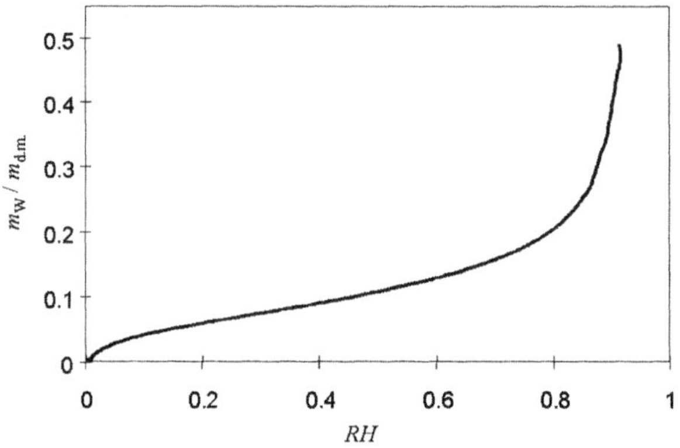

Fig. 16 Isothermal (25°C) dehydration of a bread crumb sample during a Knudsen TG run. The DTG trace was scaled with respect to that of pure water to obtain the corresponding RH, while the TG trace allowed evaluation of the water/dry matte mass ratio

Other applications of TG in food science and technology concern the monitoring of the thermal decomposition of many compounds, like carbohydrates, proteins and fats [34]. Guar and Xanthan gums decompose above 250°C with formation of carbonyl compounds; sugars undergo caramelization and, in the presence of proteins, the Maillard reaction, as in the case of honey. The decomposition of amino acids mainly produce CO, CO_2, alkylamines and HCN, as can be verified from the relevant mass spectrometry patterns.

The thermal degradation of lipids in the presence of air deserves a particular attention, since the early uptake of oxygen by the unsaturated bonds produces a mass gain after an induction period, the length of which may be referred to as a measure of the resistance of the fat to oxidation. The mass uptake is a direct measure of the chemical stability to the peroxides formed and the maximum oxidation rate depends on the exposed surface area (which means that oxygen diffusion determines the reaction rate [34]). The maximum of mass gain is followed by the decomposition of the product and the consequent mass loss, which mainly corresponds to the release of aldehydes, ketones, alcohols and esters.

The combination of TG with FTIR allows identification of volatile compounds released during a heating run. The cavity of the thermobalance is in connection with an IR spectrophotometer to which an inert gas flux conveys the volatile compounds stripped away from the sample cell. Some instruments provide the simultaneous record of TG, DTG, Heat Flow and IR spectrum. This combination is recommended to check the formation of simple volatile compounds which are produced in relatively large amounts, as in the case of main degradation processes (see above) during the TG run.

Other available instruments combine TG with a mass spectrometer (TG-MS). This approach is more sensitive and is therefore recommended when the volatiles produced have a rather low concentration (few ppm). The volatiles released by the sample at the atmospheric pressure are stripped by a flow of inert gas (He, N_2, Ar) and conveyed through a suitable orifice toward a chamber where the pressure is lower (about 10 Pa). Because of the pressure drop between the TG furnace and this chamber, the molecules attain a very high speed. A fraction of these molecules is collected through a sampling cone, under the action of a turbo-molecular pump that produces a vacuum of about 10^{-3} Pa, and conveyed to the ionisation chamber of the mass spectrometer. The connection tube must be heated in order to avoid condensation of the volatiles.

Mechanical and rheological properties

The stiffness and storage modulus, G' and G'', are of fundamental importance in the characterization of food systems. They change with temperature, water content and RH, and can be related to transitions or chemical modifications of the main components of a given food, those of polymer nature playing a major role. The system which behaves like a rigid brittle solid for $T < T_g$ becomes rubbery just above the T_g threshold as a result of the increased polymer segmental mobility. The rheological properties of the material are connected with these physico-chemical states. In recent years significant advances have been made in both the theory of viscoelasticity and the related instrumentation for Dynamic Mechanical Analysis (DMA) and Thermal Mechanical Analysis (TMA) became rather popular in food science investigations. These techniques can be combined with thermo-dielectric analysis and nuclear magnetic resonance, and other methods [31] to investigate glass transitions and their effect on mechanical properties, as well as molecular mobility and diffusivities in food ingredients and products.

A very common attitude is to look at the shape of tanδ – *vs.* –*T* trace ($\tan\delta = G' / G''$) which shows a maximum in the vicinity of T_g. There are good reasons to be very careful in such an approach [12] since the maximum of tanδ may not correspond to the maximum inflection point of G' and, above all, can occur at a different temperature when a different oscillation frequency is employed in the test. For this reason the combination with other techniques is highly recommendable [33].

Some instruments are more suitable to the study of 'solid' materials, while others are equipped with test heads that are designed for viscous fluids. Because of the main use in assessing the mechanical stiffness of some foods, TMA instruments can be referred to as 'texture meters' equipped to allow a temperature scan at a given heating rate. DMA instruments that involve oscillating stresses or strains can instead be referred to as rheometers that allow both frequency and temperature sweeping: this peculiarity make them suitable to predict, through the Boltzmann superposition principle, the behaviour of a given material in a frequency range that is not actually accessible. This information is of outmost importance to get some information of the mechanical properties at the molecular level.

An example of well characterized food systems is represented by bread doughs. The viscoelastic behaviour of a dough is non linear at all, except for small deformations. The G'–*vs.*–T trace shows an upward bending in the temperature range where starch gelatinization takes place [34] possibly because of major changes at the molecular level. This finding can be combined with those relevant to the dielectric constant and ^{1}H relaxation time T2 [31], which are related to the mobility of water molecules and changes of their short range surroundings. The combination of these different experimental approaches leads to the conclusion that water mobility decreases during the gelatinisation inducing a similar effect on the displacements of the other dough components.

References

1 Sá, M. M. and Sereno, A. M.: Glass transitions and state diagrams for typical natural fruits and vegetables, *Thermochim. Acta*, 246 (1994) 285–297.

2 Slade, L. and Levine, H.: Glass transition and water-food interactions, *Adv. Food Nutr. Res.*, 38 (1995) 103–269.

3 Roos, Y. H.: Phase transitions in foods, *Academic Press,* New York (1995).

4 Slade, L. and Levine, H.: Beyond water activity: recent advances based on an alternative approach to the assessment of food quality and safety, *Crit. Rev. Food Sci. Nutr.*, 30 (1991) 115–360.

5 Roos, Y. H.: Glass transition-related physico chemical changes in foods. Food Technology, Overview outstanding symposia in Food science and Technology (1995), October issue, 97–102.

6 Fessas, D. and Schiraldi, A.: State diagrams of arabinoxylan-water binaries, *Thermochim. Acta,* 370 (2001) 83–89.

7 Slade, L. and Levine, H.: Water and the glass transition.Dependence of the glass transition on composition and chemical structure: special implications for flour functionality in cookie baking, *J. Food Eng.,* 22 (1994) 143–188.

8 Kalentunç, G. and Breslauer K. J.: Glass transition of extrudates: relationship with processing-induced fragmentation and end-product attributes, *Ceral Chem.*, 70 (1993) 548–552.

9 Roudaut, G. Maglione, M. van Dusschoten, D. and Le Meste, M.: Molecular mobility in glassy bread: a multispectroscopy approach, *Cereal Chem.*, 76 (1999) 70–77.

10 Bizot, H. Le Bail, P. Leroux, B. Davy, J. Roger, P. and Buleon, A.: Calorimetric evaluation of the glass transition in hydrated, linear and branched polyanhydroglucose compounds, *Carbhydr. Polym.*, 32 (1997) 33–50.

11 Chinachoti, P. Kim-Shin, M. Mari, F. and Lo, L.: Gelatinization of wheat starch in the presence of sucrose and sodium chloride: correlation between gelatinizaton temperature and water mobility as determined by oxyen-17 nuclear magnetic resonance, *Cereal Chem.*, 68 (1991) 245–248.

12 Peleg, M.: A note of the tan $\delta(T)$ peak as a glass transition indicator in biosolids, *Rheol. Acta,* 34 (1995) 215–220.

13 Cocero, A. M. and Kokini, J. L.: The study of the glass transition of glutenin using small amplitude oscillatory rheologcal measurements and differential scanning calorimetry. *J. Rheol.*, 35 (1991) 257–270.

14 Johnson, J. M. Davis, E. A. and Gordon, J.: Interactions of starch and sugar water measured by electron spin resonance and differential scanning calorimetry, *Cereal Chem.*, 67 (1990) 286–291.
15 Shah, N. K. and Ludescher, R. D.: Phosphorescence probes of the glassy state in amorphous sucrose, *Biotechnol. Prog.*, 11 (1995) 540–544.
16 Russel, P. L.: Gelatinization of starches of different amylose/amylopectin content. A study by differential scanning calorimetry, *J. Cereal Sci.*, 6 (1987) 133–145.
17 Goff, H. D. Montoya, K. and Sahagian, M. E.: The effect of microstructure on the complex glass transition occurring in frozen glucose model systems and foods. In Amorphous food and pharmaceutical Systems. H. Levine Ed., Royal Soc. Chemistry, Cambridge, UK, 145–157.
18 Tester, R. F. and Debon, S. J. J.: Annealing of starch: a review, *Int. J. Biol. Macromol.*, 27 (2000) 1–12.
19 Le Bail, P. et al.,: Monitoring the crystallization of amylose-lipid complexes during maize starch melting by synchrotron X-ray diffraction, *Biopolymers,* 50 (1999) 99–110.
20 Loisel, C. Keller, G. Lecq, G. Bourgaux, C. and Ollivon, M.: Phase transitions and polymorphism of cocoa butter, *J. Am. Oil Chem. Soc.,* 75 (1998) 425–439.
21 Keller, G. Lavigne, F. Forte, L. Andrieux, K. Dahim, M. Loisel, C. Ollivon, M. Bourgaux, C. and Lesieur, P.: DSc and X–ray diffraction coupling: specifications and applications, *J. Therm. Anal. Cal.,* 51 (1998) 783–791.
22 Zobel, H. F.: Starch crystal transformations and their industrial importance, *Starch/Staerke,* 40 (1988) 1–7.
23 Schiraldi, A. Piazza, L. Fessas, D. and Riva, M.: Thermal Analyses In Foods And Food Processes. in 'Handbook of Thermal Analysis and Calorimetry'. R. Kemp Editor, Elsevier Publ. Amsterdam, The Netherlands, Vol. 4 'From Macromolecules to Man' chapter 16 (1999) 829–921.
24 Shiotsubo, T.: Changes in enthalpy and heat capacity associated with the gelatinisation of potato starch as evaluated from isothermal calorimetry, *Carbohydr. Res.,* 158 (1986) 1–6.
25 Riva, M. Piazza, L. and Schiraldi, A.: Starci gelatinization in pasta cooking: differential flux calorimetry investigations, *Cereal Chem.,* 68 (1991) 622–627.
26 Riva, M. Schiraldi, A. and Piazza, L.: Characterization of rice cooking: isothermal differential calorimetry investigations, *Thermochim. Acta,* 246 (1994) 317–328.
27 Schiraldi, A.: Microbial growth and metabolism: modelling and calorimetric characterization, *Pure Appl. Chem.,* 67 (1995) 1873–1878.
28 Fessas, D. and Schiraldi, A.: Water properties in wheat flour dough I: classical thermogravimetry approach, *Food Chemistry.,* 72 (2001) 237–244.
29 Schiraldi, A.: Water Partition in Starch Products: Thermophysical Methods and Nuclear Magnetic Resonance Applications, in Starch and Starch Containing Origins – Structure, Properties and New Technologies, Ed. V.P. Yuryev, A. Cesaro, W. Bergthaler, Nova Science Publisher, NY (2002) 287–295.
30 Richardson, S. J. Baianu, I. C. and Steinberg, M. P.: Mobility of water in wheat flour suspensions as studied by ^{1}H and ^{17}O NMR, *J. Agr. Food Chem.,* 34 (1986) 17–23.
31 Y-Ro Kim and Cornillon, P.: Effects of Temperature and Mixing Time on Molecular Mobility in Wheat Dough, *Lebensm. Wiss. Technol.,* 34 (2001) 417–423.

32 Schiraldi, A. and Fessas, D.: Classical and Knudsen Thermogravimetry to check States and Displacements of Water in Food Systems, *J. Therm. Anal. Cal.*, 71 (2003) 221–231.
33 Laaksonen, T. J. and Roos, Y. H.: Thermal, dynamic-mechanical and dielectric analysis of phase and state transitions of frozen wheat doughs, *J. Cereal Sci.,* 32 (2000) 281–292.
34 Dreese, P. C. Faubion, J. M. and Hoseney, R. C.: Dynamic rheological properties of flour, gluten, and gluten-starch doughs. II. Effect of various processing and ingredient changes, *Cereal Chem.,* 65 (1988) 354–359.

Chapter 3

Recrystallisation of starch studied with MDSC

P. De Meuter[1], H. Rahier[2*], B. Van Mele[2]

[1]*Cerestar R&D Centre, Havenstraat 84, 1800 Vilvoorde, Belgium*
[2]*Vrije Universiteit Brussel, Dept. Polymer Science and Structural Chemistry Pleinlaan 2, 1050 Brussels, Belgium*

Introduction

In starch-based biodegradable polymers, the crystallisation of starch has an influence on the texture of the product [1, 2]. For use as a material a controlled crystallisation can be a benefit since it can improve the mechanical properties. In food the retrogradation is undesirable as it is one of the reasons of firming. Investigation of the parameters that have an influence on the crystallisation behaviour of starch are thus of great interest, since they enable to understand and to control the structure development of the material.

The native semi-crystalline morphology of starch granules is destroyed in the so-called gelatinisation process, by applying heat in the presence of water or other additives [3]. The amorphous starch system can be (re)crystallised in a consecutive thermal process. The accompanying physical events are termed retrogradation [4].

Different analytical techniques have been used to study the crystallisation of starch. X-ray being one of the most common techniques [5]. Colwell *et al.* were the first to investigate the ageing of wheat starch gels by differential thermal analysis [6]. Other techniques used are differential scanning calorimetry (DSC) [7–10], nuclear magnetic resonance [11], rheometry [12], microscopy [4], Raman [13] , infra-red spectroscopy [14] and isothermal micro-calorimetry [15].

Crystallisation studies at high starch concentrations have only been reported occasionally [8, 16–19]. A majority of research has been carried out on diluted systems, containing 30w% of starch or less. For such diluted starch systems, turbidity measurements are often used to determine retrogradation [20]. Crystallisation rates are mostly determined at two different temperature ranges of interest in the daily application: 1°C to 5°C (refrigerator) and 20°C to 25°C (room temperature). It is however impossible to get a global overview of the influence of different parameters on the crystallisation kinetics, by analysing data

* hrahier@vub.ac.de

D. Lörinczy (ed.), The Nature of Biological Systems as Revealed by Thermal Methods, 49–68.

of which the crystallisation rates have been determined only at these temperatures. However some trends have been proposed and various models to describe the (re)crystallisation mechanism of starch have been formulated. Although a lot of information is available on the crystallisation of starch, the influence of parameters like the crystallisation temperature, the glass transition temperature, the concentration, and the type of starch and additives still needs to be better understood. More recently, the importance of the glass transition temperature in understanding the crystallisation process, has led to the study of more concentrated systems [9, 10].

The knowledge of the exact value of the glass transition, T_g, is very important in the choice of appropriate ageing conditions, because crystallisation will occur between T_g and the melting temperature, T_m. In excess of water, below the maximal freeze concentration, C_g' (73 w%) [7, 18, 21], starch and water phase separate. T_g of the system was suggested to remain constant [1] instead of following the curve towards T_g of water. The minimum T_g is reported to be that of the maximal freeze concentrated phase, called T_g'. Under these conditions the crystallisation and melting of phase separated water is overlapping with T_g of plasticized starch. This makes it more difficult to measure T_g for these systems [18].

Remark that is very difficult to measure T_g of carbohydrates very precisely [22], for several reasons. First of all, T_g is very sensitive to moisture content as explained above. Evaporation of water during the analysis should therefore be prevented. To obtain these conditions, changes to the existing equipment are sometimes necessary.

Secondly, most products degrade at rather low temperatures [23–25]. Heating the samples to high temperatures during the analysis should therefore be avoided. T_g of dry starch can only be estimated by extrapolation [18, 21, 26–29] since the thermal degradation of starch at temperatures below T_g precludes investigation of the dry material [24–30].

Another difficulty is the fact that T_g of many products is smeared out over a broad temperature range, due to the high polydispersity of many natural materials [31]. It is also suggested that, as moisture content is lowered, the distribution of water in starch becomes very heterogeneous (distribution of T_g's), which also broadens the T_g range [32].

Several analytical techniques have been used to investigate the glass transition of carbohydrates. Most commonly, Differential Scanning Calorimetry (DSC) has been utilised [18, 26, 30, 33–35]. Since T_g is sometimes difficult to identify with DSC for the reasons mentioned above and due to the fact that the change in C_p at T_g is small [30], a combination with for instance Dynamic Mechanical Analysis [36] (DMA) may be necessary. Many other methods to determine Tg values have been used in literature, like Nuclear Magnetic Resonance [37] (NMR), Thermal Mechanical Analysis [38] (TMA), Dielectric Analysis [39] (DEA), and Mechanical Analysis [27]. All these techniques have advantages and disadvantages.

The extension of DSC to Modulated DSC (MDSC) has facilitated the determination of very weak and broad T_g's [19, 40–45]. This is possible since MDSC improves sensitivity and resolution simultaneously. Another advantage of (M)DSC is that evaporation of water from the sample during an experiment can be avoided by using hermetically sealed pans.

For DMA the set-up of the equipment can be adapted to reduce moisture loss [19]. A cup filled with a small amount of water was placed around the probe. This adaptation made it possible to measure samples up to 85°C without drying them out during the experiment. At higher temperatures water loss still occurs, meaning that DMA is limited to the measurement of low T_g values.

A model system composed of pregelatinised waxy starch (containing over 99% amylose) and a plasticizer is used to facilitate the understanding of the effect of different parameters on T_g. Indeed, in a more complex mixture of different ingredients the glass transition of carbohydrates is even more difficult to identify. In a first step, the influence of different plasticizers on T_g and how they determine the mechanical properties of starchy materials is studied. In a second step, the influence of the molecular weight on T_g is investigated via a series of monodisperse maltooligosaccharides. An estimation for T_g of dry starch is obtained.

The benefits of MDSC to analyse the crystallisation of concentrated amorphous starch systems, containing at least 69w% of starch and less than 31w% of water, will be explored. The influence of the concentration and the type of starch on the crystallisation rate will be studied as a function of temperature using quasi-isothermal MDSC. The melting behaviour after retrogradation will be discussed briefly.

Experimental

MATERIALS AND SAMPLE PREPARATION

The crystallisation behaviour of amorphous concentrated starch systems was studied using pregelatinised waxy corn starch with an amylopectin content of more than 99 w% (C*Gel-Instant 12410), supplied by Cerestar Belgium.

The MDSC study was performed on compressed pellets of the freeze dried starch, fitting in the reusable high-pressure stainless steel (HPS) DSC pans from PerkinElmer. The starch concentration was adjusted by adding water to the sample with a micro syringe. The samples were equilibrated overnight, at room temperature, in hermetically closed HPS pans to obtain a homogeneous distribution of water in the sample. Small errors in the water content during sample preparation are unavoidable with this method, causing small variations in the initial values of the glass transition temperature of each sample. Dextrose up to maltoheptaose, with a degree of polymerisation (DP) from 1 to 7, were obtained from Sigma. Their respective molecular mass is 180, 352, 504, 666, 828, 990 and 1152 $gmol^{-1}$.

ANALYTICAL TECHNIQUES

(Modulated) differential scanning calorimetry

(Modulated) Differential Scanning Calorimetry ((M)DSC) measurements were performed on a DSC 2920 of TA Instruments, with modulated DSC option (MDSC™) and equipped with a Refrigerated Cooling System (RCS).

The purge gas was helium (25 mlmin^{-1}). Temperature was calibrated with indium and cyclohexane. Enthalpy was calibrated with indium. Water at 35°C was used for the calibration of heat capacity, with the same modulation parameters as for the experiments.

The weights of the empty sample pan and reference pan were matched. The initial sample mass varied between 20 and 30 mg. No mass loss was observed for all thermal treatments applied.

The choice of modulation conditions and heating rate are of importance to obtain reliable results especially with HPS pans. A period of 100 s and an amplitude of 0.5°C were chosen. In non-isothermal conditions, an underlying heating rate of 1°Cmin^{-1} was never exceeded.

All MDSC measurements were started at –60°C. The samples were first heated to 170°C to erase the thermal history. After measuring the glass transition temperature of the amorphous materials during the second heating, the samples were instantaneously cooled to the (quasi)isothermal crystallisation temperature and kept (quasi)isothermally for a predetermined crystallisation time. Depending on the crystallisation temperature used, the end of crystallisation was reached in about 1500 to 5000 min. After the isothermal step, the samples were immediately cooled to –60°C. A subsequent heating to 200°C was performed under the same conditions. Some samples, however, were heated without modulation (conventional DSC), at a higher heating rate of 5°Cmin^{-1}.

For studying the crystallisation with conventional DSC, partial crystallisation experiments were carried out with the temperature program as described above (except the modulation). After partial isothermal crystallisation, the amount of crystallised material is calculated from the enthalpy of melting measured in a subsequent heating at 5°Cmin^{-1}.

Thermogravimetry (TG)

The water content of all starch samples was determined by TG, measuring the mass loss at 125°C in dry helium. The TG experiments were performed on a PerkinElmer TGA-7.

Results and discussion

CHARACTERISATION OF THE TG REGION WITH MDSC

MDSC has several advantages compared to conventional DSC for the determination of T_g of starchy systems. The total heat flow signal can be deconvoluted into heat capacity and non-reversing heat flow (Fig. 1). In the total heat flow signal the T_g can hardly be detected for this sample. T_g can be calculated much easier from the heat capacity curve. The increased sensitivity, resulting from the high instantaneous heating rate, together with the increased resolution, due to the low average (underlying) heating rate, are major benefits of MDSC for the determination of weak and broad thermal transitions, such as the glass transition of starch. This improves the accuracy of the determination of T_g and the change in heat capacity at T_g, ΔC_p [46].

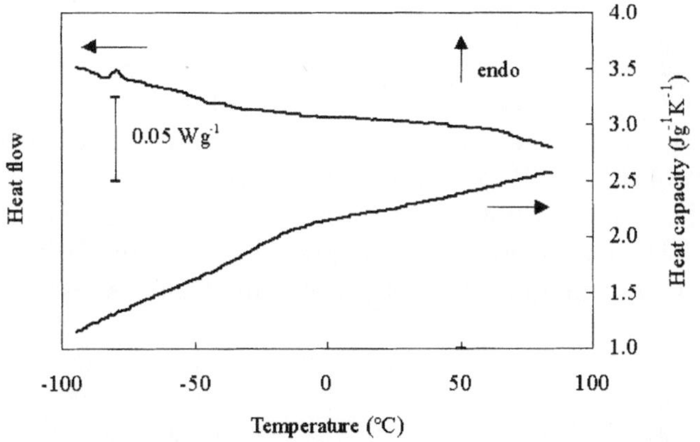

Fig. 1 Determination of T_g with the MDSC heat capacity signal (lower curve). Upper curve: total heat flow

It is however necessary to apply the correct modulation conditions like amplitude, period and heating rate to obtain reliable results. These parameters have been optimised for the pan types used [19].

TG AS A FUNCTION OF THE WATER CONTENT

The influence of the moisture content on T_g of starch is depicted in Fig. 2. The results of this work are in good agreement with literature data [22, 34, 37], also shown for comparison.

T_g decreases by the addition of water. For concentrated starch systems, the addition of one percent moisture decreases T_g with about 15°C.

Several approaches have been proposed to account for the effect of plasticizer on the position of T_g. The Couchman and Karasz approach leads to the following expression for T_g of polymer-diluent mixtures [47]:

$$T_g = \frac{W_1 \Delta C_{p1} T_{g1} + W_2 \Delta C_{p2} T_{g2}}{W_1 \Delta C_{p1} + W_2 \Delta C_{p2}} \quad (1)$$

with T_{g1}: the glass transition temperature of the plasticizer (in K); T_{g2}: the glass transition temperature of the polymer (in K); ΔC_{p1}: the change in heat capacity at T_{g1} (in $Jg^{-1}K^{-1}$); ΔC_{p2}: the change in heat capacity at T_{g2} (in $Jg^{-1}K^{-1}$); W_1: the weight fraction of the plasticizer, W_2: the weight fraction of the polymer.

Equation 1 was tried out for starch-water systems, to estimate T_g of dry starch (T_{g2}) which cannot be determined experimentally. T_{g1} and ΔC_{p1} of water are difficult to measure. Literature values are 134K and 1.94 $Jg^{-1}K^{-1}$, respectively [48]. ΔC_{p2} of dry starch was taken 0.38 $Jg^{-1}K^{-1}$ (extrapolated from own MDSC measurements). Applying the above mentioned constants and T_{g2} of starch as the only parameter to fit the Couchman-Karasz equation to the experimental results of this work (see Fig. 2), an optimised T_{g2} value for dry waxy corn starch of 250°C is obtained. The fit for pregelatinised waxy corn starch is shown in Fig. 2. It should be noticed, however, that a ΔC_{p2} value as high as 0.47 $Jg^{-1}K^{-1}$ is reported [30]. Lower values are also reported: 0.42 $Jg^{-1}K^{-1}$ for amylopectin containing 17% moisture [35], 0.30 $Jg^{-1}K^{-1}$ for high molecular weight maltodextrins [18], and 0.295 $Jg^{-1}K^{-1}$ calculated for waxy corn starch casted at 90°C [34]. The highest value reported for ΔC_{p2}, 0.47$Jg^{-1}K^{-1}$, gives a T_{g2} value for dry starch of 217°C. If 0.30$Jg^{-1}K^{-1}$ is applied a T_{g2} of about 300°C is found. Note however that by using this latter ΔC_{p2} value, the fitting is less good.

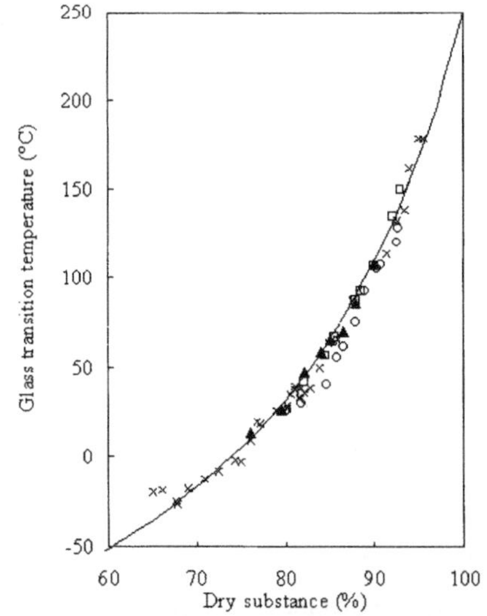

Fig. 2 T_g values of starch-water mixtures measured with DSC and MDSC: × this work, ○ K. J. Zeleznak and R. C. Hoseney [22], ■ H. Bizot *et al.*[34], ▲ M. T. Kalichevsky *et al.* [37] The line is a fit of Eq. 1 (Couchman-Karasz)

In literature, a wide range of values are reported for T_g of dry starch, obtained by extrapolation of materials containing various amounts of water: 125°C [28], 150°C [49], 243°C [18], 240–250°C [29], 285°C [34], 330°C (amylose) [27].

Below a certain concentration, called the maximal freeze concentration (C_g'), phase separation between plasticized starch and water occurs. In DSC, a melting endotherm of the water-rich phase becomes visible which overlaps with the glass transition of the plasticized starch-rich phase. In this work, the phase separation was found to occur at a waxy corn starch concentration of 73w%. The same value for C_g' has been reported by other researchers [10, 18, 21].

For concentrations near the maximal freeze concentration (C_g'), the decrease of T_g by the addition of water levels off (Fig. 2). The minimum T_g value of starch, called T_g', is reported to be 5°C [21] but other researchers have published values between 6°C [18] and 11°C [10].

In this study, the minimum value for T_g of plasticised starch is measured using MDSC. MDSC enables an accurate measurement of T_g during cooling (see Fig. 3), even if phase separation should occur thermodynamically. With MDSC the glass transition of the starch-rich phase can be separated from other thermal events during cooling. This is shown for a 68w% waxy corn starch sample in Fig. 3.

The total heat flow is separated into a non-reversing heat flow and a heat capacity signal. T_g calculated out of the C_p signal for this water content is –21°C. This determination would not be possible using only the total heat flow signal (see Fig. 3). These samples are in a non-equilibrium state during cooling and phase separation will occur during the subsequent heating. The 'cold-crystallisation', T_c, of the water-rich phase is observed in the total heat flow curve during

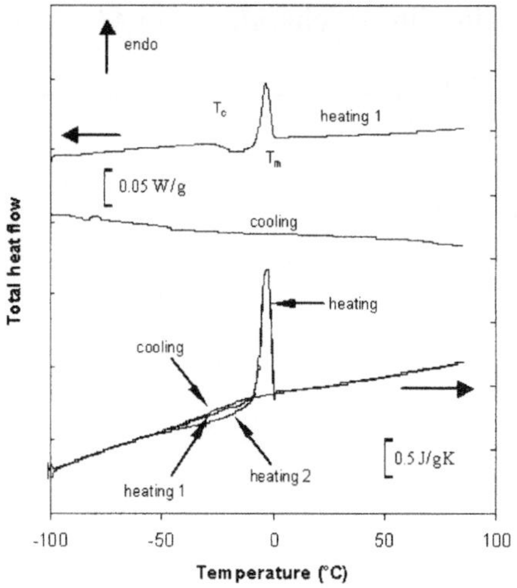

Fig. 3 Total heat flow and heat capacity measured during cooling and subsequent heating of a 68w% starch-water system. Heating 1: cold crystallisation during heating. Heating 2: reheating after cold crystallisation. T_g' is at –5°C

heating just above T_g (see Fig. 3: heating 1). The melting of this phase, at T_m, which is also seen in the heat capacity curve, follows immediately afterwards. The same effect is described for maltohexaose [46]. Note that, after phase separation, T_g of the starch-rich phase has increased in relation to the increased dry substance of this phase. By reheating the sample after cold crystallisation, T_g is shifted to T_g', which overlaps with the melting of the water-rich phase (see Fig. 3: heating 2).

A minimum value for T_g as low as –25°C is measured for a starch concentration of 66 w%. This means that even though a separate water-rich phase should be created, T_g continues to decrease, indicating that part of the additional water is still acting as a plasticizer, as long as the phase separation did not occur.

TG AS A FUNCTION OF MOLECULAR MASS

T_g of low molecular mass carbohydrates, dextrose (DP1) up to maltoheptaose (DP7), has been determined by (M)DSC analysis. In contrast with the large polydispersity of starch, the molecular weight of these maltooligosaccharides is monodisperse. The narrow distribution of the molecular mass, together with the increased ΔC_p at T_g, increase the accuracy of the measurements.

T_g of these maltooligosaccharides was measured at different concentrations (Fig. 4). For the clarity of the graph, only maltotriose (DP3), maltotetraose (DP4) and maltoheptaose (DP7) are compared with the results for starch. The decrease in T_g with increasing water content is similar as for starch.

In literature large differences are found in the reported T_g values for dry maltooligosaccharides [21, 30]. This discrepancy is probably caused by the fact that some samples were not completely dry or that degradation occurred. For this last reason, T_g of dry maltooligosaccharides with M_w above DP5 (T_g = 174°C) is

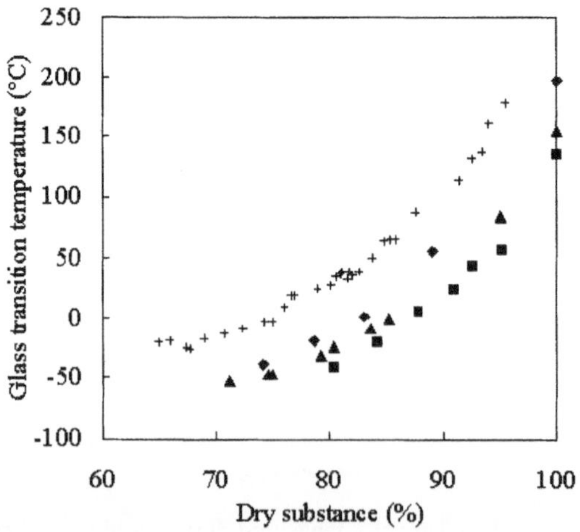

Fig. 4 T_g of maltotriose (■), maltotetraose (▲), maltoheptaose (◆) and starch (+) as a function of ds

difficult to measure without simultaneous degradation of the product. Therefore, these experiments were also done with conventional DSC at 10°Cmin^{-1} in this work, to minimise the time spent at high temperatures.

It is commonly found that the variation of T_g with degree of polymerisation (DP) can be described by an equation of the form:

$$T_g = T_g^{\infty} - \frac{A}{\mathrm{DP}} \qquad (2)$$

where T_g^{∞} is the high molecular mass limit of T_g and A is a constant.

Based on our measurements for dry maltooligosaccharides the extrapolated value to infinite molecular weight is 240°C. This value is comparable with the value of 250°C, obtained via extrapolation of T_g values of plasticized starch to 0 w% water.

TG AS A FUNCTION OF THE STARCH TYPE

The plasticizing properties of water on different kinds of starches (corn, waxy corn and potato) were investigated. The results are shown in Fig. 5. The decrease in T_g with the addition of water is similar for all starches studied, indicating that the difference in the amylose/amylopectin ratio (degree of branching) does not affect the position of T_g to a large extent. This finding is in agreement with literature [30]. Others found, however, that amylose has an increased T_g [27, 34].

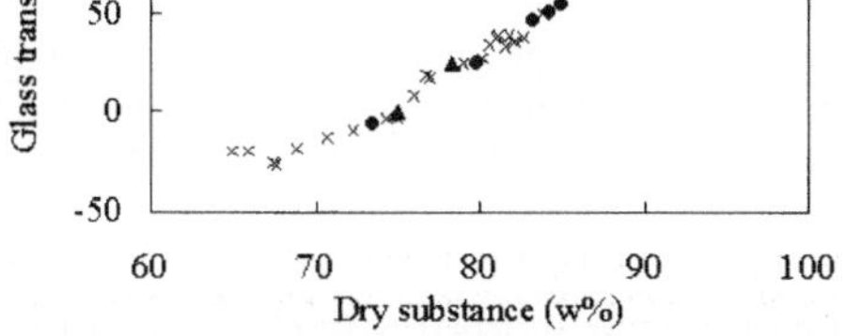

Fig. 5 Comparison of T_g of different types of starch as a function of ds: × waxy corn starch; ● corn starch; ▲ potato starch

SLOW CRYSTALLISATION OF STARCH: IN SITU MEASUREMENT WITH MODULATED DSC

Amorphous starchy materials crystallise very slowly and the crystallisation rate is strongly dependent on the starch concentration. Therefore attention has to be paid to prevent the evaporation of water (or other solvents) during the entire crystallisation. Moreover, the crystallisation process is characterised by a small exothermicity. These starch characteristics make it difficult to choose an appropriate method to measure the crystallisation behaviour of starchy systems.

The MDSC procedure used here is based on the fact that during crystallisation the heat capacity (C_p) of a material decreases [50]. With MDSC this (negative) change in heat capacity (ΔC_{pcryst}) can be measured continuously, even in (quasi) isothermal conditions. From Fig. 6, it is clear that the decrease in heat capacity occurs on the same time scale as the increase of the heat of fusion. The heat of fusion was measured after partial crystallisation at the same temperature. It is obvious that analogous but continuous information on the same time scale is available from the (quasi) isothermal MDSC heat capacity signal. It was shown previously that the crystallinity evolution seen in the heat capacity signal coincides with the one measured with X-ray, DMA and Raman spectroscopy [19].

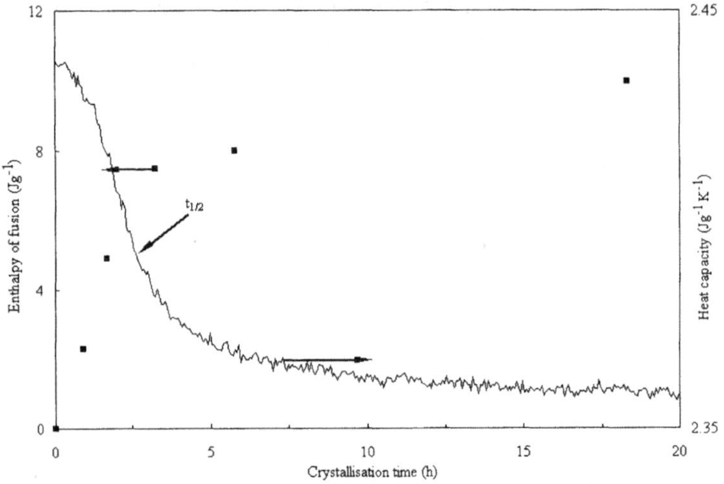

Fig. 6 Crystallisation at 60°C of 69w% pregelatinised waxy corn starch, measured as a function of crystallisation time (*h*). The MDSC heat capacity evolution (in arbitrary units) is compared to the heat of fusion after partial crystallisation

This reproducibility in a time span of 40h (and more) can only be achieved if several experimental conditions are fulfilled. The most important ones are; the fully amorphous nature, the exact concentration and homogeneous distribution of water in the starch sample before crystallisation, the constancy of temperature and water content in the starch system throughout (quasi) isothermal crystallisation. MDSC experiments with high-pressure stainless steel pans meet these experimental constraints. No water loss was noticed for all thermal treatments applied, and the deviation from the average temperature during (quasi) isothermal crystallisation was ± 0.01°C for the total time interval studied.

The MDSC method enables a quantification of the starch crystallisation process. The half conversion time ($t_{1/2}$) is defined as the time to reach half of the decrease in C_p (1/2 ΔC_{pcryst}; see Fig. 6). The reciprocal of this time ($1/t_{1/2}$) can be used as a measure for the rate of isothermal crystallisation.

The heat flow phase signal of MDSC might also be interesting in the study of starch crystallisation, and reflects the change in heat capacity (ΔC_{pcryst})[51]. However, in the experimental conditions used, the change in the heat flow phase is not always reproducible and a small effect of the crystallisation process might be superimposed [52].

The total heat flow signal (not shown), equivalent with the conventional DSC signal cannot be used to determine the crystallisation behaviour of starch. The heat released is so small and is spread out over such a long crystallisation time, that the exothermal signal is no longer reliable due to baseline drift and noise. This technique, however, is very useful to follow the crystallisation behaviour of systems that crystallise fast, like the low molecular weight components lactose and sucrose [53].

MDSC is further used to study the crystallisation rate as a function of the crystallisation temperature, concentration, starch type and the effects of crystallisation on the thermal transitions (T_g and T_m).

INFLUENCE OF THE CRYSTALLISATION ON T_g

After a first heating to 170°C in HPS pans, the initial glass transition temperature (T_{g0}) of 76w% pregelatinised starch samples was measured with MDSC. These amorphous samples were then crystallised in MDSC at temperatures between 45°C and 100°C. After the (quasi) isothermal step, the samples were immediately

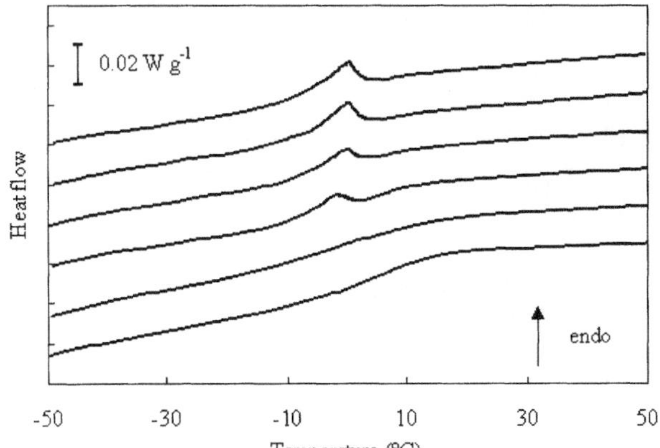

Fig. 7 Glass transition region (conventional DSC at 5°Cmin^{-1}) for 76w% pregelatinised waxy corn starch after crystallisation at 60°C for different crystallisation times t (from bottom to top: t = 0, 8, 12, 15, 24 and 48h)

cooled below T_g. In a consecutive heating, the thermal properties of the semi-crystalline material were measured (T_g, the temperature range of melting and the enthalpy of fusion).

The glass transition region of 76w% starch was measured before and after (partial) crystallisation (Fig. 7).

T_g decreases continuously as a function of crystallisation time, from 9°C for 76w% amorphous starch to –4°C for semi-crystalline starch. This is explained by the fact that water is expelled from the crystals. The remaining amorphous phase of the semi-crystalline sample will, therefore, contain more water than initially in the fully amorphous sample. A final T_g of – 4°C corresponds to about 72 wt% starch. Note that for the highest crystallisation times, a small endothermic melting peak of ice is superimposed on the glass transition signal. In these later stages of crystallisation, a water-rich phase is segregating from the plasticized starch-rich amorphous phase. In literature, this phenomenon of water expulsion from the crystals, syneresis, is only described for diluted starch systems [20, 54]. Note that for the diluted systems the accompanying decrease in T_g could never be measured, because T_{g0} already reached the minimum value for those diluted systems.

INFLUENCE OF THE ISOTHERMAL CRYSTALLISATION TEMPERATURE ON THE CRYSTALLISATION RATE

Both the melting temperature (region) and the glass transition temperature (region) of starch are important parameters controlling the rate of crystallisation. Crystallisation can only take place at temperatures between T_g and T_m [55]. Both thermal transitions are influenced by the concentration of water in the starch sample.

In Fig. 8, the half conversion time, $t_{1/2}$ defined as the time to reach half the decrease in C_p (see Fig. 6), measured with MDSC, is depicted as a function of the crystallisation temperature, giving rise to a bell-shaped curve. The bell-shaped

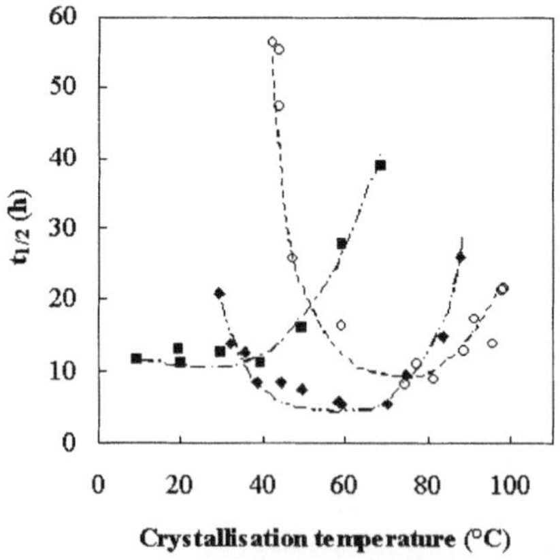

Fig. 8 Crystallisation rate as a function of the crystallisation temperature for different concentrations (× 78%, ○ 76%, ◆ 70%, ■ 60%). The lines are drawn as a guide to the eye

curve is well-known for many polymers [21, 56]. For concentrated starch-water systems, however, experimental data are scarce, but predictions were made by modelling [9, 55]. Our data are consistent with these predictions.

For 76w% waxy corn starch, the bell-shaped curve goes through a minimum at 75°C (T_{cmax}). At T_{cmax}, approximately 65°C above T_{g0}, the maximum crystallisation rate is obtained. At crystallisation temperatures below T_{cmax}, the crystallisation rate decreases ($t_{1/2}$ and t_{max} increase). Due to the higher viscosity at lower temperatures, transport of starch chains to the boundary of the starch crystal is restricted, and the crystallisation rate gets diffusion controlled. At temperatures above T_{cmax}, the crystallisation rate decreases as well ($t_{1/2}$ and t_{max} increase again), since the thermodynamic driving force for crystallisation (primary and secondary nucleation) decreases.

INFLUENCE OF THE CONCENTRATION

In Fig. 8 the bell-shaped curves for the different waxy corn starch concentrations (60, 70, 76 and 78%) are shown.

The crystallisation rate for 60w% samples was only measured at temperatures above 0°C. Therefore only part of the bell-shaped crystallisation rate curve was measured. The difference between 76w% and 78w% is small and difficult to measure accurately. The problem is that only the low end of the bell-shaped curve can be measured for the 78w% samples, since degradation of the material occurs if samples are kept for a long time (days) at temperatures above 100°C.

The temperature of maximum crystallisation rate (minimum in the bell-shaped curve), T_{cmax}, decreases with decreasing starch concentration. For 60w% samples, T_{cmax} is at about 25°C, for 70w% samples at about 60°C, and for 76w% samples at about 75°C. The scatter on the data for 60w% samples is very large. It is shown in literature that the retrogradation process could be monitored using X-ray diffraction and it was modelled by a physical formulation developed by Lauritzen and Hofmann [16, 17]. For 60 and 70w% samples, T_{cmax} values of 65°C and 80°C, respectively, were reported. No data were given for higher concentrations. These literature results are not in agreement with the findings of this work. Especially for the lowest starch concentration, a much lower T_{cmax} value is obtained.

Figure 8 also shows that the starch concentration affects the value of the maximum crystallisation rate at T_{cmax}, v_{max} (determined as $t_{1/2}$ at T_{cmax}). The scatter on the data points, however, makes it difficult to accurately establish v_{max}. The value of v_{max} is highest at a concentration of 70w%; a value for $t_{1/2}$ of about 330min is measured. By decreasing the concentration to 60w%, $t_{1/2}$ is doubled (rate of crystallisation reduced to 50%). This could be explained either by the increasing difficulty to form stable nuclei at increased solvent concentrations or by the decreasing probability for chains to meet. For a starch-water system of 76w% starch, $t_{1/2}$ increases to about 500min. This decrease in v_{max} is probably due to increasing viscosity.

Expulsion of water from the crystals during crystallisation has an effect on the crystallisation rate. For a sample with an initial concentration of 76%, T_g decreases from 9°C to –5°C thus the final concentration is approximately 73%. From Fig. 8 can be deduced that the crystallisation rate decreases with increasing water content for crystallisation temperatures well above T_{cmax}. For example from 76% to 70% at 90°C the rate decreases by a factor of about two. Since the change in concentration during crystallisation is less and since it occurs gradually, the decrease in rate will also be somewhat less than a factor of two, but the influence will anyhow be important.

In a similar way can be seen that for the crystallisation temperatures below T_{cmax} the decreasing T_g during crystallisation will increase the crystallisation rate. This auto-catalytic effect can explain why the crystallisation rate around $t_{1/2}$ is relatively high for low crystallisation temperatures, whereas for high crystallisation temperatures, the crystallisation retards itself. The effect might still be enhanced by an inhomogeneous water distribution [57].

UNIVERSAL CRYSTALLISATION RATE CURVE

An attempt was made to combine the results of Fig. 8 in order to obtain a universal crystallisation rate curve. All $t_{1/2}$ values were normalised against the corresponding $t_{1/2}$ at T_{cmax} (= v_{max}). These normalised rates were plotted as a function of $(T_c–T_g)/(T_e–T_g)$, with T_g, the glass transition temperature and T_e, the end-temperature of melting (see page 64). In this way, the crystallisation temperature T_c is normalised too, ranging between 0 ($T_c = T_g$) and 1 ($T_c = T_e$). This makes sense, as both T_g and T_e are limiting the temperature range for crystallisation. The same procedure was applied for lactose and sucrose samples [58].

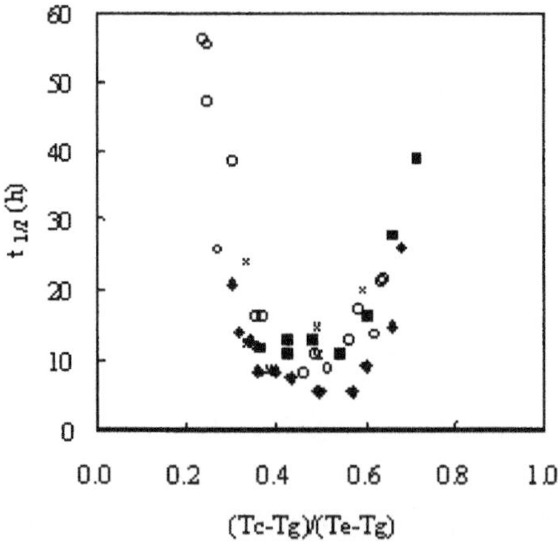

Fig. 9 Universal crystallisation rate curve for different concentrations of pregelatinised waxy corn starch (■ 60w%, ◆ 70w%, ○ 76w%): $t_{1/2}$, normalised against $t_{1/2}$ at T_{cmax}, as a function of $(T_c-T_g)/(T_e-T_g)$

T_g values for totally miscible starch-water systems were taken calculated from equation 1, even for concentrations where phase separation might occur. In this case, the crystallisation rate is related to the calculated T_g for an ideal, homogeneous starch-water system without phase separation for the whole concentration range [59]. Figure 9 clearly indicates that with this procedure a universal bell-shaped crystallisation rate curve for starch can be obtained. The value of T_e was taken as the melting temperature of the most stable crystals or the end-set temperature. The values for T_g and T_e used in both approaches are listed in Table 1.

Table 1 T_e and T_g values (°C), used in Fig. 9

	60w%	70w%	76w%
T_e	115	136	149
T_g	–51	–17	10

Figure 9 clearly shows that a universal crystallisation rate curve for starch, independent of the starch concentration, can be obtained. Differences in T_{cmax} for different concentrations are explained by changes in T_g and T_e, which in turn are explained by differences in water content. The normalised value of T_{cmax} for all concentrations is at ca. 0.5 on the normalised temperature scale, meaning a T_{cmax} value almost equal to $(T_g + T_e)/2$.
Figure 10 shows the starch-water state diagram. T_e and T_{cmax} decrease almost linearly in the concentration range studied.

A linear extrapolation of these results enables an estimation of T_{cmax} for other concentrations. Extrapolation shows that for concentrations below 50w% (food

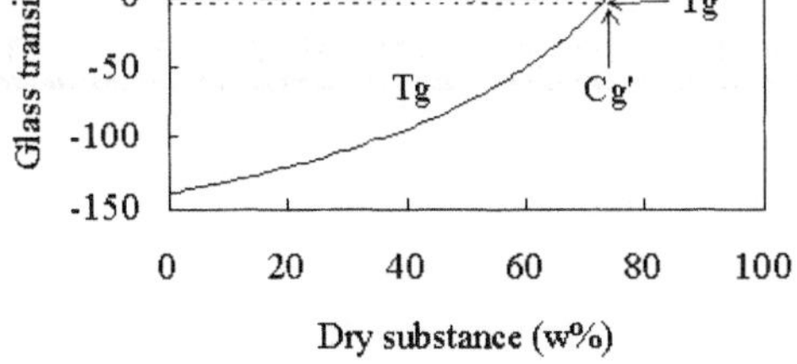

Fig. 10 State diagram showing temperature of maximum crystallisation rate, T_{cmax}, end -temperature of melting, T_e, and glass transition temperature, T_g, as a function of the starch concentration. T_g is calculated according to eq.1. $T_g' = -5$°C at $C_g' = 73$w%

systems), the calculated T_{cmax} is below T_g'. As explained before, phase separation of a water-rich phase and a starch-rich phase occurs during the crystallisation of the water-rich phase upon cooling (or reheating). Due to this interfering phase separation, the highest crystallisation rates for starch concentrations below 50w% will be found around 0°C, since crystallisation below the glass transition of the starch-rich phase ($T_g' = -5$°C) is prohibited by diffusion limitations.

INFLUENCE OF THE STARCH-TYPE [57]

Next to pregellatinised waxy corn starch, the crystallisation of pregelatinised corn starch and pregelatinised amylose extender waxy starch (aewx) was also performed in MDSC. The comparison between waxy corn starch, aewx starch and corn starch was made for 76w% starch samples.

The crystallisation rate of aewx and corn starch is much higher than the crystallisation rate of waxy corn starch. $t_{1/2}$ at T_{cmax} is less than 45min for aewx and 60min for corn compared to 500min for waxy corn.

MELTING OF STARCH [60]

The samples crystallised in MDSC conditions are crystallised quasi-isothermally, but no difference can be observed in their melting profile when compared to isothermally crystallised samples. The melting profiles of starch samples prepared at different concentrations are shown in Fig. 11. All melting profiles are very broad and at least bimodal. The broad melting range indicates a great heterogeneity of starch crystals.

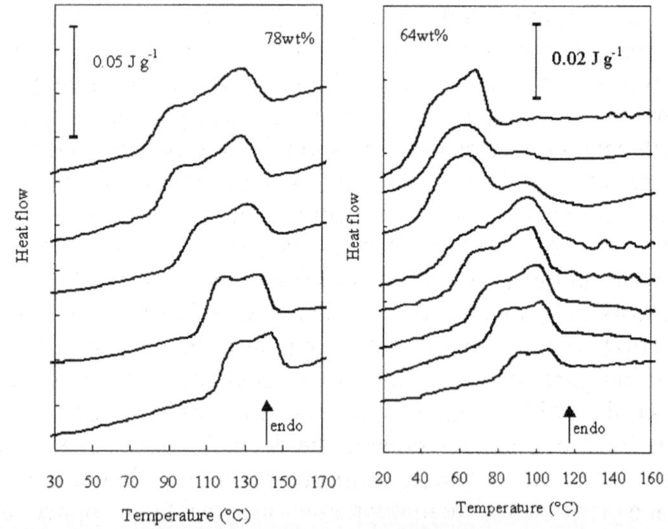

Fig. 11 Influence of T_c (indicated) on the melting endotherms for 78 and 60w% pregelatinised waxy corn starch (conventional DSC at 5°C/min)

The start of melting, T_o, shifts with the isothermal crystallisation temperature, T_c. Melting starts about 16°C above T_c for low values of T_c, whereas at high values of T_c this interval is reduced to about 5°C. This finding is independent of the concentration.

The end temperature, T_e, of melting is only slightly influenced by T_c. For 78w% samples the increase is only about 5°C for an increase in T_c of 40°C. The fact that T_e remains almost constant whereas T_o increases substantially with T_c results in a narrowing of the melting range with increasing T_c.

T_e changes largely with the concentration, from 115°C for 60w% to 150°C for 78w% starch.

Conclusions

The crystallisation window of plasticised starch is bordered by its glass transition and melting temperature. Both transitions depend on the water content. The plasticizing properties of water on starch were investigated using MDSC. Phase separation occurs from a certain concentration. However, it was remarked that during cooling, phase separation does not occur immediately and T_g still decreases, even beyond the concentration where phase separation should occur thermodynamically. The minimum T_g measured in this study with MDSC is therefore −25°C at 66w% ds.

To overcome the problem that T_g of dry starch cannot be measured without degradation, T_g is estimated using two different approaches. The first way is via extrapolation of T_g data of a series of maltooligosaccharides with increasing molecular mass. The second way is via extrapolation of T_g of starch samples with decreasing water content. A T_g value for dry starch of 240°C and 250°C, respectively, is obtained.

The glass transition temperatures at different water contents are comparable for different types of starch studied (corn, waxy corn and potato).

It is shown that Modulated Differential Scanning Calorimetry (MDSC) enables to follow in situ the slow isothermal crystallisation process of concentrated amorphous starch systems. The accurate, reproducible, and continuous measurement of the heat capacity change during (quasi) isothermal crystallisation can be related to the crystallisation process, as confirmed by other techniques, like X-ray, DMA and Raman spectroscopy.

The MDSC method with high-pressure stainless steel pans enables a systematic study of the slow crystallisation of amorphous starch in the presence of small amounts of water. The major benefits of this procedure are the combination of (*i*) easy preparation of amorphous samples with a homogeneous water distribution before crystallisation; (*ii*) excellent control of temperature and water content, even for extended crystallisation times (several days) in combination with high crystallisation temperatures (up to 100°C); (*iii*) easy measurement of the evolution of the glass transition region using the same experimental set-up, enabling to establish relations between crystallisation and other thermal transitions.

The crystallisation rate was quantified as a function of isothermal crystallisation temperature, giving rise to a bell-shaped curve.

For waxy corn starch it was found that the temperature of maximum crystallisation rate decreases with the starch concentration, from 75°C for 76w% starch to 22°C for 60w% starch. The maximum crystallisation rate depends on the starch concentration. The highest crystallisation rate is obtained for a concentration of 70w% ($t_{1/2}$ = 330min).

To be able to estimate the temperature of maximun crystallisation rate and the influence of the crystallisation temperature on the crystallisation rate for a chosen concentration, a universal crystallisation rate curve was proposed by plotting normalised rate data as a function of $(T_c-T_g)/(T_e-T_g)$. The temperature of maximum crystallisation rate for this concentrated starch systems lies at approximately $(T_g+T_e)/2$.

Crystallisation rates of different types of starch (waxy corn, corn and aewx) were compared. Retrogradation rates were found to increase with the amylose content and the size of the chain length of the amylopectin fraction. Therefore, aewx crystallises faster than corn starch, which in turn crystallises faster than waxy corn starch. Nevertheless, the maximum crystallisation rate was obtained at the same temperature (75°C at 76w% starch).

The melting endotherms of recrystallised starches are at least bimodal. The onset of melting starts only about 10°C above the crystallisation temperature. The end temperature of melting is almost independent of the crystallisation temperature.

References

1 van Soest J. J. G., Hulleman S. H. D., de Wit D., and Vliegenthart J. F. G. (1996) *Carbohydrate Polymers,* 29, 225.
2 Rindlav A., Stading M., Hermansson A., and Gatenholm P. (1998) *Carbohydrate Polymers,* 36, 217.
3 Zobel H. F. (1984) in Starch: chemistry and technology, Whistler R. L., BeMiller J. N. and Paschall E. F., Eds., Academic Press Inc., Orlando, Florida, Chapter 9, p. 285.
4 Slade L. and Levine H. (1991) Critical Rev. In *Food Sci. and Nutrition,* 30, 115.
5 Liu H., Lelievre J. and Ayoung-Chee W. (1991) *Carbohydrate Research*, 210, 79.
6 Colwell K. H., Axford D. W. E., Chamberlain N. and Elton G. A. H. (1969) *J. Sci. Fd. Agric.*, 20, 550.
7 Nakazawa F., Noguchi S., Takahashi J. and Takada M. (1984) *Agric. Biol. Chem.*, 48, 201.
8 Zeleznak K. J. and Hoseney R. C. (1986) *Cereal Chemistry*, 63, 407.
9 Slade L. and Levine H. (1989) Frontiers in Carbohydrate Research, Millane R. P., BeMiller J. N., and Chandrasekaran R., Eds., Elsevier Applied Science, New York, p. 215.
10 Jouppila K. and Roos K. (1997) *Carbohydrate Polymers*, 32, 95.

11 Farhat I. A., Blanshard J. M. V., Melvin J. L. and Mitchell J. R. (1997) Starch Structure and Functionality, Frazier P. J., Donald A. M. and Richmond P., Eds., The Royal Society of Chemistry, Cambridge, p. 86
12 Wong R. B. K. and Lelievre J. (1982) *Starch*, 34, 231.
13 Bulkin B. J., Kwak Y. and Dea I. C. M. (1987) *Carbohydrate Research,* 160, 95.
14 Wilson R. H. and Belton P. S. (1996) *Carbohydrate Research*, 180, 339.
15 Silverio J., Svensson E., Eliasson A.-C. and Olofsson G. (1996) *J. Therm. Anal.*, 47, 1179.
16 Blanshard J. M. V. and Farhat J. A., Proceedings 25th Anniversary Euro Research and development CPC Europe, 35th Anniversary Institute for Research and Development CPC Deutschland, CPC Deutschland, Heilbronn, November 1998, p. 187–200.
17 Jouppila K., Kansikas J. and Roos Y. H. (1998) *Carbohydrate Polymers*, 36, 143.
18 Roos Y. and Karel M. (1991) *Journal of Food Science*, 56 (6), 1676.
19 De Meuter P., Amelrijckx J., Rahier H., Van Mele B. (1999) *J. Pol Sc. B Pol Phys.*, 37, 2881.
20 Ring S. G., Colonna P., I'Anson K. J., Kalichevsky M. T., Miles M. J., Morris V. J., and Orford P. D. (1987) *Carbohydrate Research*, 162, 277.
21 L. Slade and H. Levine (1995) Advances in Food and Nutrition Research, 38, Glass Transitions and water-food structure interactions, Academic Press, San Diego, Chapter 2, p. 103–269.
22 Zeleznak K. J. and Hoseney R. C. (1987) *Cereal Chemistry*, 64(2), 121.
23 Aggarwal P., Dollimore D. and Heon K. (1997) *J. Therm. Anal. Cal.*, 50, 7.
24 Ciesielski W. and Tomasik P. (1996) *Carbohydrate Polymers*, 31, 205.
25 Aggarwal P. and Dollimore D. (1996) *Talanta*, 43, 1527.
26 van den Berg C. (1992) *Carbohydrates in the Netherlands*, 8, 23.
27 Nakamura S. and Tobolsky A. V. (1967) *Journal of Applied Polymer Science,* 11, 1371.
28 van den Berg C. in Concentration and drying of Foods (1986) D. MacCarthy, Ed., Elseviers Applied Science, London, p.11–36
29 Biliaderis C. G. (1990) Thermal Analysis of Foods, V. R. Harwalkar and C.-Y. Ma, eds., Elsevier Applied Science, London, Chapter 7, p. 168–220.
30 Orford P. D., Parker R., Ring S. G. and Smith A. C. (1989) *Int. J. of Biol. Macromolecules*, 11, 91.
31 Ellis T. S. (1988) *Journal of Applied Polymer Science*, 36, 451.
32 Schenz T. W., Courtney K., and Israel B. (1993) *Cryo-Letters*, 14, 91.
33 Orford P. D., Parker R. and Ring S. G. (1990) *Carbohydrate Research*, 196, 11.
34 Bizot H., Le Bail P., Leroux B., Davy J., Roger P. and Buleon A. (1997) *Carbohydrate Polymers*, 32, 33.
35 Noel T. R. and Ring S. G. (1992) *Carbohydrate Research*, 227, 203.
36 Vodovotz Y. and Chinachoti P. (1996) *Journal of Food Science*, 61(5), 932.
37 Kalichevsky M. T., Jaroszkiewicz E. M., Ablett S., Blanshard J. M. V. and Lillford P. J. (1992) *Carbohydrate Polymers*, 18, 77.
38 Yuryev V. P., Nemirovskaya I. E. and Maslova T. D. (1995) *Carbohydrate Polymers*, 26, 43.
39 Nishinari K. and Fukada E. (1980) *Journal of Polymer Science*, 18, 1609.
40 Aubuchon S. R., Thomas L. C., Theuerl W. and Renner H. (1998) *J. Therm. Anal. Cal.*, 52, 53
41 Bell L. N. and Touma D. E. (1996) *Journal of Food Science*, 61(4). 807.

42 Izzard M. J., Ablett S., Lillford P. J., Hill V. L., and Groves I. F. (1996) *J. Therm. Anal. Cal.*, 47, 1407.
43 Knopp S. A., Chongprasert S. and Nail S. L. (1998) *J. Therm. Anal.Cal.*, 54. 659.
44 Urbani R., Sussich F., Prejac S. and Cesaro A. (1997) *Thermochim. Acta*, 304/305, 359.
45 Wang G. M. and Haymet A. D., (1998) *J. Phys. Chem.,* B, 102, 5341.
46 De Meuter P., Rahier H. and Van Mele B. (1999) *International Journal of Pharmaceutics*, 192, 77.
47 Couchman P. R. and Karasz F. E., (1978) *Macromolecules*, 11, 117.
48 Kouchi A., (1987) *Nature*, 330, 550.
49 Blanshard J. M. V. (1988) Food structure – Its creation and evaluation, Blanshard J. M. V. and Mitchell J. R. eds., Butterworths, London, p. 313–330
50 Mathot V. B. F., (1994) Calorimetry and Thermal Analysis of Polymers, Hanser publishers, Munich, Chapter 5, p. 105
51 Weyer S., Hensel A. and Schick C. (1997) *Thermochim. Acta*, 304–305, 267.
52 Toda A., Oda T., Hikosaka M. and Saruyama Y. (1997) *Thermochim. Acta,* 293, 47.
53 Kedward C. J., MacNaughtan W., Blanchard J. M. V., and Mitchell J. R. (1998) *Journal of Food Science*, 63, 192.
54 Hoover R., (1995) F*ood Reviews International*, 11, 331.
55 Blanshard J. M. V. and J. A. Farhat J. A., Proc. 25th Anniversary Euro Research and development CPC Europe, 35th Anniversary Institute for Research and Development CPC Deutschland, CPC Deutschland, Heilbronn, November 1998, pp. 187–200
56 Wunderlich B., (1976) Macromolecular Physics, Vol 2, Academic Press, New York, .
57 De Meuter P., Rahier H., Van Mele B., to be published
58 Kedward C. J., MacNaughtan W., Blanshard J. M. V., and Mitchell J. R., (1998) *Journal of Food Science*, 63(2), 192.
59 Marsh R. D. L. and Blanshard J. M. V. (1988) *Carbohydrate Polymers*, 9, 301.
60 de Meuter P., Rahier H., van Mele B. to be published

Chapter 4

Calorimetric information about food and food constituents

A. Raemy*, P. Lambelet and Ph. Rousset

Nestlé Research Centre, Nestec LTD, Vers-chez-les-Blanc, CH-1000 Lausanne 26, Switzerland

Introduction

Thermal analysis and calorimetric techniques, which include principally differential scanning calorimetry (DSC), differential thermal analysis (DTA), thermogravimetry (TG), thermomanometry and adiabatic calorimetry have been widely used to investigate physico-chemical properties of foods and food ingredients as well as to determine optimal and safe food processing parameters.

In food science and technology, thermal analysis and calorimetric techniques have been extensively applied to determine specific heat values, transition enthalpies, glass transition temperatures (T_g), induction periods for oxidation phenomena or crystallization and to determine safe process conditions by detecting exothermic phenomena and determining self-ignition temperatures. They have also been used to predict thermal and structural behaviour of lipids (polymorphism) during phase transition and to control parameters (gelatinisation or retrogradation of polysaccharides, denaturation of proteins) which are essential to maintain product quality. As foods are multicomponent systems, these techniques have also helped to elucidate interactions between food constituents (macro-nutrients), *i.e.* lipid-polysaccharide, protein-polysaccharide or lipid-protein interactions.

In addition to improvement of product quality and process safety, new reasons to perform calorimetric studies of food, as well as physico-chemical studies of food in general, have appeared recently, namely:

- food aspect (surfaces, foams, etc.) has acquired more importance due to increased visual sensitivity of the consumer and to improvements in packaging,
- processing technologies have often to be adapted to entrap gases or active ingredients; in this context the glass transition phenomenon is today so important that it is sometimes called glass transition technology,

* alois.raemy@rdls.nestle.com

D. Lörinczy (ed.), The Nature of Biological Systems as Revealed by Thermal Methods, 69–98.

- new issues have appeared such as the presence of acrylamide in food products.

There is a renewed interest in the Maillard reaction from which acrylamide may arise.

The present chapter systematically analyses applications of thermal analysis and calorimetry in the field of food science and technology. It summarizes and completes preceding papers with the purpose of giving the state of the art (Raemy and Lambelet, 1991; Raemy *et al.*, 2000; Schenz, 2003).

The thermal behaviour of foods strongly depends on their composition. Therefore, we first present thermal characteristics of food constituents: carbohydrates, lipids, proteins, water, air, minor constituents and ingredients, and then we consider raw and reconstituted foods.

Both endothermic and exothermic phenomena occur in foods. Some exothermic reactions present a hazard in industrial operations or during storage. They can lead either to self-ignition and to fire or even to dust explosions in open systems such as spray-dryers, or to pressure increase and bursting of closed vessels such as extraction cells. Accordingly, use of thermal analysis and calorimetry in process safety is also briefly discussed.

Instruments, methods and procedures

The principles of thermal analysis and calorimetry are described in a recent book (Haines, 2002).

All thermal analysis and calorimetry equipment available (*e.g.* high pressure DTA, heat flow or Calvet type calorimeters, power compensation DSC instruments) give valuable information. Even the recently introduced modulated (or alternating) differential scanning calorimetry (MDSC or ADSC) can provide useful data, especially related to T_g.

Either scanning (heating and cooling) or isothermal modes are applied. In the scanning mode, two runs are often performed to clarify whether the observed phenomena are reversible. A heating scan showing melting transitions is often followed by a cooling scan under the same conditions to evidence any hysteresis between crystallization and melting temperatures. In order to obtain measurements with high sensitivity, a Micro-DSC (with high sample amount) and/or fast scanning rates should be used. In contrast, a standard DSC (with small sample amount) and/or low heating rates is generally selected to obtain measurements with high resolution.

For our studies on food materials we have used high pressure DTA analysis (Netzsch-Gerätebau 404H, Selb, Germany), heat flow or Calvet type calorimeters, a heat flow Micro-DSC (Micro-DSC III, Setaram, Caluire, France) and standard power compensation DSCs (DSC7 and Pyris 1, Perkin Elmer, Norwalk, USA). Oxidation of lipids was studied isothermally by DSC under oxygen flow in order to have an excess of oxygen. The measurements in relation with process safety were performed on a DTA under pressure (25 bar of oxygen for example) or with heat flow calorimeters using sealed cells. These sealed cells may be fitted with a pressure sensor in order to perform thermomanometry. For process safety research, adiabatic calorimetry (accelerating rate calorimetry or ARC, Columbia Scientific Industries, Milton Keynes, UK) was also used as it represents the worst situation if one considers thermal conditions.

Since calorimetric and thermal analysis techniques alone do not give complete information about the physicochemical properties of foods and food constituents, they are often used jointly with other analytical techniques such as X-ray diffraction (XRD), optical methods, rheological techniques such as dynamical mechanical analysis (DMA) or dynamical mechanical thermal analysis (DMTA), as well as high- and low-field nuclear magnetic resonance (NMR) spectroscopy.

Food processing technologies investigated

Many technologies used in the food industry may be investigated by thermal analysis and calorimetry: roasting, extraction, extrusion, freezing, freeze-drying, spray-drying, fermentation, interesterification, fractionation, hydrogenation as well as the more recent ones such as high hydrostatic pressure treatment and turbo-drying.

Thermal behaviour of major food constituents

In food many physicochemical effects can be observed in the temperature range between –50°C and 300°C. These thermal phenomena may be either endothermic, such as melting, gelatinisation, denaturation, evaporation, or exothermic such as crystallization, oxidation, fermentation. Glass transitions are observed as a shift in the baseline; this information is associated with water content and water activity determinations.

Specific heats (c_p) of foods can be calculated (Gekas, 1992) on the basis of the specific heats of the main constituents (by proportional additivity of the respective masses), but they can also be determined by DSC. The basic principles of such measurements have been described (Raemy and Lambelet, 1982) and many values of food specific heats are given in the literature (Mohsenin, 1980).

THERMAL BEHAVIOUR OF CARBOHYDRATES

The main phenomena observed with carbohydrates are release of crystallization water, melting, decomposition, gelatinisation of starch in the presence of water, retrogradation of the gel as well as glass transition, relaxation and crystallization of amorphous samples (Raemy and Schweizer, 1983; Raemy *et al.*, 1993; Roos, 1995; Blanshard and Lillford, 1993). Tables with melting and decomposition temperatures as well as enthalpies are given in the literature (Raemy and Schweizer, 1983).

Figure 1 shows a typical calorimetric curve of amorphous sucrose with glass transition and relaxation at 50°C, crystallization at 90°C as well as melting above 150°C. Figure 2 presents calorimetric curves of amorphous sucrose at increasing water activities: T_g and crystallization temperature diminish with increasing water activities.

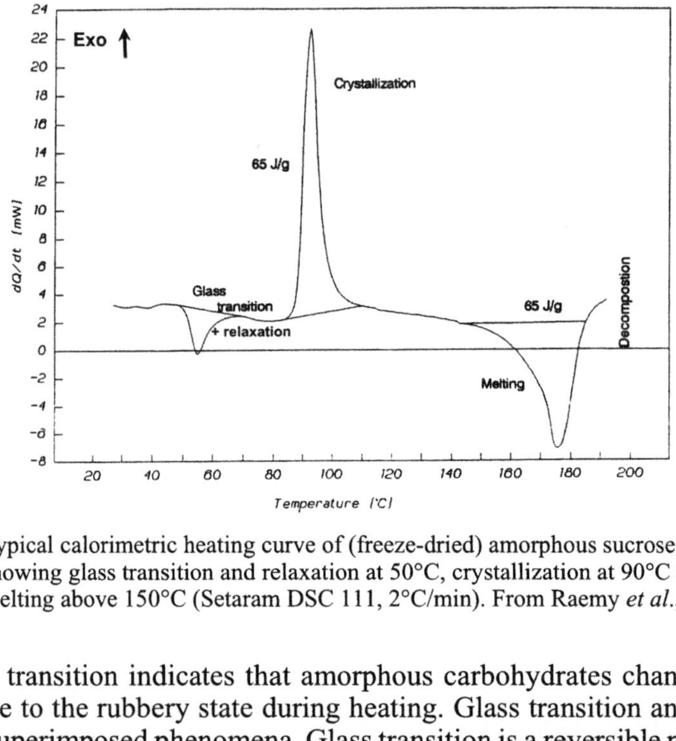

Fig. 1 Typical calorimetric heating curve of (freeze-dried) amorphous sucrose showing glass transition and relaxation at 50°C, crystallization at 90°C and melting above 150°C (Setaram DSC 111, 2°C/min). From Raemy *et al.*, 1993

Glass transition indicates that amorphous carbohydrates change from the glassy state to the rubbery state during heating. Glass transition and relaxation are often superimposed phenomena. Glass transition is a reversible phenomenon observed in DSC experiments as a change in baseline level, whereas relaxation is a non-reversible endothermic transition. Thus, performing two consecutive DSC experiments on the same sample can distinguish between these two phenomena (with MDSC only one scan is necessary as it separates reversible from non-reversible transitions). Glass transition is of particular interest in relation to storage of frozen products and food powders, and also for gas retention in powders foreseen to foam when dissolved. The gas, normally nitrogen, is for example introduced into an amorphous disaccharide matrix at a temperature above T_g, *i.e.* when the disaccharide are in a rubbery state, and encapsulated below T_g when the disaccharide are in the glassy state (Vuataz, 2002; Schoonman *et al.*, 2002). Glass transitions are more difficult to observe by DSC in starch than in mono- or disaccharides, but accurate T_g values as a function of water content can be found in the literature (Zeleznak and Hoseney, 1987; Parker and Ring, 2001).

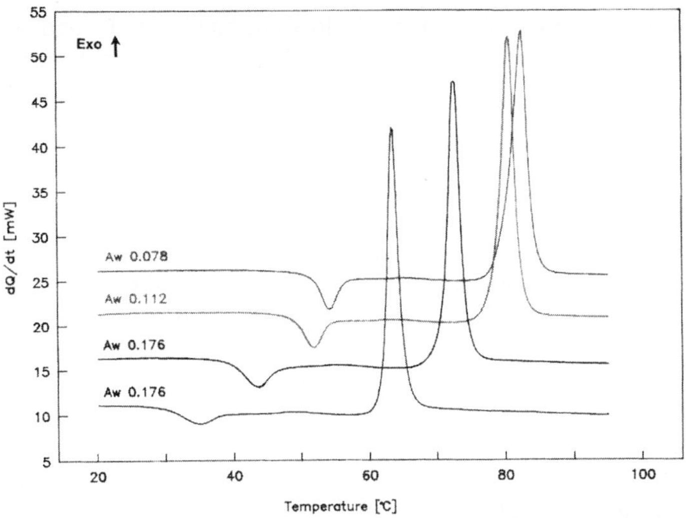

Fig. 2 Calorimetric heating curves of amorphous sucrose at different water activities (Micro-DSC III, 1°C/min) showing the effect of increasing water activities on glass transition + relaxation and on crystallization. From Raemy *et al.*, 1993

Thermogravimetry can be very useful for studying release of crystallization water, by indicating which endothermic transition observed by calorimetry corresponds to a mass loss.

Gelatinisation of starch-water systems is an endothermic non-reversible phenomenon easily observed by DSC (Ghani *et al.*, 1999). Retrogradation, which is a slow and low energy recrystallization process, can be followed by isothermal microcalorimetry (Raemy *et al.*, 1990; Silverio *et al.*, 1996), but is more often characterized after a storage period by measuring the melting transition of the retrogradated gels (Karim *et al.*, 2000). Penetrometry, DMA or DMTA provide complementary information on gelatinisation and retrogradation, which are associated with rheological modification of the products (Roulet *et al.*, 1988).

THERMAL BEHAVIOUR OF LIPIDS

Calorimetry (DTA, DSC) has been a method of choice to characterize the thermal properties of triacylglycerols (TAGs) for over 50 years, in particular their polymorphic behaviour.

Overall we can define three types of applications for lipids: determination of the thermodynamic parameters of the liquid/solid phases, monitoring of crystallization kinetics, determination of lipid quality and oxidative stability.

Thermodynamic parameters

The most common use of DSC in the lipid field has been to identify the polymorphism of TAGs and fat (Arishima *et al.*, 1991; Dimick and Manning, 1987; Garti and Sato, 1988; Huyghebaert and Hendrickx, 1971; Loisel *et al.*, 1998; Lovegren *et al.*, 1976; Merken and Vaeck, 1980; Minato *et al.*, 1997; Rousset, 1997; Rousset and Rappaz, 1996; Sato *et al.*, 1989; Sato, 1996; Spigno *et al.*, 2001; Wille and Lutton, 1966). This is done by measuring the melting enthalpy (between 50 and 200 J/g) and the melting temperature (pure components) or temperature range (complex mixtures like fat) of the phases present. As the polymorphism of TAGs is monotropic, only one form is thermodynamically stable. Thus, observing all polymorphs is not always easy (see next paragraph). A solution is to crystallize the fat using a wide range of cooling rates (*e.g.*, between 0.5°C/min and 50°C/min). Figure 3 presents the melting curves of the five most stable phases of cocoa butter; the least stable form I could not be observed due to the limited cooling capacity of the apparatus used. Changes in melting temperature and enthalpy have also been correlated to fat composition (Chaiseri and Dimick, 1989; Tan *et al.*, 2000).

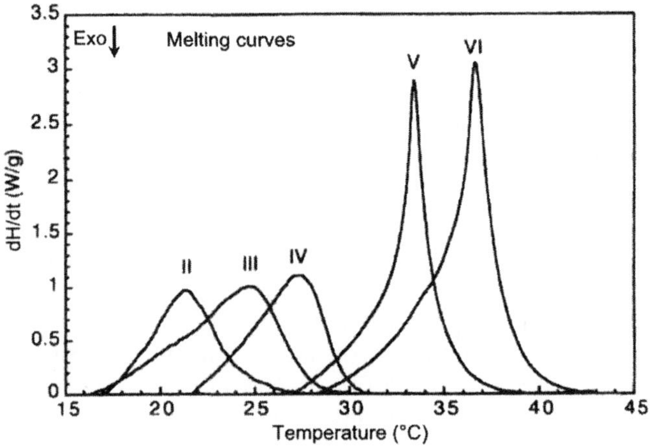

Fig. 3 DSC heating curves of 5 polymorphs of cocoa butter: II(α), III(β'_2), IV(β'_1), V(β_2) and VI(β_1) (Mettler FP900, 5°C/min). From Rousset, 1997

For binary or ternary mixtures of TAGs or fats, DSC has been used to determine phase diagrams or iso-solid diagrams, by identifying the temperature stability domains of the various phases formed (Ali and Dimick, 1994a; Ali and Dimick, 1994b; Culot, 1994; Elisabettini *et al.*, 1998; Knoester, 1972; Koyano *et al.*, 1992; Loisel *et al.*, 1998; Muhammad and Dimick, 1994; Rousset *et al.*, 1998; Timms, 1980; Timms, 1994). As previously mentioned, DSC is often used in combination with XRD for unambiguous identification.

The solid fat content (SFC) curve represents the ratio of solid to liquid in a partially crystallized lipid as a function of temperature. SFC curves are currently used in the industry for fat selection and quality control. They can be obtained from the calorimetric melting curve by sequential peak integration (Lambelet *et al.*, 1986; Kaiserberger, 1989; Bhaskar, 1998). This determination requires precise knowledge of the melting enthalpy of each phase for each fraction present in the sample, which is very difficult to assess for most fats.

As already mentioned, specific heat is another parameter that can be determined by DSC. It has been measured for various TAGs and fats (Roberts and Pearce, 1983; Rousset, 1997).

Kinetic parameters

A second domain where DSC is useful is the study of the crystallization kinetics of TAGs and fats, and of the formation and stability of their various polymorphs as a function of time and temperature (Rousset *et al.*, 1996; Rousset *et al.*, 1997). For these experiments, the lipid sample has first to be heated to at least 20°C above the melting temperature of its stable polymorph to erase all memory effects. Crystallization is then achieved either isothermally after quenching at the desired temperature or at constant cooling rate. Kinetic information has been obtained by achieving measurements either isothermally (Koyano *et al.*, 1989; Koyano *et al.*, 1991; Metin and Hartel, 1998; Rousset, 1997; Ziegleder, 1985b; Ziegleder, 1990), or under various cooling conditions (Cebula and Smith, 1991; Kawamura, 1980). Complex thermal paths like tempering stages were also studied by calorimetry to understand precisely the mechanisms that induce the formation of stable crystalline forms (Rousset and Rappaz, 1997).

Precise kinetic parameters can be determined from isothermal crystallization experiments. The variation of SFC as a function of time is obtained by sequential integration of the crystallization peak. This SFC function is used to estimate parameters of crystallization with the help of the Avrami model or more complex ones (Foubert *et al.*, 2002; Kloek *et al.*, 2000; Rousset, 2002). Nucleation induction times can also be determined from isothermal crystallization experiments. This is the time needed before nucleation can appear and is a useful indicator of the nucleation rate, being inversely proportional to it (Rousset and Rappaz, 2001). These kinetic parameters are necessary for the modelling and prediction of crystallization with analytical or numerical models (Rousset, 2002). These models are tools to know how to crystallize lipids in the desired form.

In kinetic studies, DSC signal assignment may be ambiguous and need to be combined with XRD (Keller *et al.*, 1996) and even synchrotron XRD if transformations are rapid in regard to the acquisition time. Even better, new experiments simultaneously combining DSC and synchrotron XRD revolutionise the study of crystallization (Kalnin *et al.*, 2002; Lopez *et al.*, 2000; Lopez *et al.*, 2001a; Lopez *et al.*, 2002; Ollivon *et al.*, 2001). DSC can also be used simultaneously with microscopy to identify morphologies associated with polymorphs and cooling conditions (Rousset *et al.*, 1998; Rousset and Rappaz, 1996).

Kinetic studies also help to understand the effect of compositional changes (TAGs or minor components) on crystallization (Cebula and Smith, 1992; Garti *et al.*, 1988; Tan *et al.*, 2000; Vanhoutte *et al.*, 2002b; Vanhoutte *et al.*, 2002a; Wahnelt *et al.*, 1991).

However, as the samples are not agitated, results from DSC crystallization studies are often difficult to interpret in terms of process operating conditions (Rousset and Rappaz, 2001; Ziegleder, 1985a; Ziegleder, 1988b).

Quality control

DSC crystallization curves have been used to assess the quality of oils, in particular of heated oils (Gloria and Aguilera, 1998; Tan and Man, 1999). Similarly, contamination (adulteration) of fats can be detected by calorimetry either during crystallization or melting of lipid mixtures (Kowalski, 1989; Lambelet and Ganguli, 1983; Bringer *et al.*, 1991; Marikkar *et al.*, 2002).

Lipid oxidation is an exothermic phenomenon that can be followed, at least at elevated temperatures, by DSC or preferably by isothermal calorimetry (Tan and Man, 2002; Raemy *et al.*, 1987; Kowalski, 1989). Measurements can be performed under a static air atmosphere or, better, under oxygen flow or oxygen pressure. DSC isothermal experiments allow induction times of oxidation to be determined. Tables of oxidation induction times measured by isothermal heat flux calorimetry around 100°C are reported in the literature (Raemy *et al.*, 1987). For edible oils DSC induction times were shown to correlate well with corresponding values determined by traditional methods (Tan *et al.*, 2002). DSC can, therefore, be used to assess the oxidative stability of lipids (Raemy *et al.*, 1987; Kowalski, 1989; Tan and Man, 2002) or the efficiency of food antioxidants (Raemy *et al.*, 1987; Irwandi *et al.*, 2000; Tan and Man, 2002) in bulk lipids.

THERMAL BEHAVIOUR OF PROTEINS

The main phenomena observed by DSC during heating of protein solutions (egg white or dairy proteins such as β-lactoglobulin or α-lactalbumin) are endothermic transitions commonly called denaturation (Privalov and Khechinashvili, 1974; Harwalkar and Ma, 1990; Ferreira *et al.*, 1997). As an example Fig. 4 presents a Micro-DSC curve of lyzozyme-depleted egg white showing thermal denaturation.

These transitions, although often seen as a single peak in DSC experiments, are composed of data from changes in conformational state of the proteins (unfolding, denaturation) and subsequent aggregation. Whereas the native-to-denatured change in the protein state is a co-operative phenomenon that is accompanied by significant heat uptake, change in hydrophobic interactions during protein aggregation is an exothermic process (Privalov and Khechinashvili, 1974; Biliaderis, 1983; Hayakawa *et al.*, 1996; Gotham *et al*, 1992). Although very rare, an exothermic component due to protein aggregation was observed following endothermic unfolding denaturation (Rossi and Schiraldi, 1992). The endo-

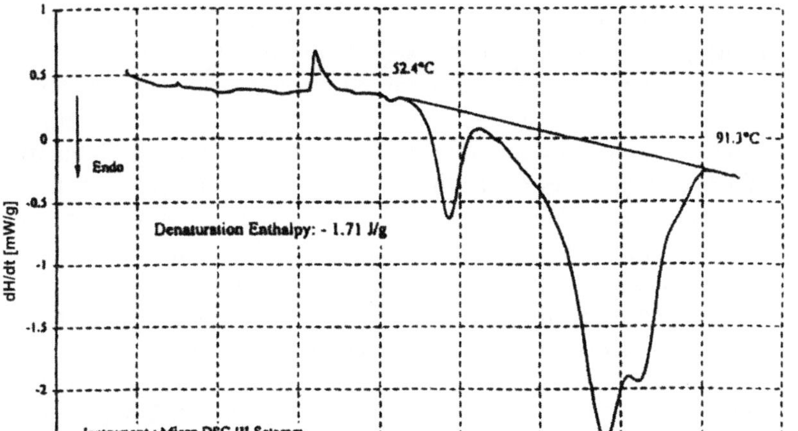

Fig. 4 Micro-DSC heating curve (1°C/min) of lyzozyme-depleted egg white (first run minus second run). From Ferreira *et al.*, 1997

thermic nature of DSC curves recorded during thermal treatments of protein solutions are an indication of the large contribution of denaturation as compared to aggregation. In fact, reported values of enthalpy changes during aggregation of proteins are small, for example 1–5 J/g for aggregation of whey proteins induced by $CaCl_2$ or proteolysis (Ju *et al.*, 1999). Though a significant underlying exothermic contribution of protein aggregation cannot always be ruled out, especially at high protein concentrations, the thermal effect due to aggregation is generally of such small amplitude in relation to the endotherm produced by denaturation that it is ignored (Donovan and Ross, 1973). In this sense, the temperature of the endothermic transition appearing in DSC analysis of protein solutions is indicative of thermal stability of the protein, and the surface of the peak corresponds to the denaturation enthalpy. For example the influence of hydration on the denaturation temperatures and enthalpies of lyzozyme has been given in the literature (Gregory, 1995). In the same way, the amount of protein that has been denaturated, e.g. during a technological process, can be determined by comparing the surface of denaturation transition to the total denaturation enthalpy (Arntfield and Murray, 1981).

Thermal denaturation of some proteins, *e.g.* egg (Ferreira *et al.*, 1997) or muscle proteins (Wright and Wilding, 1984; Togashi *et al.*, 2002) is a multi-step process. Thus, thermal denaturation of rabbit (Wright and Wilding, 1984) and walleye pollack (Togashi *et al.*, 2002) myosins was shown to occur via three cooperative endothermic processes associated with unfolding of different domains in the molecule.

DSC has also been used to study mixed protein systems. For example, the presence of gluten has been shown to shift the apparent transition temperature of whey protein towards lower temperatures; this was interpreted as gluten modification of the thermal stability of the environment of whey proteins (Lupano, 2000).

Glass transition and oxidation are primarily observed with dry proteins. As heat exchange associated with the glass transition of proteins is small, this transition is rather detected based on changes in rheological parameters obtained, for example by DMTA (Pouplin *et al.*, 1999). However, the sharp decrease in mechanical properties occurring when the sample passes through the glass transition zone depends on the frequency of the forces applied to the sample in the DMTA experiment. Thus, values of T_g obtained from DMTA are not always consistent with those obtained from DSC (Hartel, 2001).

THERMAL BEHAVIOUR OF WATER

Thermal analysis and calorimetry allow mainly the observation of crystallization (undercooling), melting (of ice) and vaporisation. Since the enthalpies corresponding to these phenomena are quite high (333 J/g for ice melting and 2255 J/g for water vaporisation) they can easily be studied by standard DSC in samples with low water content. It must here be remembered that under undercooling conditions crystallization enthalpies diminish with decreasing temperature (Franks, 1982).

THERMAL BEHAVIOUR OF AIR

In many foods (beer, ice cream, etc.) air is an important constituent (if we consider volume and not mass). But due to its low density compared to the other solid or liquid constituents, air does not change the thermal properties of foods, with exception of thermal conductivity as air is a good insulator.

Thermal behaviour of some minor constituents and ingredients

CAFFEINE, VITAMINS, MINERALS

Minor food constituents, such as caffeine or vitamins, can also be analysed by thermal analysis and calorimetry. Microcalorimetry has been used to monitor the thermal stability of vitamins A and C (Runge and Hefer, 2000; Spigno *et al.*, 1999). For caffeine we observed a solid-solid transition around 135°C and melting around 230°C, which is in agreement with corresponding data found in the literature (Hemminger and Cammenga, 1989). The high temperature stability of caffeine explains that this substance is still available to the consumer after coffee roasting. Specific heat is generally indicated for the ashes (minerals).

EMULSIFIERS

Calorimetry has been used to characterize pure emulsifiers, in particular their crystallization. In fact, crystallization and polymorphic behaviour need to be known since the physical state of the emulsifier determines its interactions with the lipid and the aqueous phase (Krog, 1997). For example, polymorphism of monoacylglycerols (MAGs) has been studied using DSC (Kodali *et al.*, 1985; Lutton, 1971; Watanabe, 1997).

Reactions and interactions between food constituents

CONTINUOUS SYSTEMS

In addition to thermal phenomena of isolated constituents, reactions and interactions between food constituents can also be evidenced by calorimetric techniques.

The browning reactions between proteins and reducing sugars, part of the Maillard reaction, are exothermic reactions that can be followed by DSC. They are associated with relatively small enthalpies (less than 100 J/g) and take place at temperatures above the T_g of the involved ingredients. Investigations of the Maillard reaction by calorimetry are rare. Examples are studies of Maillard reactions occurring between lactose and casein (Raemy *et al.*, 1983), as well as between starch and amino acids (Kapusniak, 1999). Figure 5 presents the calorimet-

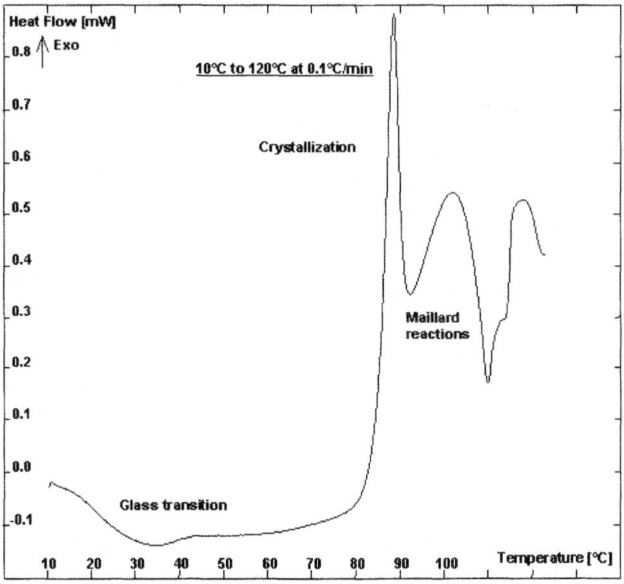

Fig. 5 Calorimetric heating curve of whey protein concentrate (WPC) showing glass transition around 30°C, crystallization around 88°C followed immediately by Maillard reactions (Micro-DSC III at low (0.1°C/min) heating rate)

ric curve of whey protein concentrate (WPC) showing glass transition around 30°C, crystallization around 88°C followed immediately by an exothermic peak due to Maillard reactions. These phenomena depend strongly on water activity.

Interactions between proteins and polysaccharides have indirectly been shown by DSC. Thus, interaction between ovalbumin and glutamate glucan was demonstrated throughout the reduction of temperature-induced precipitation of ovalbumin in the presence of calcium ions during the addition of glutamate glucan (Delben and Stefanchich, 1998); interaction between sodium caseinate with κ-carrageenan was evidenced by the observation of a progessive broadening of the DSC transition peak of carrageenan during the addition of sodium caseinate (Oakenfull *et al.*, 1999); complexation of a small globular protein by anionic polysaccharides was shown by a decrease in the temperature of protein denaturation (Tolstoguzov, 1993).

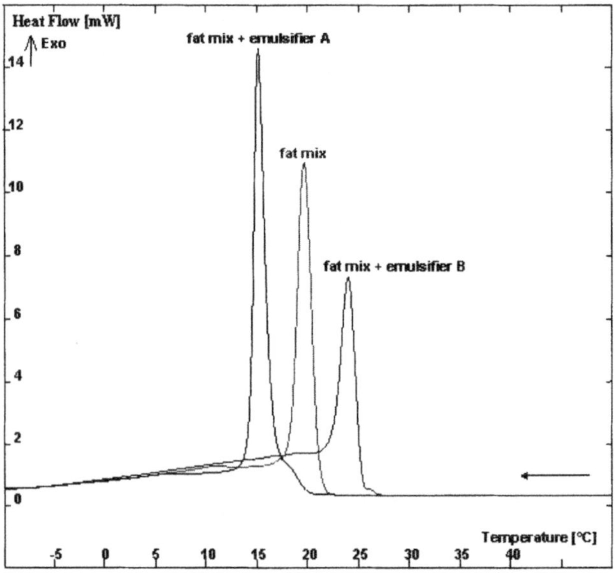

Fig. 6 Micro-DSC cooling curves of a mix of lauric and palm based fat (Butao ICE NG, Aarhus United, DK) in the absence of emulsifier, in the presence of emulsifier A (Datem, Panodan, Danisco, DK) or emulsifier B (mix of saturated MAG and diacylglycerol, Emuldan HA40, Danisco, DK) showing their effect on fat crystallization (Micro-DSC III at low (0.2°C/min) cooling rate). From Raemy, 2003

Lipid-polysaccharide interactions are commonly studied by DSC. For example DSC associated with high energy XRD has been proven to be very fruitful for studying starch-lipid complexes (Lebail *et al.*, 1999). The presence of inclusion complexes of amylose with lipids was revealed by transitions in the temperature range of 95–130°C (Eliasson, 1994; Villwock *et al.*, 1999; Lebail *et al.*, 1999;

Chien *et al.*, 1999; Ozcan and Jackson, 2002). This complex formation is the basis of an analytical method for measuring amylose content in starchy raw materials using DSC (Mestres *et al.,* 1996). The method is based on the enthalpy change which occurs during the exothermic formation of complexes between amylose and phospholipids during cooling. Formation of an amylopectin-lipid complex has been shown indirectly, by the decrease in gelatinisation enthalpy recorded for a waxy maize starch in the presence of lipids, and the reduced retrogradation of this maize-starch lipid mixture (Eliasson, 1994; Villwock *et al.*, 1999).

Binding of lipids to proteins has been evidenced by calorimetry throughout a shift in protein denaturation and enthalpy. For instance such modifications in DSC curves were observed with mixtures of lipids and either β-lactoglobulin (Puyol et al., 1994) or ovalbumin (Grinberg *et al.*, 2002).

In lipid systems, minor components can interact with TAGs and affect crystallization properties (Elisabettini *et al.*, 1996; Rao *et al.*, 2001; Siew and Ng, 2000; Smith *et al.*, 1994, Yuki, A. *et al.*, 1990). Figure 6 presents the influence of two emulsifiers on the crystallization of a fat during a cooling ramp. Due to different interactions, emulsifier A decreases the crystallization temperature of the fat whereas emulsifier B increases it. Emulsifier A is probably incorporated into the TAGs crystal structure and due to distortion retards nucleation and/or growth. Emulsifier B crystallizes before the main fat (small exothermal shoulder) and seeds crystallization of the fat.

DISPERSED SYSTEMS

Emulsifier-water systems

DSC has proven to be an excellent technique for studying the thermal behaviour of lipid-water systems that can be regarded as models of the lipid matrix of cell membranes (Blume, 1991). These systems involve emulsifiers that may exhibit highly ordered self-assembly structures, which are liquid crystalline phases. DSC has been applied to the study of endothermic phase transitions appearing in lipid-water systems when they transform from the gel to the liquid crystal phase (Chapman *et al.*, 1974; Tölgyesi *et al.*, 1985) and for determining thermotropic and lyotropic behaviour of these systems (Qui and Caffrey, 1999; Briggs *et al.*, 1996). DSC provides direct thermodynamic information such as transition temperatures and enthalpies, without revealing the identity of the transforming phases (Chung and Caffrey, 1992). Identification and structure characterization of liquid crystalline phases have to be achieved by another technique, e.g. XRD (Chung and Caffrey, 1992). Similarly, DSC has been applied to investigate the thermal behaviour of several emulsifier-water systems modified by changing the pH-value, the ionic composition of the environment or by chemical agents (Tölgyesi *et al.*, 1985; Forte *et al.*, 1998; Fournier *et al.*, 1998).

Temperature-composition phase diagrams for emulsifier-water systems are either described in the literature (Krog and Larsson, 1968; Krog and Borup, 1973; Qiu and Caffrey, 1999; Qiu and Caffrey, 2000; Misquitta and Caffrey,

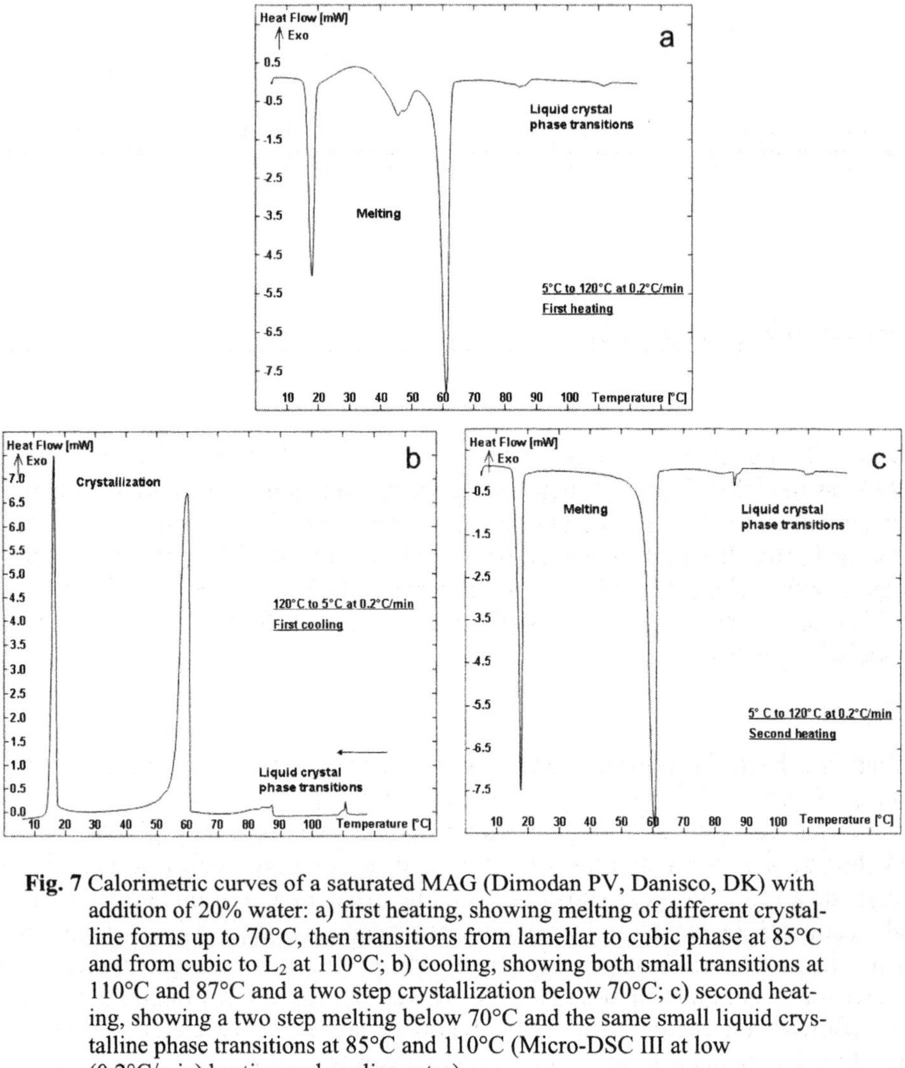

Fig. 7 Calorimetric curves of a saturated MAG (Dimodan PV, Danisco, DK) with addition of 20% water: a) first heating, showing melting of different crystalline forms up to 70°C, then transitions from lamellar to cubic phase at 85°C and from cubic to L_2 at 110°C; b) cooling, showing both small transitions at 110°C and 87°C and a two step crystallization below 70°C; c) second heating, showing a two step melting below 70°C and the same small liquid crystalline phase transitions at 85°C and 110°C (Micro-DSC III at low (0.2°C/min) heating and cooling rates)

2001) or given by the emulsifier suppliers. Even if XRD remains the most widely used technique, calorimetric techniques like DSC and particularly Micro-DSC can be of great help for establishing temperature-composition phase diagrams of emulsifier-water systems (Grabielle-Madelmont and Perron, 1983; Thoen, 1995; Demus *et al.*, 1999).

Some graphs are presented here to demonstrate the performance of Micro-DSC used at low heating and cooling rates. Figure 7a presents the calorimetric curve of a saturated MAG with 20% water showing melting of different crystalline forms up to 70°C and weak liquid crystal transitions at 85°C and 110°C. Comparison with following cooling curve (Fig. 7b) shows that there is practically no hysteresis between the temperatures of the phenomena; however, the crystalline form melting at 45°C has disappeared. The second heating curve (Fig. 7c) confirms the reversible character of most transitions.

Emulsions

Thermal behaviour of lipids in a dispersion or emulsified form is quite different from that of the same fat in bulk. DSC is useful as it is sensitive enough to catch transformations in dispersed phases, in particular when used simultaneously with synchrotron XRD (Kalnin *et al.*, 2002). Fat crystallization in oil-in-water emulsions has thus been followed by DSC in combination with either NMR spectroscopy (Özilgen *et al*, 1993) or XRD (Awad *et al.*, 2001; Lopez *et al.*, 2001b; Lopez *et al.*, 2001c; Ollivon *et al.*, 2001).

Thermal behaviour of raw and reconstituted foods

GENERAL CONSIDERATIONS

Most of the physico-chemical effects observed with food constituents are also found in the calorimetric curves of raw and reconstituted foods; examples are coffee beans, chicory roots, cereals or milk powders and infant formulas (Raemy, 1981; Raemy and Lambelet, 1982; Raemy and Löliger, 1982; Raemy *et al.*, 1983). The thermal phenomena observed with pure minor constituents will, however, not be observed once these constituents are dispersed in a food matrix. Many raw and reconstituted foods contain water. Therefore, measurements of such products in sealed cells above 100°C must only be performed with great precaution because of pressure increase due to water vapour and gas release during decomposition.

In addition to these phenomena, some interactions between food constituents, such as the Maillard reactions which occur between proteins and reducing sugars, may be observed, for example, as an exothermic phenomenon in calorimetric curves of milk powders or infant formulas.

Considering the emulsifier-water systems, calorimetric techniques can be applied in the structuring of food products (Heertje *et al.*, 1998) to obtain low calorie foods for instance or in encapsulating or creating new flavour cocktails (Leser *et al.*, 2003; Vauthey *et al.*, 2000). These new situations will require from the scientist or the food technologist new physicochemical measurements to determine the involved phases.

The melting of ice is often used to determine freezable water, which is considered as 'free water' by opposition to 'bound water' which is not freezable. As the enthalpy of vaporization of water at 100°C is high, boiling can easily be observed with open cells, even in foods containing small amounts of water. However, other exothermic effects such as carbohydrate crystallization or decomposition cannot be clearly observed under these conditions; measurements in relation to process safety are, therefore, achieved under inert gas pressure or in sealed cells.

DSC crystallization curves have also been used to determine the oil content in dry food products. The amount of oil is calculated from the area of the peak recorded during cooling of the product (Iannotta *et al.*, 2001; Aguilera and Gloria, 1997).

THE CASE OF CHOCOLATE

Chocolate is a suspension of particles (sugar, cocoa and milk powders) in a continuous fat matrix. Crystallization of fat determines important textural properties of chocolate, like firmness at room temperature and melting sensation in the mouth. Fat in chocolate is mainly cocoa butter. It has a complex polymorphism with six forms (I to VI). To obtain a good-quality product that is stable during storage, cocoa butter must be crystallized in the form V. Less stable forms have a lower melting point; they are thus less pleasant in the mouth; in addition, their recrystallization into more stable phases will induce fat bloom (white colour caused by big fat crystals reflecting light) during storage. Therefore, before being moulded or enrobed and cooled, chocolate must be tempered, *i.e.,* submitted to a specific thermal process. This consists in keeping the chocolate at a suitable temperature to form crystal nuclei followed by a second isothermal stage at a higher temperature to melt unstable phase nuclei and keep the stable ones. At the end of this process around 1% of form V crystals are present. After cooling, there is a mixture of phases IV and V, but after a few days of storage, all crystals have transformed into the form V. DSC has been used for many years to control and study the crystallization of fat during the production of chocolate (Wagner *et al.*, 1997), in particular to understand and follow:

- the tempering stage (Adenier *et al.*, 1984; Merken *et al.*, 1982; Schuster and Ziegleder, 1992; Yella *et al.*, 1996; Yella *et al.*, 1997; Ziegleder *et al.*, 1988; Ziegleder and Kegel, 1989); Figure 8 presents the typical melting peak of fat crystals of form V created after a tempering stage in chocolate;
- the cooling stage (Hausmann *et al.*, 1993a; Hausmann *et al.*, 1993b; Ziegleder *et al.*, 1988);
- the storage stage (Hausmann *et al.*, 1994), with specific studies on bloom formation (Kleef, 1995) or fat migration from an inner filling (Walter and Cornillon, 2002; Ziegleder and Schwingshandl, 1998) under accelerating conditions (temperature cycles).

In addition, similarly to what was described in paragraph 4.2 for pure lipids, DSC has been used to measure various properties of fat directly in the chocolate:

- thermodynamic properties: melting range and melting enthalpy, polymorphism (Hausmann *et al.*, 1994). DSC is the main method used to monitor polymorphism, which cannot easily be studied by XRD directly in chocolate due to the presence of crystalline sucrose that hides fat peaks;
- kinetics of crystallization (Stapley *et al.*, 1999; Stapley and Fryer, 2001; Ziegleder, 1988a);

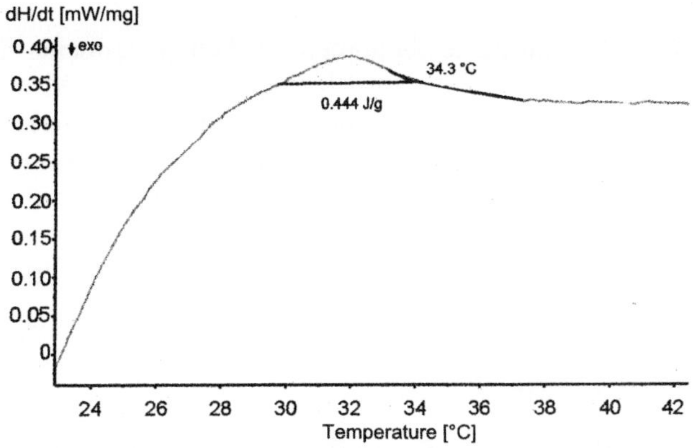

Fig. 8 Calorimetric heating curve of a chocolate after tempering, showing the melting of form V cocoa butter crystals created during this process stage. DSC allows to estimate the quantity of crystals formed, in this case about 0.9% (Netzsch DSC 200, 10°C/min). From Wagner *et al.*, 1997.

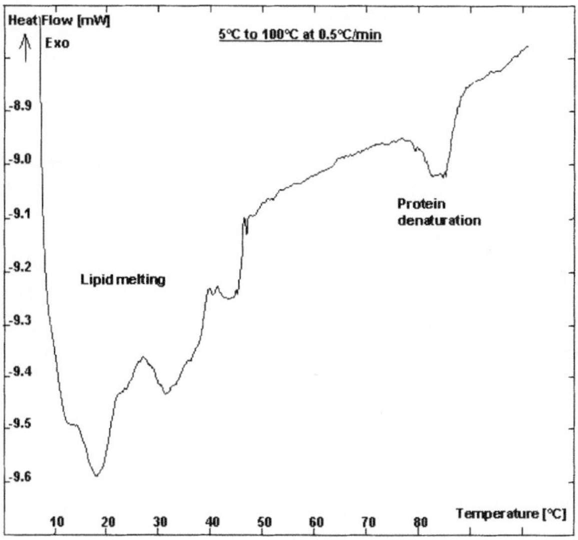

Fig. 9 Calorimetric heating curve of a foamed emulsion showing melting of different crystalline forms of the lipids below 60°C and protein denaturation around 85°C (Micro-DSC III, 0.5°C/min)

- effect of minor components of chocolate such as lecithin on the crystallization of cocoa butter (Hachiya *et al.*, 1989a; Hachiya *et al.*, 1989b; Hachiya *et al.*, 1989c; Hachiya *et al.*, 1989d; Savage and Dimick, 1995).

DSC can also be used to study physicochemical properties of other components, *e.g.*, crystallinity of sugars in chocolate products (Gloria and Sievert, 2001).

THERMAL BEHAVIOUR OF FOAMS

Foamed products present the same phenomena as their constituents. To demonstrate this, Micro-DSC can be of interest as the large sample size used with such instruments allows the foamed sample to be representative of the material. This is exemplified in Fig. 9 where the heating curve of a foamed emulsion (β-lactoglobulin, acacia gum, palm oil) shows melting of different crystalline forms of the lipid followed by protein denaturation.

Logically, because of the very low density of air, the specific heat of a foamed product is the same as that of the corresponding bulk sample.

Microbiological studies

Calorimetric techniques used in the isothermal mode allow the growth of micro-organisms under aerobic or anaerobic conditions to be followed (Gustafsson, 1991; von Stockar *et al.*, 1997; Riva *et al.*, 1998). Figure 10 presents isothermal calorimetric curves of malted yeast extract showing an important aerobic micro-organism growth under air, a weak growth under nitrogen and a very weak growth under strictly anaerobic conditions.

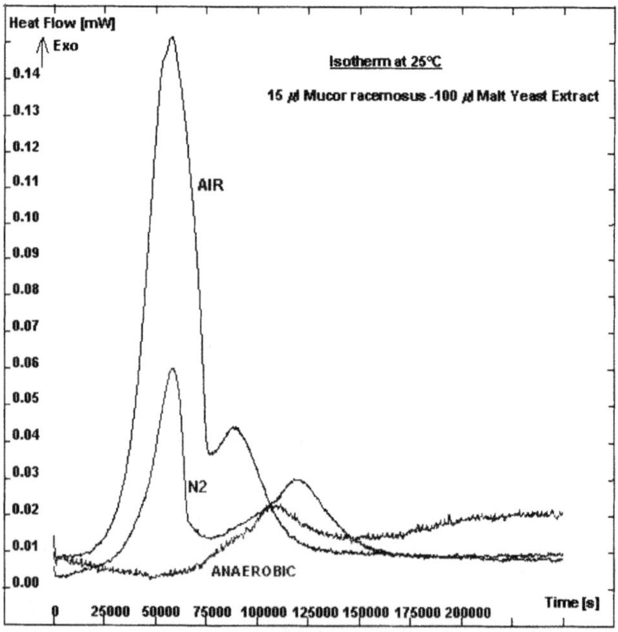

Fig. 10 Isothermal calorimetric curves of malt yeast extract under air, under nitrogen and under strict anaerobic conditions showing important micro-organism growth under air, weak growth under nitrogen and even weaker under strict anaerobic conditions (Micro-DSC III at 25°C). Courtesy of the Microbiology team at the Nestlé Research Centre

Process safety

Carbohydrate decomposition, which sometimes immediately follows melting, lipid oxidation (especially if oil is present at the surface of the product) and protein oxidation may present a hazard in industrial operations (roasting, high temperature drying, etc.).

The role of thermal analysis and calorimetry for determining safe conditions of industrial processes has already been explained elsewhere (Raemy and Löliger, 1985; Raemy *et al.*, 1985; Raemy, 1988; Raemy and Gardiol, 1987; Raemy, 2001). The application of adiabatic calorimetry to the study of cellulose decomposition has been described in detail (Raemy and Ottaway, 1991). Thermomanometry allows the pressure increase due to water vapour pressure, evolved roasting or decomposition gases and air compression (the pressure increase due to the dilatation of the pressure sensor has to be deduced) to be monitored.

In the case of safety studies, thermal analysis and calorimetric techniques must sometimes be applied unconventionally as measurements have to be carried out under conditions close to those of the process to be studied (Raemy, 1992; Raemy *et al.*, 2000).

Conclusion

This systematic approach demonstrates that thermal analysis and calorimetric techniques are increasingly applied in food science and food technology. The development of highly sensitive calorimeters has also opened up new fields of application. The various thermal transformations which can be observed by these techniques either during heating/cooling, or isothermally (for example after a temperature quench), allow food constituents as well as raw and reconstituted foods to be characterized. In the same way, reactions and interactions between food constituents can be studied with these techniques. Information provided by thermal analysis and calorimetric techniques is thus useful to food scientists and helps food technologists design processes allowing to prepare foods with optimal quality. Calorimetric information on foods is also useful for those dealing with food authenticity as well as for safety of industrial processes.

There is a large choice of instruments available today. All these instruments allow high performance measurements if one chooses the best instrument for a given problem. Each scientist has to determine the optimal parameters according to the studied problem, the amount of sample available and the equipment; it is often a question of compromise. The strategy in a laboratory devoted to thermal analysis and calorimetry is to buy complementary equipment, for example a DSC instrument for small samples and high heating rates and a Micro-DSC for large samples and low heating rates.

Although powerful, thermal analysis and calorimetric techniques have limitations. For example assignments of thermal transitions are not always unambiguous. Also they do not give a complete picture of a food. For these reasons these techniques are often used in combination with other physical techniques such as XRD (for unambiguous identification), optical methods, rheology, NMR spectroscopy as well as with chemical methods such as chromatography and electrophoresis. Thanks to these additional techniques, coupled (for example TG coupled to mass spectrometry) or not, thermal analysis and calorimetric techniques are even more powerful when applied to food science and food technology.

For the future, we think that isothermal microcalorimeters will give interesting results in microbiology, oxidation and new topics. The new microplate microcalorimetry instrument has to find his place here. Considering scanning mode, the Nano-DSC developed on the basis of the Privalov calorimeter will be useful mainly for studying proteins in solution. For the time being, instruments with very fast (100°C/min and more) heating rates leave us sceptical, as food materials are generally poor conductors: there will therefore be great temperature gradients inside the sample. In addition the rates of heating or cooling for large food quantities are normally not as high (microwave heating and liquid nitrogen cooling are rarely used industrially for food); there is thus a risk of obtaining results of no practical interest in this context. Among the related techniques, Thermally Stimulated Current (TSC) spectroscopy has certainly a role to play in food studies as well as Micro Thermal Analysis (μTA) which uses a tiny resistive thermal probe within an atomic force microscope (AFM) to collect local images related to sample topography and thermal conductivity. Considering modeling and kinetics, calorimetric information is often used (Schwarzberg, 2002) with Arrhenius' or more sophisticated equations to describe industrial processes; the developments based on finite element techniques presented recently (Roduit, 2002) have certainly a promising future.

References

Adenier, H., Chaveron, H., and Ollivon, M. (1984) Control of chocolate tempering with the aid of simplified differential thermal analysis, *Sciences des Aliments,* 4, 213–231.

Aguilera, J.M., and Gloria, H. (1997) Determination of oil in fried potato products by differential scanning calorimetry, *J. Agr. Food Chem.,* 45, 781–785.

Ali, A.R.M., and Dimick, P.S. (1994a) Thermal analysis of palm mid-fraction, cocoa butter and milk fat blends by DSC, *J. Am. Oil Chem. Soc.*, 71, 299–302.

Ali, A.R.M., and Dimick, P.S. (1994b) Melting and solidification characteristics of confectionery fats: anhydrous milk fat, cocoa butter and palm kernel stearin blends, *J. Am. Oil Chem. Soc.*, 71, 803–806.

Arishima, T., Sagi, N., Mori, H., and Sato, K. (1991) Polymorphism of POS. I. Occurrence and polymorphic transformation, *J. Am. Oil Chem. Soc.*, 68, 710–715.

Aronhime, J.S. (1988) Applications of thermal analysis (DSC) in the study of polymorphic transformations, *Thermochim. Acta,* 134, 1–14.

Arntfield, S.D., and Murray, E.D. (1981) The influence of processing parameters on food protein functionality. I. Differential scanning calorimetry as an indicator of protein denaturation, *Can. Inst. Food. Sci. Technol. J.,* 14, 269–275.

Awad, T., Hamada, Y., and Sato, K. (2001) Effects of addition of diacylglycerols on fat crystallization in oil-in-water emulsion, *Eur. J. Lipid Sci. Technol.*, 103, 735–741.

Bhaskar, A.R., Rizvi, S.S.H., Bertoli, C., Fay, L.B., and Hug, B. (1998) A comparison of physical and chemical properties of milk fat fractions obtained by two processing technologies, *J. Am. Oil Chem. Soc.*, 75, 1249–1264.

Biliaderis, C.G. (1983) Differential scanning calorimetry in food research. A review, *Food Chem.*, 10, 239–265.

Blanshard, J.M.V., and Lillford, P.J. (1993) *The Glassy State in Foods*, Nottingham University Press, Nottingham.

Blume, A. (1991) Biological calorimetry: membranes, *Thermochim. Acta,* 193, 299–347.

Briggs, J., Chung, H., and Caffrey, M. (1996) The temperature–composition phase diagram and mesophase structure characterization of the monoolein/water system, *J. Phys.* II (France), 6, 723–751.

Bringer, R., Rudzik, L., Weber, T., and Wüst, E. (1991) Detection of foreign fat in milk fat by means of differential calorimetry, *Milchwissenschaft,* 46, 304–307.

Cebula, D.J., and Smith, K.W. (1991) Differential scanning calorimetry of confectionery fats. Pure triglycerides: effects of cooling and heating rate variation, *J. Am. Oil Chem. Soc.,* 68, 591–595.

Cebula, D.J., and Smith, K.W. (1992) Differential scanning calorimetry of confectionery fats. II. Effects of blends and minor components, *J. Am. Oil Chem. Soc.,* 69, 992–998.

Chaiseri, S., and Dimick, P.S. (1989) Lipid and hardness characteristics of cocoa butters from different geographic regions, *J. Am. Oil Chem. Soc.*, 72, 1491–1496.

Chapman, D., Urbina, J., and Keough, K.M. (1974) Biomembranes phase transitions, *J. Biol. Chem.,* 249, 2512–2521.

Chien, J.T., Lien, Y.Y., and Shoemaker, C.F. (1999) Effect of polarity of complexing agents on thermal and rheological properties of rice starch gels, *Cereal Chem.*, 76, 837–842.

Chung, H., and Caffrey, M. (1992) Direct correlation of structure changes and thermal events in hydrated lipid established by simultaneous calorimetry and time-resolved x-ray diffraction, *Biophysical J.*, 63, 438–447.

Culot, C. (1994) *Modélisation du Comportement Polymorphique des Triglycérides*, Thesis, Université Notre Dame de la Paix, Namur.

Delben, F., and Stefancich, S. (1998) Interaction of food polysaccharides with ovalbumin, *Food Hydrocolloids,* 12, 291–299.

Demus, D., Goodby, J., Gray, G. W., Spiess, H. W., and Vill, V. (1999) *Physical properties of liquid crystals*, Wiley–VCH, Weinheim.

Dimick, P. S. and Manning, D. M. (1987) Thermal and compositional properties of cocoa butter during static crystallization, *J. Am. Oil Chem. Soc.*, 64, 1663–1669.

Donovan, J. W., and Ross, K. D. (1973) Increase in the stability of avidin produced by binding of biotin. A differential scanning calorimetric study of denaturation by heat, *Biochem.*, 12, 512–517.

Eliasson, A. C. (1994) Interactions between starch and lipid studied by DSC, *Thermochim. Acta,* 246, 343–356.

Elisabettini, P., Desmedt, A., Durant, F. (1996), Polymorphism of stabilized and nonstabilized tristearin, pure and in the presence of food emulsifiers, *J. Am. Oil Chem. Soc.*, 73 (2), 187–192.

Elisabettini, P., Lognay, G., Desmedt, A., Culot, C., Istasse, N., Deffense, E., and Durant, F. (1998) Synthesis and physicochemical characterization of mixed diacid triglycerides that contain elaidic acid, *J. Am. Oil Chem. Soc.*, 75, 285–291.

Ferreira, M., Hofer, C., and Raemy, A. (1997) A calorimetric study of egg white proteins, *J. Thermal Anal.*, 48, 683–690.

Forte, L., Andrieux, K., Keller, G., Grabielle-Madelmont, C., Lesieur, S., Paternostre M., Ollivon, M., Bourgaux, C., and Lesieur, P. (1998) Sodium taurocholate-induced lamellar-micellar phase transitions of DPPC determined by DSC and X-ray diffraction, *J. Therm. Analysis Cal.*, 51, 773–782.

Foubert, I., Vanrolleghem, P. A., Vanhoutte, B., and Dewettinck, K. (2002) Dynamic mathematical model of the crystallization kinetics of fats, *Food Res. Int.*, 35, 945–956.

Fournier, I., Barwicz, J., and Tancrède, P. (1998) The structuring effects of amphotericin B on pure and ergosterol- or cholesterol-containing dipalmitoylphosphatidylcholine bilayers : a differential scanning calorimetry study, *Biochim. Biophys. Acta,* 1373, 76–86.

Franks, F. (1982) The properties of aqueous solutions at subzero temperatures, in Franks, F. (ed.) *Water: a comprehensive treatise*, Plenum, New York.

Garti, N., and Sato, K. (1988) *Crystallization and Polymorphism of Fats and Fatty Acids*, Marcel Dekker, New York.

Garti, N., Schlichter, J., and Sarig, S. (1988) DSC studies concerning polymorphism of saturated monoacid triglycerides in the presence of food emulsifiers, *Fett Wiss. Technol.*, 90, 295–299.

Gekas, V. (1992) *Transport phenomena of foods and biological materials*, CRC Press, London.

Ghani, M. A., Che Man, Y., Ali A., and Hashim D. M. (1999) DSC: gelatinisation of sago starch in the presence of sucrose and sodium chloride, *J. Sci. Food Agric.*, 79, 2001–2009.

Gloria, H., and Aguilera, J.M. (1998). Assessment of the quality of heated oils by Differential Scanning Calorimetry, *J. Agr. Food Chem.*, 46, 1363–1368.

Gloria, H., and Sievert, D. (2001) Changes in the physical state of sucrose during dark chocolate processing, *J. Agr. Food Chem.*, 49, 2433–2436.

Gotham, S. M. Fryer, P. J., and Pritschard, A.M. (1992) α-lactoglobulin denaturation and aggregation reactions and fouling deposit formation: a DSC study, *Int. J. Food Sci. Technol.*, 27, 313–327.

Grabielle-Madelmont, C., and Perron, R. (1983) Calorimetric studies on phospholipid-water systems, *J. Colloid and Int. Sci.*, 95, 471–482.

Gregory, R. B. (1995) Protein hydration and glass transition behavior, in R.B. Gregory (ed.), *Protein-solvent interactions*, Marcel Dekker, New York, pp. 191–264.

Grinberg, V. Ya., Grinberg N. V., Mashkevich A. Ya., Burova, T. V., Tolstoguzov, V. B. (2002) Calorimetric study of interaction of ovalbumin with vanillin, *Food Hydrocolloids,* 16, 333–343.

Gustafsson, L. (1991) Microbiological calorimetry, *Thermochim. Acta,* 193, 145–171.

Hachiya, I., Koyano, T., and Sato, K. (1989a) Crystallization behavior of dark chocolate seeded with crystal of fat materials, *Yukagaku,* 38, 699–704.

Hachiya, I., Koyano, T., and Sato, K. (1989b) Observation of seeding effects on fat bloom of dark chocolate, *Food Microstructure,* 8, 257–261.

Hachiya, I., Koyano, T. and Sato, K. (1989c) Seeding effects on solidification behavior of cocoa butter and dark chocolate. I. Kinetics of solidification, *J. Am. Oil Chem. Soc.*, 66, 1757–1762.

Hachiya, I., Koyano, T., and Sato, K. (1989d) Seeding effects on solidification behavior of cocoa butter and dark chocolate. II. Physical properties of dark chocolate, *J. Am. Oil Chem. Soc.*, 66, 1763–1770.

Haines, P. J. (2002) *Principles of Thermal Analysis and Calorimetry*, Royal Society of Chemistry, Cambridge.

Hartel, R. W. (2001) *Crystallization in Foods*, Aspen Publishers, Gaithersburg.

Harwalkar, V. R., and Ma, C. Y. (1990) *Thermal Analysis of Foods*, Elsevier Applied Science, London.

Hausmann, A., Tscheuschner, H. D., and Tralles, I. (1993a) Influence of cooling conditions on crystallization of chocolate. II, *Zucker- u. Süsswaren Wirtschaft,* 46, 492–498.

Hausmann, A., Tscheuschner, H. D., Tralles, I., and Zscheile, H. (1993b) Influence of cooling conditions on the formation of new crystals in chocolate, *Zucker- u. Süsswaren Wirtschaft,* 46, 65–74.

Hausmann, A., Tscheuschner, H. D., and Tralles, I. (1994) Influence of storage conditions on the quality of chocolate. III, *Zucker- u. Süsswaren Wirtschaft,* 47, 118–123.

Hayakawa, I., Linko, Y., and Linko, P. (1996) Mechanism of high pressure denaturation of proteins, *Lebensm.-Wiss. u.-Technol.*, 29, 756–762.

Heertje, I., Roijers, E. C., and Hendricks, H. A. (1998) Liquid crystalline phases in the structuring of food products, *Lebensm.-Wiss. u.-Technol.*, 31, 387–396.

Hemminger, W. F., and Cammenga, H. K. (1989) *Methoden der Thermischen Analyse*, Springer-Verlag, Berlin, 11, 278–289.

Higami, M., Ueno, S., Segawa, T., Iwanami, L., and Sato, K. (2003) Simultaneous synchrotron radiation X-ray diffraction-DSC analysis of melting and crystallization behavior of trilauroylglycerol in nanoparticles of oil-in-water emulsion, *J. Am. Oil Chem. Soc.*, 80, 731–739.

Huyghebaert, A., and Hendrickx, H. (1971) Polymorphism of Cocoa Butter, shown by Differential Scanning Calorimetry, *Lebensm.-Wiss. u.-Technol.*, 4, 59–63.

Iannotta, N., Oliviero, C., Ranieri, G. A., and Uccella, N. (2001) Determination of the oil content in olives by the DSC technique, *Eur. Food Res. Technol.*, 212, 240–243.

Irwandi, J., Man, Y. B., Kitts, D. D., Bakar, J., and Jinap, S. (2000) Synergies between plant antioxidant blends in preventing peroxidation reactions in models and food oil systems, *J. Am. Oil Chem. Soc.*, 77, 945–950.

Ju, Z. Y., Hettiarachchy, N., and Kilara, A. (1999) Thermal properties of whey protein aggregates, *J. Dairy Sci.*, 82, 1882–1889.

Kaiserberger, E. (1989) DSC investigations of the thermal characterization of edible fats and oils, *Thermochim. Acta,* 151, 83–90.

Kalnin, D., Garnaud, G., Amenitsch, H., and Ollivon, M. (2002) Monitoring fat crystallization in aerated food emulsions by combined DSC and time-resolved synchrotron X-ray diffraction, *Food Res. Int.*, 35, 925-934.

Kapusniak, J., Ciesielsky, W., Koziol, J. J., and Tomasik, P. (1999) Reaction of starch with amino α-acids, *Eur. Food Res. Technol.*, 209, 325–329.

Karim, A. A., Norziah, M. H., and Seow, C. C. (2000) Methods for the study of starch retrogradation, *Food Chem.*, 71, 9–36.

Kawamura, K. (1980) The DSC thermal analysis of crystallization behavior in palm oil. II, *J. Am. Oil Chem. Soc.*, 57, 48–52.

Keller, G., Lavigne, F., Loisel, C., Ollivon, M., and Bourgaux, C. (1996) Investigation of the complex thermal behavior of fats, *J. Thermal Anal.*, 47, 1545–1565.

Kleef, F. V. (1995) Crystallization and fat bloom of chocolate, *Voedingsmiddelentechn.*, 28, 11–13.

Kloek, W., Walstra, P., and van Vliet, T. (2000) Crystallization kinetics of fully hydrogenated palm oil in sunflower oil mixtures, *J. Am. Oil Chem. Soc.*, 77, 389–398.

Knoester, M. (1972) Solid-liquid equilibrium of binary mixtures of triglycerides with stearic and palmitic chains, *Chem. Phys. Lipids*, 9, 309–319.

Kodali, D. R., Redgrave, T. G., Small, D. M., and Atkinson, D. (1985) Structure and polymorphism of 3-acyl-sn-glycerols, *Biochem.*, 24, 519–525.

Kowalski., B. (1989) Determination of oxidative stability of edible vegetable oils by pressure differential scanning calorimetry, *Thermochim. Acta*, 156, 347–358.

Kowalski., B. (1989) Sub-ambient differential scanning calorimetry of lard and lard contaminated by tallow, *Int. J. Food Sci. Technol.*, 24, 415–420.

Koyano, T., Hachiya, I., Arishima, T., Sagi, N., and Sato, K. (1991) Polymorphism of POS. II. Kinetics of melt crystallization, *J. Am. Oil Chem. Soc.*, 68, 716–718.

Koyano, T., Hachiya, I., Arishima, T., Sato, K., and Sagi, N. (1989) Polymorphism of POP and SOS. II. Kinetics of melt crystallization, *J. Am. Oil Chem. Soc.*, 66, 675–679.

Koyano, T., Hachiya, I., and Sato, K. (1992) Phase behavior of mixed systems of SOS and OSO, *J. Phys. Chem.*, 96, 10514–10520.

Krog, N., and Larsson, K. (1968) Phase behaviour and rheological properties of aqueous systems of industrial distilled monoglycerides, *Chem. Phys. Lipids*, 2, 129–143.

Krog, N., and Borup, A. P. (1973) Swelling behaviour of lamellar phases of saturated monoglycerides in aqueous systems, *J. Sci. Food Agric.*, 24, 691–701.

Krog, N. (1997) Food emulsifiers and their chemical and physical properties, in S. Friberg and K. Larsson (eds.), *Food Emulsions*, Marcel Dekker, New York, pp. 141–188.

Lambelet, P., Desarzens, C., and Raemy, A. (1986) Comparison of NMR and DSC methods for determining the solid fat content of fats, *Lebensm.-Wiss. u.-Technol.*, 19, 77–81.

Lambelet, P., and Ganguli, N.C. (1983) Detection of pig and buffalo body fat in cow and buffalow ghees by differential scanning calorimetry, *J. Am. Oil Chem. Soc.*, 60, 1005–1008.

Le Bail, P., Bizot, H., Ollivon, M., Keller, G., Bourgaux, C., and Buléon, A. (1999) Monitoring the crystallization of amylose-lipid complexes during maize starch melting by synchrotron X-ray diffraction, *Biopolymers*, 50, 99–110.

Leser, M. E., Michel, M., and Watzke, H. J. (2003) Food goes nano-new horizons for food structure research, in E. Dickinson, T. van Vliet (eds.), *Food Colloids, Biopolymers and Materials*, Royal Society of Chemistry, Cambridge, pp. 3–13.

Loisel, C., Keller, G., Lecq, G., Bourgaux, C., and Ollivon, M. (1998) Phase transitions and polymorphism of cocoa butter, *J. Am. Oil Chem. Soc.*, 75, 425–439.

Lopez, C. Lesieur, P. Keller, G. and Ollivon, M. (2000) Thermal and structural behavior of milk fat - 1. Unstable species of cream, *J. Colloid Int. Sci.*, 229, 62–71.

Lopez, C. Lavigne, F. Lesieur, P. Bourgaux, C. and Ollivon, M. (2001a) Thermal and structural behavior of milk fat. Unstable species of anhydrous milk fat, *J. Dairy Sci.*, 84, 756–766.

Lopez, C., Lesieur, P., Keller, G., and Ollivon, M. (2001b) Crystallization in emulsion: application to thermal and structural behavior of milk fat, in N. Widlak, R. Hartel and S. S., Narine (eds.) *Crystallization and Solidification Properties of Lipids*, AOCS Press, Champaign, pp.190–199.

Lopez, C., Riaublanc, A., Lesieur, P., Bourgaux, C., Keller, G., and Ollivon, M. (2001c) Definition of a model fat for crystallization-in-emulsion studies, *J. Am. Oil Chem. Soc.*, 78, 1233–1244.

Lopez, C., Bourgaux, C., Lesieur, P., and Ollivon, M. (2002) Crystalline structures formed in cream and anhydrous milk fat at 4 C, *Lait,* 82, 317–335.

Lovegren, N. V., Gray, M. S., and Feuge, R. O. (1976) Polymorphic changes in mixtures of confectionery fats, *J. Am. Oil Chem. Soc.*, 53, 83–88.

Lupano, C. E. (2000) Gelation of mixted systems whey proteins concentrate-gluten in acidic solutions, *Food Res. Int.*, 33, 691–696.

Lutton, E. (1971) The phases of saturated 1-monoglycerides C_{14}-C_{22}, *J. Am. Oil Chem. Soc.*, 48, 778–781.

Marikkar, J. M. N., Lai, O. M. Ghazali, H. M., and Man, Y. B. C. (2002) Compositional and thermal analysis of RBD palm oil adulterated with lipase-catalyzed interesterified lard, *Food Chem.*, 76, 249–258.

Merken, G. V., and Vaeck, S. V. (1980) Study of polymorphism of cocoa butter by differential scanning calorimetry, *Lebensm.-Wiss. u.-Technol.*, 13, 314–317.

Merken, G. V., Vaeck, S. V., and Dewulf, D. (1982) Determination of the technological properties of cocoa butter by means of differential scanning calorimetry, *Lebensm.-Wiss. u.-Technol.*, 15, 195–198.

Mestres, C., Matencio, F., Pons, B., Yajid, M., and Fliedel, G. (1996) A rapid method for the determination of amylose content by using differential scanning calorimetry, *Starch,* 48, 2–6.

Metin, S., and Hartel, R. W. (1998) Thermal analysis of isothermal crystallization kinetics in blends of cocoa butter with milk fat or milk fat fractions, *J. Am. Oil Chem. Soc.*, 75, 1617–1624.

Minato, A., Ueno, S., Yano, J., Smith, K., Seto, H., Amemiya, Y., and Sato, K. (1997) Thermal and structural properties of sn-1,3-dipalmitoyl-2-oleoylglycerol and sn-1,3-dioleoyl-2-palmitoylglycerol binary mixtures examined with synchrotron radiation X-ray diffraction, *J. Am. Oil Chem. Soc.*, 74, 1213–1220.

Misquitta, Y., and Caffrey, M. (2001) Rational design of lipid molecular structure, *Biophysical J.*, 81, 1047–1058.

Mohsenin, N. N., (1980) *Thermal Properties of Foods and Agricultural Materials*, Gordon and Breach, New York

Muhammad, A. A., and Dimick, P. S. (1994) Melting and solidification characteristics of confectionery fats: anhydrous milk fat, cocoa butter and palm kernel stearin blends, *J. Am. Oil Chem. Soc.*, 71, 803–806.

Oakenfull, D. Miyoshi, E. Nishinari, K. and Scott, A. (1999) Rheological and thermal properties of milk gels formed with k-carragenan. I. Sodium caseinate, *Food Hydrocolloids,* 13, 525–533.

Ollivon, M., Loisel, C., Lopez, C., Lesieur, P., Artzner, F., and Keller, G. (2001) Simultaneous examination of structural and thermal behaviors of fats by coupled X-ray diffraction and differential scanning calorimetry techniques: application to cocoa butter polymorphism, in N. Widlak, R. Hartel and S. S. Narine (eds.), *Crystallization and Solidification Properties of Lipids*, AOCS Press, Champaign, pp. 34–41.

Ozcan, S., and Jackson D. S. (2002) The impact of thermal events on amylose–fatty acid complexes, *Starch,* 54, 593–602.

Özilgen, S., Simoneau, C., German, J. B., McCarthy, M. J., and Reid, D. S. (1993) Crystallization kinetics of emulsified triglycerides, *J. Sci. Food Agric.* 61, 101–108.

Parker, R., and Ring, S. G. (2001) Aspects of the physical chemistry of starch, *J. Cereal Sci.,* 34, 1–17.

Pouplin, M., Redl, A., and Gontard, N. (1999) Glass transition of wheat gluten plasticized with water, glycerol or sorbitol, *J. Agric. Food Chem.,* 47, 538–543.

Privalov, P. L., and Khechinashvili, N.N. (1974) A thermodynamic approach to the problem of stabilization of globular protein structure: a calorimetric study, *J. Mol. Biol.,* 86, 665–684.

Puyol, P., Perez, M. D., Peiro, J. M., and Calvo, M. (1994) Effect of binding of retinal and palmitic acid to bovine β-lactoglobulin on its resistance to thermal denaturation, *J. Dairy Sci.,* 77, 1494.

Qiu, H., and Caffrey, M. (1999) Phase behavior of the monoerucin/water system, *Chem. Phys. Lipids,* 100, 55–79.

Qiu, H., and Caffrey, M. (2000) The phase diagram of monoolein/water system: metastability and equilibrium aspects, *Biomaterials,* 21, 223–234.

Raemy, A. (1981) Differential thermal analysis and heat flow calorimetry of coffee and chicory products, *Thermochim. Acta,* 43, 229–236.

Raemy, A., and Loeliger, J. (1982) Thermal behavior of cereals studied by heat flow calorimetry, *Cereal Chem.,* 59, 189–191.

Raemy, A., and Lambelet, P. (1982) A calorimetric study of self-heating in coffee and chicory, *J. Food Technol.,* 17, 451–460.

Raemy, A., and Schweizer, T. F. (1983) Thermal behaviour of carbohydrates studied by heat flow calorimetry, *J. Thermal Anal.,* 28, 95–108.

Raemy, A., Hurrell, R., and Löliger, J. (1983) Thermal behavior of milk powders studied by differential thermal analysis and heat flow calorimetry, *Thermochim. Acta,* 65, 81–92.

Raemy, A., and Löliger, J. (1985) Self-ignition of powders studied by high pressure differential thermal analysis, *Thermochim. Acta,* 85, 343–346.

Raemy, A., Lambelet, P., and Löliger, J. (1985) Thermal analysis and safety in relation to food processing, *Thermochim. Acta,* 95, 441–446.

Raemy, A., and Gardiol, M. (1987) Paramètres thermodynamiques et sécurité des opérations industrielles, Association Scientifique Internationale du Café (ASIC), 12e Colloque, Montreux (CH), 320–330.

Raemy, A., Frölicher, I., and Loeliger, J. (1987) Oxidation of lipids studied by isothermal heat flux calorimetry, *Thermochim. Acta,* 114, 159–164.

Raemy, A. (1988) Une méthodologie d'investigation des réactions exothermiques, de l'auto-inflammation et de l'explosion de poussières adaptée aux produits alimentaires, C.A.T., Lille (F), 19, pp. C3.1–C3.3.

Raemy, A., Kaabi, C., and MacInnes, W. M. (1990) Mise en évidence de la rétrogradation de l'amidon par microcalorimétrie isotherme, in AFCAT (ed.) *Calorimetrie et Analyse Thermique*, Clermont-Ferrand (F), 20–21, pp. 73–78.

Raemy, A., and Lambelet, P. (1991) Thermal behaviour of foods, *Thermochim. Acta,* 193, 417–439.

Raemy, A., and Ottaway, M. (1991) The use of high pressure DTA, heat flow and adiabatic calorimetry to study exothermic reactions, *J. Thermal Anal.,* 37, 1965–1971.
Raemy, A. (1992) From thermal analysis to safety science, *J. Thermal Anal.,* 38, 437–443.
Raemy, A., Kaabi, C., Ernst E., and Vuataz, G. (1993) Precise determination of low level sucrose amorphism by microcalorimetry, *J. Thermal Anal.,* 40, 437–444.
Raemy, A., Lambelet, P., and Garti, N. (2000) Thermal behavior of food and food constituents, in N. Garti (ed.), *Thermal Behavior of Dispersed Systems*, Marcel Dekker, New York, pp. 477–505.
Raemy, A. (2001) La mesure des réactions exothermiques des aliments par analyse thermique différentielle sous pression et calorimétrie différentielle programmée, in AFCAT (ed), *Calorimetrie et Analyse Thermique*, Hammamet (TN), 32, pp. 63–64.
Raemy, A. (2003) Behavior of foods studied by thermal analysis : introduction, *J. Thermal Anal.,* 71, 273–278.
Rao, R., Sankar, K. U., Sambaiah, K., and Lockesh, B. R. (2001) Differential scanning calorimetry studies on structured lipids from coconut oil triglycerides containing stearic acid, *Eur. Food Res. Technol.,* 212, 334–343.
Riva, M., Fessas, D., Franzetti, L., and Schiraldi, A. (1998) Calorimetric characterization of different yeast strains in doughs, *J. Thermal Anal.,* 52, 753–764.
Roberts, R. T., and Pearce, J. (1983) *Determination of the specific heats of confectionery products and commercial oils and fats*, Leatherhead Food Research Association
Roduit, B. (2002) Prediction of the progress of solid-state reactions under different temperature modes, *Thermochim. Acta,* 388, 377–387.
Roos, Y. (1995) *Phase Transition of Foods*, Academic Press, New York
Rossi, M., and Schiraldi, A. (1992) Thermal denaturation and aggregation of egg proteins, *Thermochim. Acta,* 119, 115–123.
Roulet, P. ,MacInnes, W. M., Würsch, P., Sanchez, R. M., and Raemy, A. (1988) A comparative study of the retrogradation kinetics of gelatinized wheat starch in gel and powder form using X-rays, DSC and dynamic mechanical analysis, *Food Hydrocolloids,* 2, 381–396.
Rousset, P., and Rappaz, M. (1996) Crystallization kinetics of the pure triacylglycerols glycerol-1,3-dipalmitate-2-oleate, glycerol-1-palmitate-2-oleate-3-stearate, and glycerol-1,3-distearate-2-oleate, *J. Am. Oil Chem. Soc.,* 73, 1051–1057.
Rousset, P. (1997) *Etude Expérimentale et Modélisation de la Cristallisation de Triacylglycérols et du Beurre de Cacao*, Thesis 1718, EPFL, Lausanne, Switzerland
Rousset, P., and Rappaz, M. (1997) Alpha-melt-mediated crystallization of 1-palmitoyl-2-oleoyl-3-stearoyl-sn-glycerol, *J. Am. Oil Chem. Soc.,* 74. 693–697.
Rousset, P., Rappaz, M., and Minner, E. (1998) Polymorphism and solidification kinetics of the binary system POS-SOS, *J. Am. Oil Chem. Soc.,* 75. 857–864.
Rousset, P., and Rappaz, M. (2001) Experimental study and computer modeling of the dynamic and static crystallization of cocoa butter, in N. Widlak, R. Hartel and S. Narine (eds.), *Crystallization and Solidification Properties of Lipids*, AOCS Press, Champaign, Ill, pp. 96–109.
Rousset, Ph. (2002) Modeling crystallization kinetics of triacylglycerols, in A.G. Marangoni and S.S. Narine (eds.), *Physical Properties of Lipids*, Marcel Dekker, New York, pp. 1–36.
Runge, F. E., and Hefer, R. (2000) Use of microcalorimetry in monitoring stability studies. Example: Vitamin A esters, *J. Agr. Food Chem.,* 48, 47–55.

Sato, K. (1996) Polymorphism of pure triacylglycerols and natural fats, in F.B. Padley (ed.), *Advances in Applied Lipid Research, volume 2*, JAI Press, London, pp. 213–268.

Sato, K., Arishima, T., Wang, Z. H., Ojima, K., Sagi, N., and Mori, H. (1989) Polymorphism of POP and SOS. I. Occurrence and polymorphic transformation, *J. Am. Oil Chem. Soc.*, 66, 664–674.

Savage, C. M., and Dimick, P. S. (1995) Influence of phospholipids during crystallization of hard and soft cocoa butters, *Manufacturing Confectioner,* 75, 127–132.

Schenz, T. W., (2003) Thermal analysis, in S.S. Nielsen (ed.), *Food Analysis*, third edition, Kluwer Academic, Plenum publishers, New York, pp. 517–528.

Schoonman, A., Ubbink, J. B., MacInnes, W. M., Watzke, H. J., (2002) Uptake and transport of gas in microstructured amorphous matrices, in H. Levine (ed.), *Amorphous food and pharmaceutical systems*, The Royal Society of Chemistry, Cambridge, pp. 98–112.

Schuster, S. C., and Ziegleder, G. (1992) DSC measurement of the degree of tempering of fluid pre-crystallized chocolate mass under production conditions, *Zucker- u. Süsswaren Wirtschaft,* 45, 324–326.

Schwarzberg, H. G. (2002) Modeling bean heating during batch roasting of coffee beans, in J. Welti-Chanes, G. V. Barbosa-Canovas, and J. M. Aguilera (eds), *Engineering and food for the 21st Century*, CRC Press, Boca Raton, pp. 871–890.

Siew, W. L., and Ng, W. L. (2000) Differential scanning thermograms of palm oil triglycerides in the presence of diglycerides, *J. Palm Oil Research,* 12, 1–7.

Silverio, J., Svensson E., Eliasson A.-C., and Olofsson G. (1996) Isothermal microcalorimetric studies on starch retrogradation, *J. Thermal Analysis,* 47, 1179–1200.

Smith, P. R., Cebula, D. J., and Povey, M. J. W. (1994), The effect of lauric-based molecules on trilaurin crystallization, *J. Am. Oil Chem. Soc.*, 71, 1367–1372.

Spigno, G., Marco de Faveri, D., and Perego, P. (1999) Thermal stability of ascorbic acid and sodium erithorbate, *Industrie Alimentari,* 38, 538–545.

Spigno, G., Pagella, C., and Faveri, D.D. (2001) DSC characterisation of cocoa butter polymorphs, *Ital. J. Food Sci.,* 13, 275–284.

Stapley, A. G. F., Tewkesbury, H., and Fryer, P. J. (1999) The effects of shear and temperature history on the crystallization of chocolate, *J. Am. Oil Chem. Soc.,* 76, 677–685.

Stapley, A. G. F., and Fryer, P. J. (2001) The effects of shear and temperature history on the crystallization of chocolate, in *Proceedings of the 8th International Congress on Engineering and Food* (ICEF 8), Lancaster, pp. 235–240.

Tan, C. P., and Man, Y. B. (1999) Quantitative differential scanning calorimetric analysis for determining total polar compounds in heated oils, *J. Am. Oil Chem. Soc.,* 76, 1047–1057.

Tan, C. P., and Man, Y. B. C. (2000) Differential scanning calorimetric analysis of edible oils: Comparison of thermal properties and chemical composition, *J. Am. Oil Chem. Soc.,* 77, 143–155.

Tan, C.P., and Man, Y.B. (2002) Recent developments in differential scanning calorimetry for assessing oxidative deterioration of vegetable oils, *Trends Food Sci. Technol.,* 13, 312–318.

Tan, C. P., Man, Y. B. C., Selamat, J., and Yusoff, M. S. A. (2002) Comparative studies of oxidative stability of edible oils by differential scanning calorimetry and oxidative stability index methods, *Food Chem.,* 76, 385–389.

Thoen, J. (1995) Thermal investigations of phase transitions in thermotropic liquid crystals, *Int. J. Modern Physics,* B9, 2157–2218.

Timms, R. E. (1980) The phase behaviour and polymorphism of mixtures of cocoa butter and milk fat, *Lebensm.-Wiss. u.-Technol.*, 13, 61–65.

Timms, R. E. (1994) Physical chemistry of fats, in *Fats in Food Products*, D.P.J. Moran, and K.K. Rajah (eds.), Blackie, Glasgow, pp. 1–27.

Togashi, M., Kakinuma, M., Nakaya, M., Ooi, T., and Watabe, S. (2002) Differential scanning calorimetry and circular dichroism spectrometry of walleye pollack myosin and light meromyosin, *J. Agr. Food Chem.*, 50, 4803–4811.

Tölgyesi, F., Szőgyi, M., and Györgyi, S. (1985) DSC study of the influence of chemical environment on the structure of lyotropic liquid crystals, *Thermochim. Acta,* 93, 37–40.

Tolstoguzov, V. B. (1993) Functional properties of food proteins: role of interactions in protein systems, in K. D. Schwenke and R. Mothes (eds.), *Food Proteins. Structure and Functionality*, VHC, Weinheim, pp. 203–209.

Vanhoutte, B., Dewettink, K., Foubert, I., Vanlerberghe, B., and Hyughebaert, A. (2002a) The effect of phospholipids and water on the isothermal crystallization of milk fat, *Eur. J. Lipids Sci. Technol.,* 104, 490–495.

Vanhoutte, B., Foubert, I., Duplacie, F., Huyghebaert, A., and Dewettinck, K. (2002b) Effect of phospholipids on isothermal crystallization and fractionation of milk fat, *Eur. J. Lipid Sci. Technol.,* 104, 738–744.

Vauthey, S., Milo, Ch., Frossard, Ph., Garti, N., Leser, M. E., and Watzke, H. J. (2000) Structured fluids as microreactors for flavor formation by the Maillard reaction, *J. Agric. Food Chem.,* 48, 4808–4816.

Villwock, V. K., Eliasson, A. C., Silverio, J., and BeMiller, J. N. (1999) Starch-lipid interactions in common, waxy, *ae du*, and *ae su2* maize starches examined by differential scanning calorimetry, *Cereal Chem.,* 76, 292–298.

Von Stockar, U., Duboc, P., Menoud, L., and Marison, I. V. (1997) On-line calorimetry as a technique for process monitoring and control in biotechnology, *Thermochim. Acta,* 300, 225–236.

Vuataz, G. (2002) The phase diagram of milk: a new tool for optimising the drying process, *Lait,* 82, 485–500.

Wagner, K. J., Bretschneider, U., and Rampke, T. (1997) Practical use of DSC for chocolate masses. Factors affecting the production process, *Zucker- u. Süsswaren Wirtschaft,* 50, 472–475.

Wahnelt, S., Meusel, D., and Tulsner, M. (1991) Influence of isomeric diglycerides on phase transitions of cocoa butter – investigations by isothermal DSC, *Fett-Wissenschaft-Technol.,* 93, 174–178.

Walter, P., and Cornillon, P. (2002) Lipid migration in two-phase chocolate systems investigated by NMR and DSC, *Food Res. Int.,* 35, 761–767.

Wang, G., Lin, H. N., Lin, S. ,and Huang, C. H. (1995) Phosphatidylcholines with *sn*-1 saturated and *sn*-2 cis-monounsaturated acyl chains. Their melting behavior and structures, *J. Biol. Chem.,* 270, 22738–22746.

Watanabe, A. (1997) On the sub-α-form and the α-form in monoacylglycerols, *J. Am. Oil Chem. Soc.,* 74, 1569–1573.

Wille, R., and Lutton, E. (1966) Polymorphism of cocoa butter, *J. Am. Oil Chem. Soc.,* 43, 491–496.

Wright, D. J., and Wilding, P. (1984) Differential scanning calorimetric study of muscle and its proteins: myosin and its subfragments, *J. Sci. Food Agric.*, 35, 35–372.

Yella, R. S., Full, N., Dimick, P. S., and Ziegler, G. R. (1996) Tempering method for chocolate containing milk-fat fractions, *J. Am. Oil Chem. Soc.*, 73, 723–727.

Yella, R. S., Full, N., Dimick, P. S., and Ziegler, G. R. (1997) Tempering method for chocolate containing milk-fat fractions, *Kennedy's Confection,* 4, 28–31.

Yuki, A., Matsuda, K., and Nishimura, A. (1990) Effect of sucrose polyesters on crystallization behavior of vegetable shortening and margarine fat, *J. Jap. Oil Chem. Soc.*, 39, 236–244.

Zeleznak, K. J., and Hoseney, R. C. (1987) The glass transition in starch, *Cereal Chem.*, 64, 121–124.

Ziegleder, G. (1985a) Improved crystallization behaviour of cocoa butter under shearing, *Int. Zeitschrift Lebensm. Technol. Verfahrenstechnik,* 36, 412–416.

Ziegleder, G. (1985b) The measurement of the crystallization of cocoa butter. The isotherm DSC method, *Zucker- u. Süsswaren Wirtschaft,* 38, 258–263.

Ziegleder, G. (1988a) Crystallization of chocolate masses. Part 1. Seed crystal formation, *Zucker- u. Süsswaren Wirtschaft,* 41, 165–168.

Ziegleder, G. (1988b) Crystallization of cocoa butter under static and dynamic conditions (differential scanning calorimetry, rheometry) *Süsswaren,* 32, 487–493.

Ziegleder, G. (1990) DSC thermal analysis and kinetics of cocoa butter crystallization, *Fett-Wissenschaft-Technol.*, 92, 481–485.

Ziegleder, G., Becker, K., Baumann, M., and Rosskopf, O. (1988) The crystallization of chocolate masses. Part II. Precrystallization and tempering, *Zucker- u. Süsswaren Wirtschaft,* 41, 238–243.

Ziegleder, G., and Kegel, M. (1989) The crystallization of chocolate masses. Part III. DSC determination of cooling crystallization, *Zucker- u. Süsswaren Wirtschaft,* 42, 338–342.

Ziegleder, G., and Schwingshandl, I. (1998) Kinetics of fat migration within chocolate products. III. Fat bloom, *Fett/Lipid,* 100, 411–415.

Chapter 5

Using DSC for monitoring protein conformation stability and effects on fat droplets crystallinity in complex food emulsions

P. Relkin*

Ecole Nationale Supérieure des Industries Alimentaires, Unité Mixte de Recherche Science de l'Aliment et de l'Emballage 1, Avenue des Olympiades, F- 91 744, Massy-France

Introduction

Proteins and fats are used as inherent parts of the formulation of many food emulsions, such as mayonnaise, salad dressing, frozen dessert, milk whipped-or ice creams [11, 18, 66, 67]. In these complex food emulsions, besides fatty acid and triacylglycerol composition of fat droplets, many factors such as structure and concentration of emulsifiers, droplet size, droplet-droplet interactions, nature of boundary layer at the oil/water interface have effects on their physical stability and organoleptic quality [1, 2, 4, 5, 37, 46, 63]. Their processing involves heating and cooling steps, and monitoring desired or detrimental effects of time-temperature parameters can help to maximize processing steps and can constitute a way to produce emulsions with a good storage quality.

Among the group of thermal analysis methods, differential scanning calorimetry is one of the frequently used techniques to relate structural behaviour of food materials (raw ingredients or food samples) to their composition and to processing factors [25, 26, 33, 50, 58, 69]. The published calorimetric data appear to be greatly influenced by several intrinsic and extrinsic factors, including chemical differences among food sample constituents (nature and concentration of a major component, salts, cellular content or sugars, other organic or mineral solutes), and also differences among DSC instrumentation (calorimeter response time, validity of equilibrium thermodynamics, and methodologies applied for analysis of DSC curves [6, 20, 59, 61, 65].

In this chapter we focused on examples of application of DSC for evaluation of heat-induced conformation changes of globular whey proteins in model solutions and in complex systems, in relation to growing of fat droplet crystallinity in protein-stabilized emulsions and by using power compensation and/or heat flux DSC systems.

---

* relkin@ensia.inra.fr

D. Lörinczy (ed.), The Nature of Biological Systems as Revealed by Thermal Methods, 99–126.

Thermal transitions contributing to structuration of food containing globular proteins and fats

STABILITY/INSTABILITY OF GLOBULAR PROTEINS

The conformational stability of folded native structures of proteins results from a balance of attractive and repulsive forces within the polypeptide itself and also between proteins and co-solvents/co-solutes, and between proteins and surfaces [10, 43]. Fibrillar proteins, of particular concern for meat science and industry, are compact particles with internal organisation. They display extended overall shapes; but present the same types of secondary structures as globular proteins, sharing basically the same denaturation properties with globular proteins (Lőrinczy and coll., this volume.). Prolamins and caseins constitute quite different types of proteins. When considered as individual molecules, they are much less compact and organised than globular proteins [39, 29]. The concept of denaturation cannot apply fully to them, but the degree to which they can interact with globular proteins (e.g in milk systems), is related to their techno-functionality in foods [24, 31, 32].

The high ordered structure of most globular proteins is due to forces such as

- Hydrophobic interactions: repulsive interaction between water molecules and non-polar amino acids in proteins, leading to minimal hydration of the buried hydrophobic core
- H-bonds: strong dipole-dipole attractive force between covalently bonded H atoms and electronegative atoms such as oxygen and nitrogen
- Van der Waals forces: interactions between fixed or induced dipoles
- Salt bridges: interactions between ionised amino acids

Perturbation in physico-chemical parameters such temperature, pressure, pH, is accompanied by changes from the initial conformational state to other spatial arrangements classified as denatured arrangements [43, 48]. The denatured states of a protein result from the breaking up of labile linkages leaving disorganised structures with altered surfaces and a more of their hydrophobic core exposed to aqueous medium. The term 'denatured' state describes an alteration in the original native structure without hydrolysis of primary covalent bonds, following a first reversible step where multimeric proteins may dissociate and denature within another or several unfolding processes, and following the simplified scheme [54, 59]:

$$N_x \text{ (native)} \overset{K_{eq}}{\leftrightarrow} x\,D \text{ (denatured/unfolded)} \overset{k}{\rightarrow} A \text{ (aggregated)} \qquad (1)$$

Proteins in their native folded structure have well-defined primary, secondary and tertiary structures that are characteristic of each particular protein. Analysis of DSC thermogrammes, based on equilibrium thermodynamics, provide energetic data on protein denaturation following the reversible two-state model [48], and also a non-two stage unfolding mechanism [20].

Proteins used in food manufacturing are most often in powdered forms, indicating that they were submitted to extraction, purification, pasteurisation, concentration and drying procedures [13, 19]. Under these conditions, the initial conformation state of commercially available proteins used in food industry is far from an initial folded state. And, since several decades DSC is used to evaluate the degree to which structural stability of proteins may be affected by perturbation such as that of environmental conditions used in food processing. Globular proteins which are stabilized by non-covalent interaction forces (hydrogen, ionic and hydrophobic bonding) and also by covalent cross-linking [32], can have more or less conformation 'rigidity', depending on the presence or absence of disulfide/free thiol groups, and on the location of hydrophilic and ionisable amino-acid groups on the surface. Among milk proteins, whey proteins such as β-lactoglobulin, bovine serum albumin (BSA), and immunoglobulin G (IgG) contain both disulfide bonds and free sulfhydryl groups, while β-lactalbumin (α -la), another globular whey protein, contains disulfide bonds but no free thiol group (Table 1). For caseins (the major protein component of milk) only α_{S_2}- and κ-caseins contain disulfide bonds (–S–S–) but no free thiol group (–SH). Under heat-treatment used in food processing, hydrophobic amino acids initially buried in the core of globular proteins became more exposed to the aqueous phase, and this unfolding mechanism may be followed by covalent modifications such as SH/S–S interchange reactions between whey proteins themselves, and also between caseins and whey proteins [12, 24, 34].

Table 1. Some chemical parameters of globular proteins from bovine milk

Protein Abbr.	β-lg	α-la	BSA	Caseins			
				α-s_1	β-	κ-	α-s_2
Amino-acid groups	162	123	582	199	209	169	207
Molecular mass (Kg mol^{-1})	18.3	14.6	66. 2	23.6	24.0	19.0	25.2
Disulfure bonds (S–S)	2	4	17	0	0	1	1
Thiol groups (–SH)	1	0	1	0	0	0	0
Isoelectric pH (pI)	5.2	4.3	5.3	5.0	5.2	5.5	5.3
Concentration in milk (gL^{-1})	2–4	1–1.5	0.4	10	9.3	3.3	2.6

Data from reference [32]

Oil-in-water emulsions are characterized by a dispersing medium, fat droplets, and interfaces [18, 67]. Proteins, which have the ability to interact with fat molecules, orienting their hydrophobic moieties toward fat crystals or oil phase, can form a protein layer around the fat droplets that is known to determine a main factor for physical stability of O/W emulsions [7, 11, 15, 30]. Full protein coverage of fat droplets is known to be a determinant factor for resistance to coalescence in simple protein-stabilized emulsions. In complex emulsions, such as those used for preparation of milk whipped-creams, frozen desserts or ice creams, small molecular mass emulsifiers and proteins (used in combination) can compete for the oil-water interface [16, 22, 46]. And, in addition to displacement of proteins from the interface, crystallinity of colliding fat droplets is supposed to play a role in their decreasing resistance to coalescence under the cooling and shearing steps used for their preparation [4, 14, 54]. Furthermore, the temperature required for crystallisation of emulsified fat is lower than for a non-emulsified fat of the same composition and it depends on the emulsifier type [21].

Thus, monitoring changes in structure stability of food components, by means of easy-to-use methodologies, could be of great importance to search ways to monitor effects of combined factors on the structural characteristics of foods, and help to maximize their processing and storage quality.

Using DSC for determination of heat-induced calorimetric parameters

DSC PRINCIPLE

DSC is powerful for monitoring physical state changes (liquid/solid) or molecular conformation or structural perturbations through the change of one thermodynamic parameter: temperature. Commercially available calorimeters working on the basis of different measuring principles (power compensation or heat flux calorimeters) measure a temperature difference that is linked to the energy changes involved during heat-induced reactions in sample materials [28].

In DSC, the furnace provides the same temperature programme to a sample pan (containing the material to be studied) and to a reference pan (containing a non-reacting material). If T_0 is the starting temperature, the programmed temperature T at time t, and constant heating rate, $\mathrm{d}T / \mathrm{d}t = \beta$ (°C s^{-1}) is:

$$T = T_0 + \beta t \tag{2}$$

For ideally symmetrical sample and reference pans this results in equality between the heat that flows through the instrument source to the sample and reference pans. If an endothermic or exothermic reaction occurs in the sample pan, the symmetry is disturbed, and there is a temperature difference ($\Delta T = T_s - T_r$) between the reacting sample and non-reacting reference. The DSC measured signal, ΔT is proportional to the heat flow rates between the instrument source and sample (to be studied) and reference. Following Newton's law the rate at which heat is transferred from the instrument source, at temperature T, to the sample or to the reference is:

$$(\mathrm{d}q/\mathrm{d}t)_{s,r}=(T-T_{s,r})/R \tag{3}$$

$$(\mathrm{d}q/\mathrm{d}t)_s-(\mathrm{d}q/\mathrm{d}t)_r=-(\Delta T)/R \tag{4}$$

Where $T_{s,r}$ is the temperature of the sample (or reference), R is the thermal resistance (°C W^{-1}), between symmetrical samples and holders. The rate of heat flow between the furnace and the sample pan when an exothermic ($\mathrm{d}H/\mathrm{d}t < 0$) or endothermic ($\mathrm{d}H/\mathrm{d}t > 0$) reaction occurs in the sample pan with heat capacity at constant pressure equal to C_s is:

$$(\mathrm{d}q/\mathrm{d}t)_s = C_s\,\mathrm{d}T_s/\mathrm{d}t + (\mathrm{d}H/\mathrm{d}t) \tag{5}$$

$\mathrm{d}H/\mathrm{d}t$ being the instantaneous heat absorbed ($\mathrm{d}H/\mathrm{d}t > 0$) or generated ($\mathrm{d}H/\mathrm{d}t <0$) by the reaction. The rate of heat flow between the furnace and reference with heat capacity at constant pressure equal to C_r, and for which $\mathrm{d}H/\mathrm{d}t = 0$ by definition, and $\mathrm{d}T_r/\mathrm{d}t = \mathrm{d}T/\mathrm{d}t = \beta$ (steady-state heating mode) is

$$(\mathrm{d}q/d)_r = C_r\,\mathrm{d}T_r/d\mathrm{t} = C_r\,\beta \tag{6}$$

Therefore, the relation between the reaction heat flow rate ($\mathrm{d}H/\mathrm{d}t$) and the measurement signal ($\Delta T = T_s - T_r$) is

$$\mathrm{d}H/\mathrm{d}t= -(\Delta T/R) - (C_s-C_r)\,\beta - C_s\,\mathrm{d}(\Delta\,T)\,/\mathrm{d}t \tag{7}$$

This expression links the heat flow rate ($\mathrm{d}H/\mathrm{d}t$) caused by the temperature difference (ΔT) between the reacting sample and the non-reacting reference. It contains a first term assigned to this temperature difference, a second term due to differences between the heat capacities of sample and reference materials (initial deviation of the signal after reaching its quasi-steady state in scanning mode), and a third term taking into account the thermal inertia ($\tau = RC_s$) of the system when ΔT is detected.

In heat flux DSC system, operating with cylinder-type containers, thermocouples connected in series between the containers and the furnace measure the differential signal, ΔT as a voltage. In power compensation DSC, the temperature difference is 'compensated' by an additional increasing (endothermic reaction) or decreasing (exothermic reaction) heating power. In both DSC systems the measured temperatures and sample temperatures are different. There is a systematic temperature lag occurring in non steady-state condition and depending on instrument characteristics and operating conditions (temperature scanning rate, sample weight and heat capacity). The sample temperature is related to the programmed temperature and characteristic time constant (τ):

$$T_{s,r} = \tau \, \mathrm{d}T_{s,r} \,/\, \mathrm{d}t \tag{8}$$

R and $\tau = RC$ are determined by calibration measurements from melting curves of pure standards. During melting of a pure standard:

T_s is constant $\mathrm{d}T_s/\mathrm{d}t = 0$ $\quad$ $\mathrm{d}(T_s - T_r)/\mathrm{d}t = -\,\mathrm{d}T_r/\mathrm{d}t = \beta$ $\quad$ $\mathrm{d}^2(T_s - T_r)/\mathrm{d}t^2 = 0$

Therefore, the thermal conductance ($1/R$) between sample holder and sample is deduced by application of Eq. 7, from melting curves of pure materials. It is determined from the slope of the linear increase in the heat flow rate with temperature. The determination of the sample heat capacity (mC_s), before and after a thermal transition, may be performed [69] relatively to the heat capacity of a different sample with known heat capacity (mC'). Pure water or synthetic sapphire is used for determination of sample heat capacity, following a three-step method, where calorimetric heat flow rates obtained from samples of known and unknown heat capacity ($\mathrm{d}H/\mathrm{d}ts$ and $\mathrm{d}H/\mathrm{d}t'$) are determined after subtraction of the zero base line (two-empty pans):

$$mC_s(T) = \frac{mC'\left[\left(\mathrm{d}H/\mathrm{d}t_s\right) - \left(\mathrm{d}H/\mathrm{d}t_0\right)\right]}{\left(\mathrm{d}H/\mathrm{d}t'\right) - \left(\mathrm{d}H/\mathrm{d}t_0\right)} \tag{9}$$

In non steady-state conditions ($\Delta T \# 0$) the asymmetry between sample and reference sides introduces a complex dependence of heat capacities, scanning rates, and thermal resistances on the shape of DSC signal. From Eq. (7), the effects on $\mathrm{d}H/\mathrm{d}t$ of DSC experimental conditions (scan rate, sample weight, thermal inertia) can be minimized by using sample and reference as similar as possible, in regards with their heat capacities and thermal resistances, and also a scanning rate as low as possible.

DSC PARAMETERS

Calorimetric parameters associated with a thermal transition occurring in the sample pan are extracted from DSC signals after subtraction of zero DSC signal, obtained by using two pans filled with reference materials (buffer for study of transitions in protein solutions, or empty pans for fat samples). Temperature of transition mostly used is that of extrapolated peak onset (T_{extr}), or peak maximum (T_{max}), the temperature of the observed maximum deviation of the heat-flow signal, corresponding to approximatively 50% denaturation reaction. The area between the peak and a sample baseline drawn from temperatures corresponding to pre- and post-transitions (maximum amount sample materials in initial and final states, respectively), divided by the amount of reacting materials in the sample pan, is used to determine the apparent heat of reaction, $Q_{cal,}$ involved during the thermal transition. In the physico-chemical environmental conditions used for food preparation, thermal transitions observed for protein conformational changes occur without a significant shift between the pre- and post-transition region, and approximation of a straight baseline drawn by interpolation between the beginning and the end points of the transition is mostly used.

Due to its relatively low characteristic time constants (<few seconds), power compensation DSC is used for determination of the fractional completion of reaction x (T), as a function of temperature T, from the pattern analysis of the heat flow involded during the scanning mode:

$$A_{\mathrm{T}} = x_{\mathrm{T}} \int_{\mathrm{T}}^{\mathrm{T1}} \left(dH / dt \right)_{\mathrm{T}} = x_{\mathrm{T}} Q_{\mathrm{cal}} \tag{10}$$

A_{T}, is the apparent heat of reaction at temperature T, calculated from the partial area under the exothermal heat flow curve), and Q_{cal} the calorimetric heat of reaction, calculated by using a straight base-line sample drawn between the initial and final deviations of the heat flow.

In the present study, this methodology is used to compare the growing-up of fat crystals in non-emulsified fat samples and in emulsions which differed only by the nature of adsorbed proteins.

Using DSC for monitoring heat-induced conformational changes in protein solutions

PROTEIN IN SIMPLE MODEL SOLUTIONS

The DSC curves shown in Fig. 1 represent the apparent heat flow rate arising from energy uptake during the heat-induced conformational transitions of β-lactoglobulin (major bovine whey protein) as observed by using two calorimeters operating at different conditions (6.1% protein concentration, 10°Cmin^{-1}

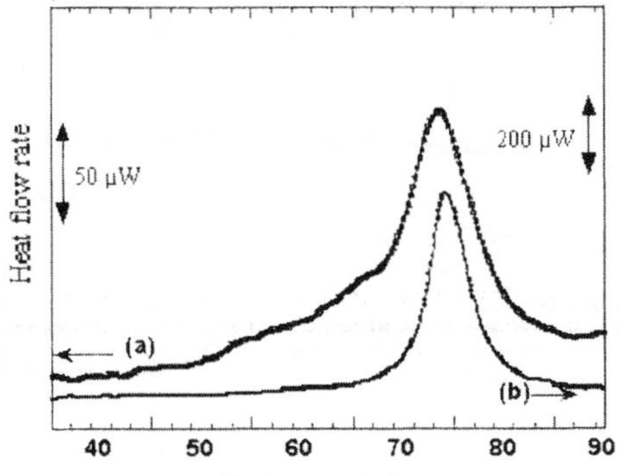

Fig. 1 Examples of DSC heating curves obtained by using heat flux (a) and power compensated (b) DSC systems to monitor heat-induced conformation stability of β-lactoglobulin dispersed in distilled water (pH6), in the following conditions: (a) 1°C.min^{-1} (scanning rate), 1% protein concentration (b) 10°C.min^{-1} (scanning rate, 6.1% protein concentration. From reference [52]

and 45 mg for compensated heat calorimeter, and 1% protein concentration, 1°Cmin^{-1} and 750 mg for heat flux calorimeter). The heat flux DSC is applied to study protein thermal denaturation in the more diluted condition and at a low heating rate ensuring thermodynamic equilibrium in a relatively high sample volume. Power compensation DSC, is applied at a higher scan rate and to a lower sample volume to quantify heat conformational stability of more concentrated protein solution. These DSC curves were obtained after subtraction of the signal obtained by re-heating. Calorimetric parameters determined for this example of irreversible heat-induced transition of β-lg at pH 6, are reported in Table 2. The results obtained by using power compensation DSC are in agreement with published results from similar β-lg solution and heating rate [13, 44, 45, 53, 54]. Calorimetric data extracted from power compensation and heat flux DSC curves show differences between the temperatures of extrapolated peak onset (T_{extr}), peak maximum (T_{max}), and overall heats of reaction (Q_{cal}). These differences arise from differences in protein concentrations, heating rates, mass samples, but slightly from DSC systems that are based on different principles (heat flux or power compensation). The curves in Fig. 2 were obtained by using the power compensation calorimeter from β-lg protein solutions at various concen-

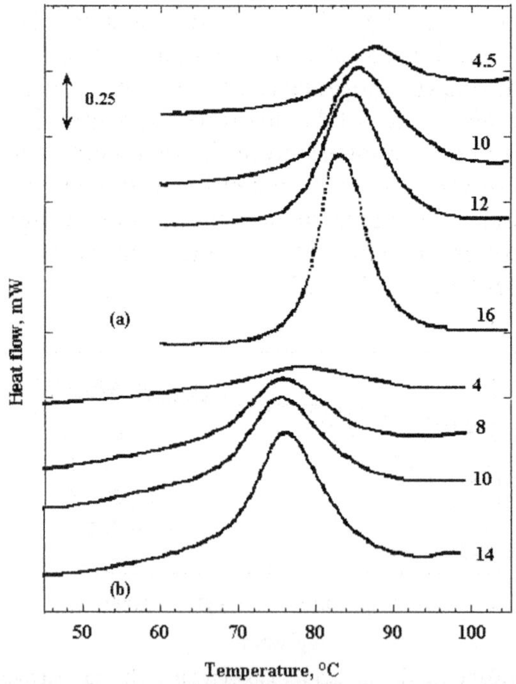

Fig. 2 Influence of β-lactoglobulin concentration in distilled water (200 mM NaCl and acidic or neutral pH) on DSC heating curves obtained at 10°Cmin^{-1}. (a) pH 3.5 and protein concentration ranging from 16.0 to 4.5 % .(b) pH 7 and protein concentration ranging from 14.0 to 4.0 %

Table 2. Calorimetric parameters extracted from signals obtained by using power compensation DSC (6.1% protein concentration) or heat flux DSC (1% protein concentration), T_{max} (temperature of maximum heat flow deviation), T_{extr} (extrapolate peak onset temperature), Q_{cal} (heat of reaction)

Concentration (wt%)	Heating rate dT/dt (°C min^{-1})	Sample mass (mg)	T_{max}(°C)	T_{extr}(°C)	Q_{cal} (Jg^{-1})
6.1	10	45	79.5±0.2	75.0±0.2	13.4±0.5
1	1	750	78.1±0.05	70.0±0.5	15.7±1.1

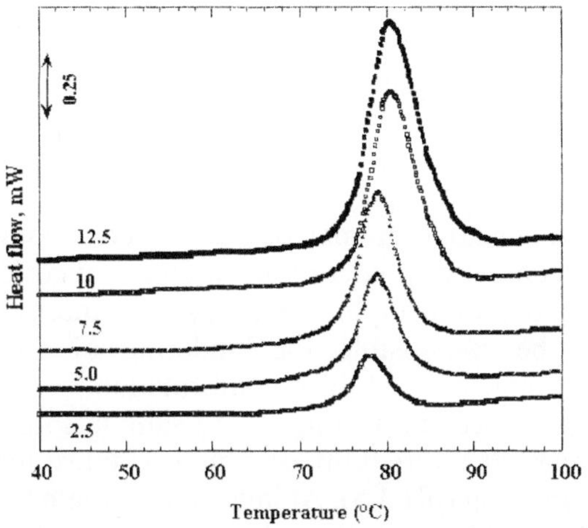

Fig. 3 Influence of heating rate (in °C.min^{-1}) on DSC curves obtained from a β-lactoglobulin solution (4.5% protein concentration in distilled water at 200 mM NaCl and pH 3.5)

trations and two different pHs (pH 3.5 for curves 2a; or pH 7 for curves 2b). They are characteristic of proteins with a higher conformational stability against heating [33] when in acidic condition (pH 3.5) than at neutral pH (pH 7). Varying the protein concentration at a constant heating rate (10 °C.min^{-1}) has effects on the location of peak maximum, T_{max} and Q_{cal}. However, the level to which protein concentration has effect on T_{max} values is dependent on pH condition. Protein concentration has a higher effect on the calorimetric parameters at pH 3.5 than at pH 7, with T_{max} varying by approx 7°C for pH 3.5 (Fig. 2a) instead of 2°C between 14 and 4% protein concentration for pH 7 (Fig. 2b). In parallel, DSC experiments performed on protein solution at pH 3.5 and 4.5% protein concentration, but by varying heating rate between 2°C.min^{-1} and 15°Cmin^{-1} (Fig. 3) indicated that the higher the heating rate, the higher T_{max} values.

Generally for small single domain globular proteins the reversible two-state model for the transition between native and denatured states is relevant, and thermodynamics of associated reaction can be monitored. In addition to transition cooperativity, there is equality between the calorimetric enthalpy change, Q_{cal}, and the modified van't Hoff expression, $\Delta_{VH}H$ which can be extracted from the DSC signal:

$$\Delta_{VH}H = 4RT_{max^2}\Delta h_{max} / \left(A\mathrm{d}T / \mathrm{d}t\right) \tag{11}$$

Where R is the gas constant (8.31 J. mol^{-1}K^{-1}), T_{max} is the peak temperature, Δh_{max} his the peak value of the heat flow (W) involved during the reaction, A (J) is the total area under the peak, and dT/dt (K s^{-1}) is the heating scan rate. In the case of multistate transitions in proteins of which the conformation is stabilised by interactions joining various domains, the ratio $Q_{cal} / \Delta_{VH}H$ is greater than 1 [6, 10, 48].

For most of the proteins used in conditions of food applications (protein concentration, ionic strength, protein samples which were prepared at industrial pilot scale), heat related denaturation is at most partly reversible and often not reversible at all. The ratio $Q_{cal} / \Delta_{VH}H$ lower than 1, may indicate the occurrence of superimposed irreversible aggregation process [20, 33, 48, 49]. Non reversibility of heat related denaturation may be due to stereo isomery or several possible mechanisms such as: deamidation of amino acid residues, hydrolysis of peptide bonds, disruption of disulphide bonds, and isomerization of proline residues which could be responsible for hindering of the refolding of the polypeptide chain into the native folded conformation.

In our examples, a second heating of the sample solution showed a peak corresponding to a partial reversibility of structural transitions, but only for the solution in acidic condition (pH 3.5). At low ionic strength (< 0.15 M NaCl), and fast heating rate (β > 2.5°Cmin^{-1}), the calorimetric heat of reaction Q_{cal} did not vary significantly, but T_{max} decreased upon increasing protein concentration. Following these results, indicating that Q_{cal} involved in these experiments could represent the enthalpy change between the protein initial and final conformational states, and could be used as an indicator index of protein conformational stability. The observed lowering of T_{max} with increased protein concentration could result from increasing rate of irreversible aggregation (D→A), following the Lumry-Eyring models [36] suggesting that dissociation of multimeric proteins into monomers (Eq. 1) does not take place before the unfolding mechanism. Upon heating hydrophilic interactions (hydrogen bonds, Van der Waals interactions, electrostatic interactions between charged groups, specific binding) are weakened, whereas hydrophobic interactions are strengthened at temperature ranging from 60 to 80°C. The hydrophobic interactions are exothermic, and the breaking of the other bonds is endothermic. When protein unfolding is followed by protein aggregation, the DSC curve is sharpened, and determination of $Q_{cal} / \Delta_{VH}H$, following a reversible two-state transition, as described above, leads to a value which is lower than 1. However, in all of our experimental conditions, the appearance of the second heating curve was changed (shift in temperature and decrease in magnitude of transition, or no peak at all). Considering the simplified scheme represented by equation 1, if $k > K$, most denatured proteins are converted irreversibly into A species (aggregates) and the thermal behaviour of the system is kinetically controlled by the rate-limiting step reaction.

The heat involved in the aggregation step (D→A) is much lower compared to that of the first denaturation step (N↔D), then the calorimetric heat of reaction could be very close to that of differences between the enthalpy of the initial and final protein conformational states [59]. β-lg exists as monomers below pH 3.5, as dimers at pH between isoelectric pH of 5.2 and alkaline pH, and as octamers between pH 3.5 and pH 5.2 [40]. In an earlier study [55], activation energy of β-lg heat-induced denaturation, determined by application of equation 10, was found to be dependent on protein concentration ($400 < E_a < 550$ kJmol^{-1} for protein concentration ranging between 3.5% and 24%). Considering this effect of protein concentration, we suggested that the reaction mechanism of denaturation could have a rate-determining step involving the interaction of two monomers. The decrease in Q_{cal} with increasing protein concentration was explained by a mechanism involving a fast initial step of partial unfolding, followed by a slow interchain hydrophobic reaction.

Application of Lumry-Eyring model, requires also the recording of thermograms at different scan rates (β), and correction of T_{max} for the thermal lag, as described above. T_{max}, temperature corresponding to maximum deviation of heat flow for a defined temperature scan rate, is linked to the activation energy (E_a) and pre-exponential factor (Z) by the relation:

$$\ln (\beta / T^2_{max}) = \ln (ZR/E_a) - E_a/RT_{max} \qquad (12)$$

The activation enthalpy ($\Delta H^{\#}$) and entropy ($\Delta S^{\#}$) may be determined according to the Lumry-Eyring theory:

$$\Delta H^{\#} = E_a - R\,T \qquad (13)$$

$$\Delta S^{\#}/R = \ln (Z\,h) - \ln (e\,kB\,T) \qquad (14)$$

where, kB is the Boltzman constant (1.38 10-23 J.K^{-1}), T temperature in K, and h is the Planck constant (6.62 10^{-34} J.s).

Results reported in Table 3, show examples of thermodynamic parameters of heat-induced denaturation of β-lg solutions at pH 3.5, determined following the Lumry-Eyring theory applied to DSC curves obtained from for protein concentration ranging between 3.5 and 24 % and heating rates between 2.5°Cmin^{-1} and 15°Cmin^{-1}. There is evidence that the highest difference between Q_{cal} and $\Delta_{HV}H$ values (tentatively calculated by application of Eq. 11), and $\Delta H^{\#}$ (Eqs 12 and 13) is observed for the highest protein concentration, in parallel with increased sharpness of the peak transition (Fig. 2a). These results, obtained according to Lumry-Eyring models, show that transition in molecular structures and reaction mechanism may be linked [59].

Table 3. Comparison between calorimetric heat of reaction (Q_{cal}) and enthalpy changes of reaction determined from Lumry-Eyring model[59] and Van't Hoff equation [48], see Eqs (12, 13) in the text

Concentration (wt%)	T_{max} (5°C min^{-1})	$\Delta H^{\#}$ (kJ mol^{-1})	Q_{cal} (kJ mol^{-1})	$\Delta_{VH}H$ (kJ mol^{-1})	$Q_{cal}/\Delta_{VH}H$
3.5	89.7	316	307	413	0.74
24.0	84.5	339	238	590	0.4

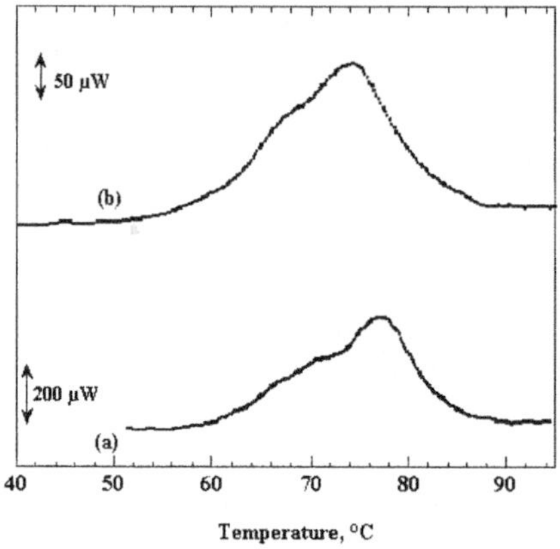

Fig. 4 Examples of DSC heating curves obtained by using powercompensated (a) and heat flux (b) systems and a of whey protein isolate dispersed in distilled water (pH 6.7), in the following conditions: (a) 5°Cmin^{-1} (scanning rate), 10 % protein concentration (b) 0.5°Cmin^{-1} (scanning rate), 3% protein concentration

The whey fraction of bovine milk, obtained after precipitation of caseins, is a widely used source of functional ingredients in food systems [42]. Whey proteins fraction is mainly composed by β-lactoglobulin, β-lg (55%), α-lactalbumin, α-la (22%) and bovine serum albumin, BSA (7%). Whey protein concentrates and isolates are commercially available in powdered-forms, they have different contents of protein, lactose and minerals, depending on the procedure used for their preparation. The protein composition of a whey protein isolate is reflected in the shape of the DSC power compensated and heat flow DSC heating curves (Fig. 4). These curves were obtained from a solution of whey protein isolate, and a power compensation DSC (10%, 45 mg, 5°Cmin^{-1}) or a heat flow DSC (3%, 750 mg, 0.5°Cmin^{-1}). In both the two curves, besides the main peak located at 78°C, there are two shoulders located at around 68°C and 73°C. These thermal events correspond to heat-induced denaturation of β-lg (the major protein component), and α-la and BSA (minor components), respectively [56, 59]. Again, this example shows a good agreement between the calorimetric parameters extracted from DSC signals obtained with calorimeters working with different conditions (high scan rate and protein concentration or low scan rate and low protein concentration).

For food applications, milk protein isolates (whey proteins and caseins) characterized by a high protein content and a low lactose content [42] are usually purchased in powdered forms. They are re-dispersed in solution of skim milk permeate to provide mineral ions and lactose to mimic equivalent milk physico-chemical environmental conditions. The curves in Fig. 5, show heat flow DSC signals (1°Cmin^{-1}) obtained from protein solutions (S 100 and S 80) consisting of a milk permeate (5% *w/w*) and a whey protein isolate weather alone (S 100) or in mixture with 20% caseins (S 80), respectively. Solutions S 100 containing only whey proteins were prepared at different concentrations 5.3% (a), 2.88% (b) and 0.9% (c). Solution S 80, containing whey proteins and caseins were prepared at 5.3% total concentration, consisting of 4.2% whey proteins and 1.1% caseins (d). The overall shape of these curves differed, depending on the total protein concentration of whey proteins when alone (a-b-c) or in presence of caseins (d). The peak temperature is the highest (82°C) for the lowest protein concentration (0.9%). For 2.88% and 5.3% total concentration of whey proteins, T_{max} values are very similar (76.2°C), but for the highest total protein concentration (5.3%) when whey proteins are in mixture with caseins, T_{max} is lowered by approx 1°C, relative to the solution without caseins (Fig. 5d). The shoulder located at 65°C in the DSC curves corresponding to 0.9% and 2.65% protein concentration, is much less visible at 5.3% total protein concentration, whatever the presence or absence of caseins. This difference, could be explained by the lack of protein dissociation, before denaturation, when at the highest protein concentration [59]. In addition, at 5.3% total concentration, DSC curves show an exothermic effect, which seems to occur at a temperature correspond-

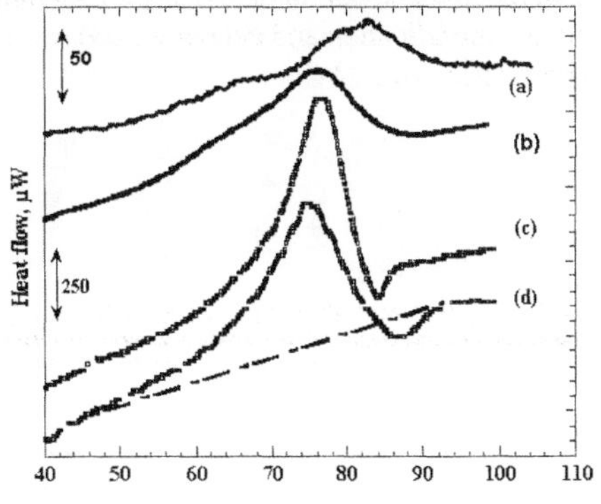

Fig 5. DSC heating curves obtained from protein solutions in 5% *w/w* skimmed milk permeate (pH 6.7) at 1°Cmin^{-1} (heat flux DSC) at different total protein concentrations in the absence of caseins (a, b, c), and at 5.3% total protein concentration of which 20% caseins (d). Total protein concentrations: 0.9% (a), 2.65% and 5.3% (b and c)

ing to the end of heat-induced denaturation. This exothermic event located at $T > T_{max}$ is observed for whey proteins at 5.3% concentration in both the absence (c) or the presence of caseins (d), and only for 1°Cmin^{-1} thermal rate, but neither for 10% whey proteins and 10°Cmin^{-1} nor for 3% protein concentration and 0.5°Cmin^{-1} (Fig. 4). The exothermic signal appearing at both a relatively high protein concentration and a low heating rate, is due to aggregation between whey proteins themselves or between whey proteins and caseins. Thus, in these physico-chemical conditions, the exothermic aggregation mechanism may be separated from the denaturation endotherm, as previously reported for other globular proteins [31, 61].

INDEX OF PROTEIN CONFORMATIONAL STABILITY

The overall heat of reaction (Q_{cal}), determined after subtraction of the sample baseline (two crucibles filled with distilled water) may be used as an indicator index of enthalpy difference between the protein initial and final conformational states. At pH 4.6, a characteristic pH value, which is close to iso-electric pH of milk proteins (pH at which global charge of milk proteins is close to zero), all denatured proteins can precipitate under centrifugation [19, 57]. This test is used to check the ability of DSC for evaluation of an index of protein conformational stability [52]. Briefly, whey protein samples from different manufacturers and sources are dispersed in distilled water (10% *w/w*), to prepare protein solutions at pH 6.7 and pH 4.6. After stirring for 2 hours, they are submitted to centrifugation (15,000 g for 30 min), then the supernatants are analysed for their protein content by using Kjeldahl method, as described in reference [52]. The loss of solubility at pH 4.6 is calculated from the following expression:

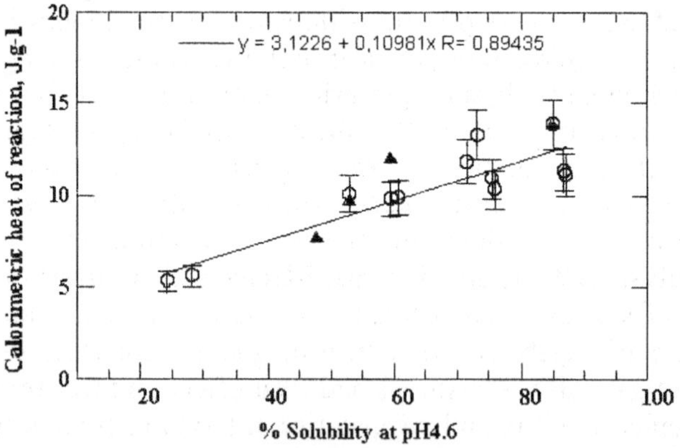

Fig. 6 Evolution of calorimetric heat of reaction, Q_{cal} (in J.g^{-1}) as a function of protein solubility at pH 4.6, determined for various commercially available whey protein powders (closed symbols correspond to data obtained from pre-heat treated whey protein isolates), from reference (52)

$$\text{Loss of solubility at pH4.6} = 100\%\left[1 - \frac{\text{Soluble proteins at pH 4.6}}{\text{Soluble proteins at pH 6.7}}\right] \qquad (15)$$

It can be compared to Q_{cal} values, calculated from the peak area under the DSC curves obtained from corresponding protein solutions at pH 6.7, before centrifugation. The results reported in Fig. 6 indicate that Q_{cal} values increases in parallel with protein solubility after precipitation at pH 4.6. Despite differences among protein sources, DSC data give valuable information in regards with protein conformation stability, as affected by technological processing steps (concentration, purification, drying) and physico-chemistry of protein environment.

Table 4. Protein composition, calorimetric heat of reaction (Q_{cal}) and peak temperature (T_{max}), and protein solubility at pH 4.6 of solutions in skim milk permeate (5% *w/w*) of whey proteins (S 100A and S 100B)) or mixtures of 80% whey proteins and 20% caseins (S 80 A and S 80B). WP A: non pre-heat treated whey proteins; WP B: pre-heat treated (80°C for 90 s) whey proteins ; CSN A: non pre-heat treated caseins; CSN B: pre-heat treated caseins

Sample	WP A	WP B	CSN A	CSN B	% solubility (pH 4.6)	Q_{cal} (Jg^{-1}) (T_{max} in °C)
S100A	5.3	0.0	0.0	0.0	86±7	9.4±0.1 (76.3)
S 80A	4.3	0.0	1.0	0.00	77±2	8.1±0.1 (75.0)
S100B	0.0	5 .3	0.0	0.00	39±3	ND
S 80B	0.0	4.3	0.0	1.0	37±2	ND

In many practical situations, proteins having different high-ordered structures (e.g. micellar or globular architectures) and having different extent of damaged initial structures may be present in the same preparation. To mimic this kind of sample heterogeneity, solutions of whey proteins alone (S 100), or in mixture with 20% *w/w* caseins (S 80) were pre-heat treated at 65°C under gentle stirring, and then pumped at 50 $Lhour^{-1}$ through stainless steel tubes heated at 80°C. By using appropriate lengths of tube coils the proteins holding time was set at 80°C for 90 s. The protein solutions (5.3%) were then cooled to 4°C, before analysis for determination of the effects of this heat treatment on protein solubility at pH 4.6, and DSC signal. Power compensation DSC working at $10°Cmin^{-1}$ heating rate, gave a very noisy signal, and heat flux DSC which has a higher sensitivity gave a large endothermic peak lying between 40 and 95°C (not shown), indicating that under the first heat-treatment (80°C or 90s) the irreversible structural change was not total. This qualitative result was confirmed by a higher partial loss of protein solubility after precipitation at pH 4.6, for solutions 100B and 80B (containing pre-heat treated proteins at 80°C for 90 s), than non pre-heat treated proteins (Table 4).

PROTEIN IN COMPLEX SOLUTIONS

Examples of non-fat solids used to prepare complex food emulsions are: proteins, lactose, and minerals from milk powders, sucrose and glucose syrup, mixture of mono-and diglycerides and stabilizers (carraghenan, locus bean, guar). DSC curves obtained from solutions containing these non-fat milk solids in mixture with either whey proteins (M 100A and M 100B) or mixtures of whey proteins and caseins (M 80A and M 80B) are shown in Fig. 7. The differences among the physico-chemistry of protein environmental conditions (due to presence of emulsifiers, sucrose and stabilizers), result in a large shift toward a higher temperature, without a no significant change in the overall heat of reaction. An additional endothermic event (shown only for M 80B solution) was observed between 45 and 65°C for all the complex protein solutions. This peak is due to phase transition of mono- and di-glycerides [21]. The other peak, located at a higher temperature, is due to heat-induced conformation change of whey proteins, which were added to the other non-fat solids, as explained above and in reference [64].

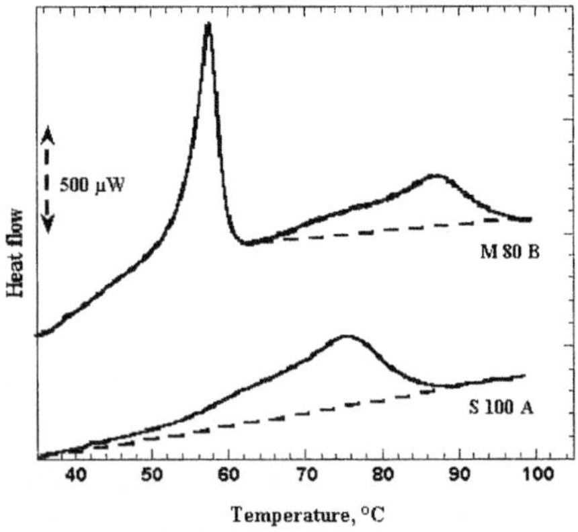

Fig 7. DSC heating crves (1°Cmin^{-1}, heat flux DSC) obtained from protein solutions (2.88% *w/w*) in 5% *w/w* skimmed milk permeate (pH 6.7), and in the absence or presence of other non-fat solids used for preparation of complex food emulsions (see text). S 100A 100% whey proteins in 5% *w/w* skimmed milk permeate M 100A 100% whey proteins in 5% *w/w* skimmed milk permeate and in mixture with non-fat solids used for preparation of complex food emulsions M 80B 80% whey proteins and 20% caseins, in 5% *w/w* skimmed milk permeate and in mixture with non-fat solids used for preparation of complex food emulsions (see text)

In comparison with the DSC curve corresponding to heat transition of whey protein (2.88%) in skim milk permeate (Fig. 5b), protein solutions in presence of additional non-fat solids show a very large increase in T_{max} (by approx. ~12.5°C for mixtures M 100, and 10°C for mixtures M 80), but a small decrease

Table 5. Protein composition, calorimetric heat of reaction (Q_{cal}) and peak temperature (T_{max}), and protein solubility at pH 4.6 of solutions in skim milk permeate (5% *w/w*) of whey proteins (M 100A and M 100B)) or mixtures of 80% whey proteins and 20% caseins (M 80A and M 80B), in presence of other non fat solids used for preparation of ice-cream model systems. WP A: non pre-heat treated whey proteins; WP B: pre-heat treated (80°C for 90 s) whey proteins; CSN A: non pre-heat treated caseins; CSN B: pre-heat treated caseins

	WP A	WP B	CSN A	CSN B	Protein solubility at pH4.6	Q_{cal}, (Jg^{-1}) T_{max}, (°C)
M 100A	2.88	0.00	0.00	0.00	79±1	7.30±0.05 (86.4)
M 80A	2.30	0.00	0.58	0.00	75±2	6.90±0.05 (86.3)
M 100B	2.03	0.85	0.00	0.00	75±3	6.90±0.05 (85.8)
M 80B	1.62	0.68	0.41	0.17	70±1	6.90±0.05(87.1)

in Q_{cal} (by less than 20%). These results indicate a stabilization effect of protein conformation against denaturation, and against a subsequent aggregation mechanism, as checked from determination of the amount of insoluble proteins at pH 4.6 (Table 5). The different solutions, consisting of non heat-treated whey proteins (M 100A) or mixture of 80% whey proteins and 20% caseins (M 80A), and solutions containing heat-treated proteins (M 100B and M 80B), were homogenized with 9% hydrogenated palm kernel oil for preparation of complex food emulsions, named (E 100A, E80A, E 100B and E80B).

Using DSC for monitoring rise of fat crystals in complex emulsified systems

In complex emulsions used for preparation of milk-whipped creams or ice creams, lipid ingredient comes mostly from milk fat (in United States) or vegetable fats (palm oil or palm kernel oil) in other countries. These lipids are constituted by a wide diversity of fatty acids and triacylglycerols (TG), each characterized by its own melting temperature [23, 38]. Their chemical and physical properties may be modified by hydrogenation or fractionation, with characteristic melting profiles and growing-up of solids.

The manufacturing process of complex emulsion consists of different steps. Briefly, in the first step, solid fat is heated above its melting temperature (50°C), and lipophilic emulsifiers are dispersed in the lipid melt. In the second step this lipid-melt phase is mixed under stirring, at the same temperature, with the aqueous phase containing water-soluble ingredients, before homogenisation, cooling down and aging at 4°C [63, 64]. Numerous authors observed that crystallization temperature of dispersed fat droplets is lowered, compared to bulk fat [26, 27, 41, 47, 51, 62, 68]. The degree of super cooling needed to initiate crystallization in milk fat globules in recombined creams, cocoa-butter or hardened palm kernel oil droplets in oil-in-water emulsions [27, 51, 68] was found in the order of 20°C, and 26°C for dispersed tristearin and tripalmitin [14, 27, 70]. The super cooling was shown to depend on the mean droplet size and on lipophilic or hydrophilic nature of emulsifiers [41, 62]. In oil-in-water emulsions, molecules such as emulsifiers are capable of partitioning between hydrophilic phases (continuous aqueous phase) or hydrophobic (interior of dispersed fat droplets) or interface, with consequence not only on the initial temperature of crystallization but also on the rise of solids content [14]. It was postulated that following the hypothesis of heterogeneous nucleation, large droplets containing the highest number of catalytic impurities will crystallize first, and as crystallization proceeds small droplets containing less impurities will crystallize at lower temperatures .

RISE OF FAT SOLIDS IN BULK AND EMULSIFIED SAMPLES

Numerous techniques such as dilatometry, ultrasonic velocity measurements, X-ray diffraction and differential scanning calorimetry, Nuclear Magnetic Resonance, Electron Spin Resonance Spectroscopy [9, 14, 26, 35, 60, 70], are used for monitoring crystallization behaviour in dispersed fat droplets. Examples of DSC curves obtained by using power compensation and heat flux DSC are shown in Figs 8–11. All the non-emulsified and emulsified palm kernel oil samples, were heated to 50°C (crystal melting) before cooling and re-heating at the same scan rate. The crystallization curves recorded during cooling of non-emulsified fat samples at 5°Cmin^{-1} or 1°Cmin^{-1} (Fig. 8a) show distinguishable exothermal events corresponding to formation of crystals (or polymorphs) composed by TG fractions having different melting-points [38]. A first peak of crystallization is clearly separated from the other one for 5°C. min^{-1} cooling rate, but it is much less marked for 1 °C.min^{-1} cooling rate (Fig. 8a), and not observable

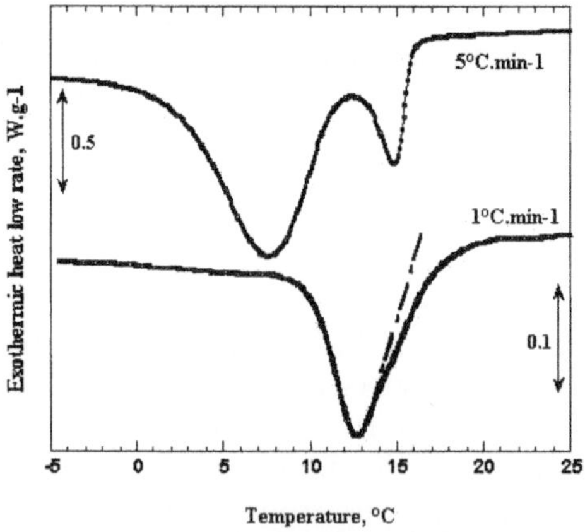

Fig. 8a DSC curves obtained from palm kernel oil samples upon cooling from 50 °C to –10°C, at 5°C min^{-1} (power compensated DSC) and 1°C min^{-1} (heat flux DSC)

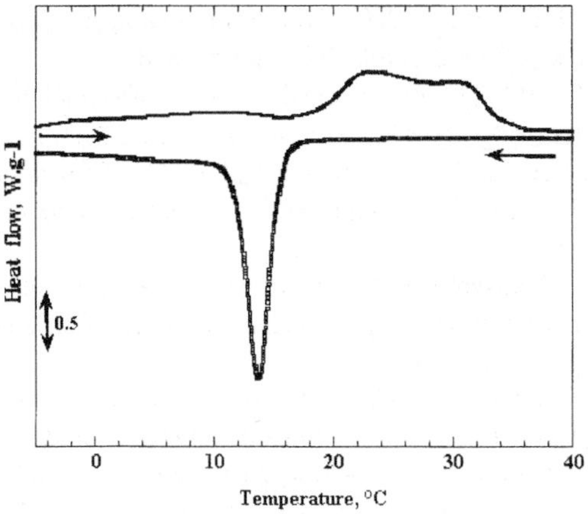

Fig. 8b DSC curves obtained from hydrogenated palm kernel oil sample which was pre-melted at 50°C, then cooled to –10°C and re-heated at 0.5°Cmin^{-1} (heat flux DSC)

at all for at 0.5°C.min^{-1} (Fig. 8b). The signals of heat flow released or absorbed under cooling and re-heating of the non-emulsified, and oil-in-water E100 or E 80 emulsions containing whey proteins alone or in mixture with caseins, respectively, are shown in Figs 9, 10 and 11. These curves obtained from emulsions differed from those obtained from bulk fat in regards with the super cooling values (Table 6), and also in regards with the rise of solid content (Fig. 12) under cooling experiments, as determined from Eq. 14. The differences among emulsified samples are not due to different cooling rates ranging between 5°Cmin^{-1} and 0.5°Cmin^{-1}, but rather due to protein types contained in the emulsions. For the

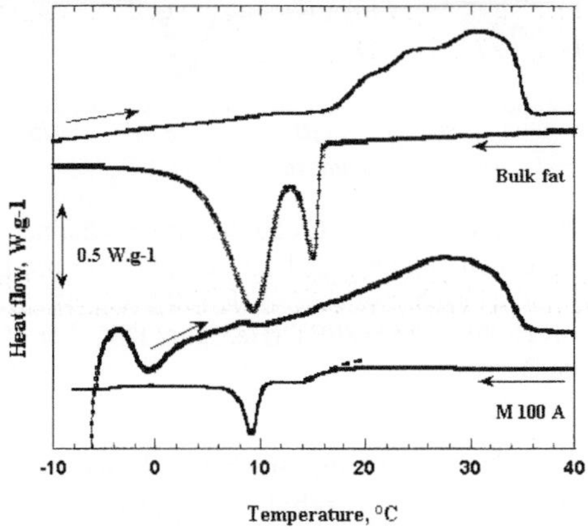

Fig. 9 DSC curves obtained from non-emulsified (bulk sample) and emulsified (E 100A) hydrogenated palm kernel oil samples, which were pre-melted at 50°C, then cooled to –10°C and re-heated at 5°Cmin^{-1} (power compensation DSC), M 100A: emulsion without caseins (see text)

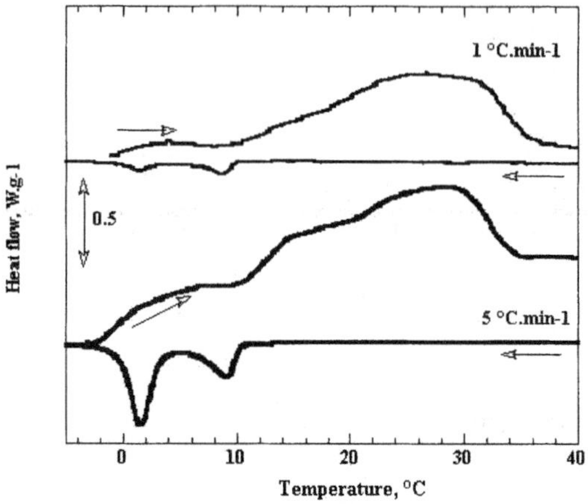

Fig. 10 DSC curves obtained from emulsified (M 80A) hydrogenated palm kernel oil, which were pre-melted at 50°C, then cooled to –10°C and re-heated at 5°Cmin^{-1} (power compensation DSC) or at 1°Cmin^{-1} (heat flux DSC), M 80A: emulsion containing milk proteins of which 80% whey proteins and 20% caseins (see text)

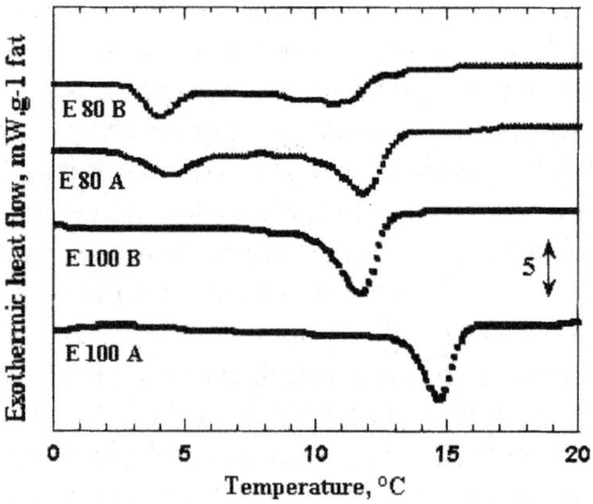

Fig. 11 Influence of milk protein types on DSC cooling curves obtained from complex food emulsions containing non pre-heat treated (A) or pre-heat treated (B) whey proteins in the absence of caseins (E 100A and B) or in the presence of 20% caseins (E 80A and E 80B). (Heat flux DSC from 50°C to –10°C, at 0.5°Cmin^{-1})

emulsions E 80 A and E 80B containing mixture of caseins and whey proteins, the cooling curves show two distinguishable separated peaks, whereas emulsions E 100 A and E 100B containing pure whey proteins show one major crystallisation peak preceded by a slight shoulder for E 100A containing non pre-heated treated whey proteins. The changes in the heat flow released during

Table 6 Calorimetric parameters of fat crystallisation obtained from non emulsified hydrogenated palm kernel oil samples (PKO) and ice-cream mixes containing different milk proteins (see Table3 for protein composition of corresponding continuous phase)

Sample	Crystallization				Supercooling
	T_i	$T_{max}1$ ±0.2°C	$T_{max}2$	$\Delta_{cal}H$ ±2Jg^{-1}	$\Delta T=T_f-T_i$ ±0.5°C
PKO					
5°Cmin^{-1}	16.6	14.4	7.5	-70.0	20.0
0.5°Cmin^{-1}	16.5	13.7	-	-76.0	19.5
E 100A					
5°Cmin^{-1}	16.7	14.1s	9.1	-23.0	19.8
0.5°Cmin^{-1}	16.1	13.7s	9.3	-23.3	20.4
E 100B					
5°Cmin^{-1}	9.7	7.7	1.0	-23.0	25.4
0.5°Cmin^{-1}	13.2	10.0	–	-24.9	23.3
E 80A					
5°Cmin^{-1}	10.5	8.8	0.9	-26.7	25.4
0.5°Cmin^{-1}	12.9	11.8	4.4	-23.0	23.6
E 80B					
5°Cmin^{-1}	9.8	8.1	0.7	-24.0	26.4
0.5°Cmin^{-1}	12.3	11.0	4.0	-21.0	24.2

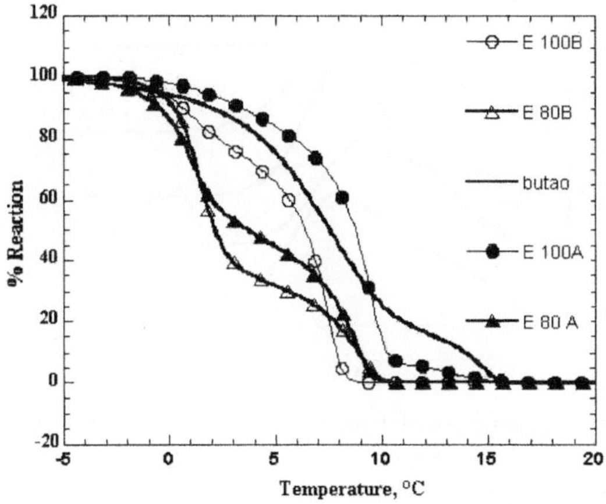

Fig. 12 Evolution of fat crystallinity as a function of cooling temperature (5°Cmin^{-1} from 50°C to –10°C) in complex food emulsions containing non pre-heat treated (A) or pre-heat treated (B) whey proteins in the absence of caseins (E 100A and B), or in the presence of 20% caseins (E 80A and E 80B)

the cooling experiments are reflected by differences in the growing-up of solids (Fig. 12), as determined from Eq. 14. Crystallization in non emulsified fat is anticipated, in comparison with emulsified fat, whereas as crystallization proceeds droplets in emulsions seem to crystallize following at least two different growing rates depending on milk protein types (native of pre-heated whey proteins, alone or in mixture with caseins).

EFFECTS OF FAT DROPLET SIZE AND ADSORBED PROTEIN TYPE ON FAT CRYSTALLINITY

The differences in the growing of fat crystallinity among the emulsions containing different milk proteins (Fig. 12) can be due to differences in globule size distributions or to different types of adsorbed molecules [21, 47, 62, 68]. To check this hypothesis, fat globule size distributions were determined after dispersion in distilled water of the four emulsions. They all indicated a bimodal shape (Fig. 13) with a shoulder located at approximately 0.8 μm and a principal peak at approx. 2 μm. Dispersion of the emulsions in SDS solutions, led to aggregate dissociation under the effect of SDS molecules and to monomodal globule size distributions, with similar average volume-surface diameters (0.81±0.02 μm for E 100 and 0.87±0.07 μm for E80). Thus, the change in the rise of fat solids, reflected by changes in the cooling curves cannot be attributed to differences in fat droplet size distributions.

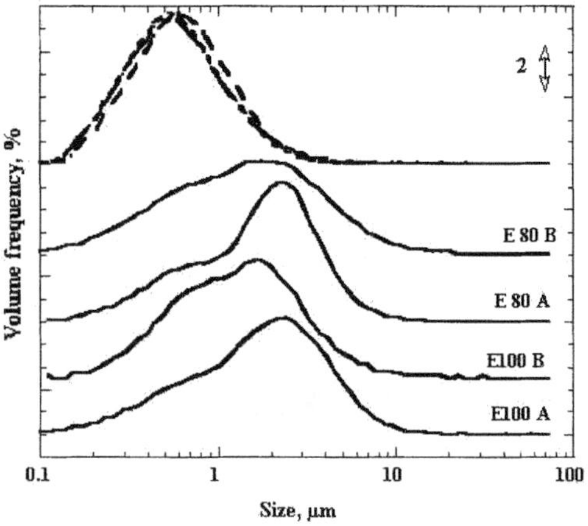

Fig. 13 Fat droplet size distributions observed after dilution in distilled water of complex food emulsions containing non pre-heat treated (A) or pre-heat treated (B) whey proteins in the absence of caseins (E 100A and B), or in the presence of 20% caseins (E 80A and E 80B), upper curves were obtained after dispersion of the emulsions in 1% SDS solutions

Evaluation of the protein surface coverage of fat droplets (amount of adsorbed proteins, in mg, divided by the specific surface area of fat droplets, in m^2) can be performed after separation of the cream layers from the aqueous phases of the emulsions and analysis of its protein and fat contents by using Kjeldahl and Mojonnier methods, respectively [63]. The results indicated that the amount of proteins adsorbed to the fat droplet surface is higher in emulsions containing caseins (13.35 mg m^{-2} for E 80A, and 13.4 mg m^{-2} for E 80B) than those containing only whey proteins (8 mg m^{-2} for E 100A, and 10.25 mg m^{-2} for E 100B). In agreement with published data [4, 22, 46, 63], caseins appear to be more surface active than whey proteins despite the presence of emulsifiers. Furthermore, by using heat-treated whey proteins, before homogenisation, the amount of adsorbed proteins increases with a slight effect on the fat droplet size distribution (Fig. 13), which cannot explain the differences observed in the evolution of fat crystallinity, depending on the milk protein types used for preparation of the ice-cream mix models (Fig 12).

RELATION BETWEEN SURFACE PROTEIN COVERAGE AND RISE OF SOLID FAT IN EMULSIONS

It is worth noting that emulsions E 80A and E 80B containing caseins presented two distinguishable exothermic peaks, whereas emulsions E 100A and E 100B without caseins showed one major exothermic peak, in DSC cooling experiments at rates ranging between 5°Cmin^{-1} and 0.5°Cmin^{-1} (Fig. 11). In the absence of caseins, evolution of the solid content is achieved in the temperature domain lying between 15°C and 5°C (5°Cmin^{-1} cooling rate) or between 13°C and 6°C (0.5°Cmin^{-1} cooling rate). In the presence of caseins the heat released upon cooling returns to the sample base line at a much lower temperature (ranging from –2°C and –5°C). Upon re-heating in scanning mode, the overall heat of melting reactions were much lower than for non emulsified samples, and it was in the following order E 100 < E 80 < bulk fat, indicating that less solids were

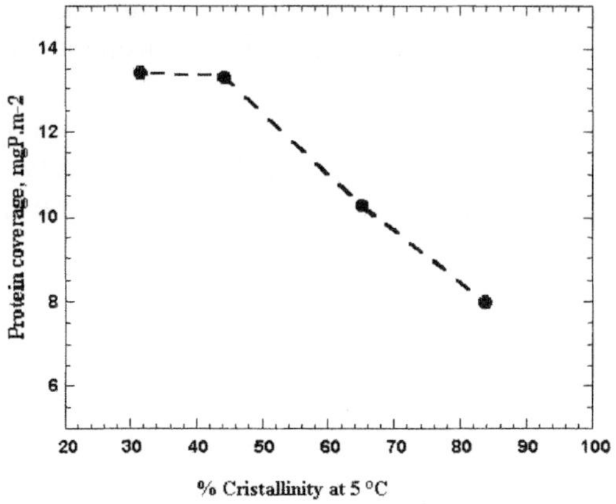

Fig. 14 Protein surface coverage in complex food emulsions containing non pre-heat treated (A) or pre-heat treated (B) whey proteins in the absence of caseins (E 100A and B), or in the presence of 20% caseins (E 80A and E 80B). The protein surface coverage is plotted as a function of fat crystallinity, as calculated by application of equation 10 (see text)

formed in emulsions containing whey proteins [51]. Evaluation of fat solids formed at 5°C upon cooling indicates that the lower the surface protein coverage (emulsions without caseins), the higher the crystallinity.

This trend could be related to accelerated growing of fat crystals in emulsions, in the following range: E 100A (8 mg m^{-2}) > E 100B (10.25 mg m^{-2}) > E 80A (13.3 mg m^{-2}) > E 80B (13.4 mg m^{-2}), as represented by the curve shown in Fig. 14.

Adsorbed small molecular weight emulsifiers act as catalyser for lipid nucleation thanks to molecular similarity [14, 17, 27, 62]. Milk proteins are also surface active, they adsorb to the surface of fat droplets and they can form relatively thicker films than emulsifiers. When in mixture with small molecular weight emulsifier, proteins may be adsorbed at fat droplet interfaces, but at a lower amount due to competition for the interface. They were shown to act as emulsion stabilizers due to steric forces [2, 4, 16, 46, 51], with consequences on the growing rate of fat crystals, depending on the milk protein types contained in the emulsions [51].

Conclusion

DSC is used to investigate heat-induced conformational or structural changes of a broad range of food ingredients (biopolymers, proteins, fats, sugars, emulsifiers) in various physico-chemical conditions, and at various weight fractions of water. Detailed description of principles and methodologies employed to obtain DSC signals and to extract information may be found in the large body of data that were published since several decades. They show a great influence of several intrinsic and extrinsic factors on calorimetric data, making difficult any generalization. In the present work, we showed that DSC can give valuable information on how examples of protein solutions and emulsions behave under various physico-chemical conditions used for preparation of complex food emulsions. Besides the well established effects of physico-chemical environmental parameters on heat-induced conformation changes of globular proteins, DSC in scanning mode can be used for evaluation of growing of fat crystals in complex food emulsions as a function of cooling temperature and in relation to other characteristics of emulsion stability.

References

1 Abd El Rahman A. M., Madkor, S. A., Ibrahim, F. S., Kilara, A. (1997) Physical characteristics of frozen desserts made with cream, anhydrous milk fat or milk fat fraction *J. Dairy Sci.*, 80, 1926–1935.

2 Barfod, N. M., Krog, N., Larsen, G., Buchheim, W. (1991) Effects of emulsifiers on protein-fat interaction in ice-cream mix during ageing. I. Quantitative analyses, *Fat Sci. Technol.*, 93, 24–29.

3 Barone, G., Del Vecchio, P., Fessas, D., Giancola, C., Graziano, G., Riccio, A., in Russo N. (1994), J. Anastasso-Poulou and Barone G. (eds), Chemistry and Properties of Biomolecular Systems, Kluwer Ac. Publ., Vol 2, p. 49.
4 Bolliger S., Goff H. D., Tharp B. W. (2000) Correlation between colloidal properties of ice cream mix and ice cream, Int. Dairy J., 10, 303–309.
5 Boode K., Walstra, P de Groot-Mostert, A.E (1993) Partial coalescence in oil-in-water emulsions.2. Influence of the properties of the fat, *Colloids and Surfaces A.*, 81, 139–151.
6 Brandts, J. F., and Lin, L. N. (1990) Study of strong to ultratight protein interactions using differential scanning calorimetry, *Biochemistry*, 29, 6927–6940.
7 Britten M., Giroux, H. J. (1993) Interfacial properties of milk protein-stabilized emulsions as influenced by protein concentration, *J. Agric. Food Chem.*, 41, 1187–1191.
8 Chen J., Dickinson, E., Iveson, G (1993) Interfacial interactions, competitive adsorption and emulsion stability, *Food Stuct.*, 12, 135–146.
9 Clausse, D., Dumas, J. P. Meijer., P. H. E. Broto, F. (1987) Phase transformation in emulsions, *J. Disp. Sci. Tech.*, 8, 1–6.
10 Cooper, A. (1999) Thermodynamics of protein folding and stability, in Protein: A comprehensive treatrise, Vol 2, pp. 217-270, Allen Geoffrey (ed) JAI Press Inc.
11 Dalgleish, D. G. (1996) Conformations and structures of milk proteins adsorbed to oil-water interfaces, *Food Res. Int.*, 29, 541–547.
12 Dalgleish, D. G., Van Mourik, L., Corredig, M. (1997) Heat-induced interactions of whey proteins and casein micelles with different concentrations of α-lactalbumin and β-lactoglobulin, *J. Agric. Food Chem.*, 45, 4806–4813.
13 De Wit, J. N., Klarenbeek, G., (1984) Effects of various treatments on structure and solubility of whey proteins, *J. Dairy Sci.*, 67, 2701–2710.
14 Dickinson, E. and McClements, D. J (1995) Fat crystallization in oil-in-water emulsions, in Advances in Food Colloids, Eds (Dickinson E. and McClements D. J), Blakie Academic & Professional, London-UK, p. 211–246.
15 Dickinson, E. (1992) Structure and composition of adsorbed protein layers and the relationship to emulsion stability, *J. Chem. Soc. Faraday Trans.*, 88, 2973–2983.
16 Dickinson, E., Gelin, J. L (1992) Influence of emulsifier on competitive adsorption of a-s-casein and β-lactoglobulin in oil-in-water emulsions, *Colloid Surface*, 63, 329–335.
17 Dickinson, E., Goller. M. I., McClements, D. J, Peasgood, S., Povey, M. J. W. (1990), Ultrasonic monitoring of crystallization in an oil-in-water emulsion, *J. Chem. Soc. Faraday Trans.*, 86, 1147–1155.
18 Dickinson, E. (1997) Properties of emulsions stabilized with milk proteins: overview of some recent developments, *J. Dairy Sci.*, 80, 2607–2619.
19 Donovan, M.and Mulvihill, D.M (1987)., Thermal denaturation and aggregation of whey proteins, *Irish J. Food Sci. Technol.*, 11, 87–100.
20 Freire, E, and Biltonen, R. L. (1978) Statistical mechanical deconvolution of thermal transitions in macromolecules, I. Theory and application to homogeneous systems. *Biopolymers,* 17, 463–479.
21 Garti, N., and Jano, J. (2001) The roles of emulsifiers in fat crystallization, in Crystallization processes in fats and lipid systems, Garti N and Sato K (eds) Marcel Dekker, New York, p. 211–250.

22 Gelin, J. L., Poyen, L., Courthaudon, J. L., Meste, M., Lorient, D. (1994) Structural changes in oil-in-water emulsions during the manufacture of ice cream, *Food Hydrocoll.,* 8, 299-308.

23 Hagemann, J. W (1988) Thermal behaviour and polymorphism of acylglycerides, *in:* Crystallization and polymorphism of fats and fatty acids, Garti N. and Sato K. (eds), Marcel Dekker, Inc., p. 9–98.

24 Haque, Z., Kristjansson, M. M., Kinsella, J. E. (1987) Interaction between k-casein and β-lactoglobulin: possible mechanisms, *J. Agric. Food Chem.,* 35, 644–649.

25 Harwalkar, V. R. and Ma, C. Y (1990). Thermal Analysis of foods, *Elsevier Appl. Sci. Publ.*, England

26 Hartel, R. W (1997) in Phase/state transitions in foods, M.A. Rao and R.W. Hartel (eds), Marcel Dekker, New York.

27 Hindle, S., Povey, M. J. I., and Smith, K. (2000) Kinetics of crystallization in n-hexadecane and cocoa butter oil-in-water emulsions accouning for droplet collision-mediated nucleation, *J. Colloid Interface Sci.*, 232, 370–380.

28 Höhne, G. W. H., Hemminger, W. and Flammersheim H.-.(1996) Differential Scanning Calorimetry. An introduction for practitioners. Springer-Verlag, Berlin, Heidelberg, NewYork

29 Holt., C., and Sawyer, L. (1988) Primary and predicted secondary structures of caseins in relation to their biological functions, *Protein Eng.,* 2, 241–259.

30 Hunt, J. A. and Dalgleish, D. G. (1994) Adsorption behavior of whey protein isolate and caseinate in Soya oil-in-water emulsions, *Food Hydrocoll.*, 8, 175–187.

31 Hunt, J. and Dalgleish, D. G. (1995) Heat stability of oil-in-water emulsions containing milk proteins: effects of ionic strength and pH, *J. Food Sci.*, 60, 1120–1123.

32 Kinsella, J. E., Whitehead, D. M.(1989) Proteins in whey: chemical, physical and functional properties, *Adv. Food Nutr. Res.*, 33, 343–438.

33 Lefèbvre, J, and Relkin, P. (1996) Denaturation of globular proteins in relation to their functional properties. In Surface activity of proteins, S. Magdassi (ed), Marcel Dekker, Inc., New York

34 Liu, T., Relkin, P., Launay, B. (1994) Thermal denaturation and heat-induced gelation of β-lactoglobulin: effects of some chemical parameters, *Thermochim. Acta*, 246, 387–403.

35 Lopez, C., Lesieur, P., Keller, G. and Ollivon, M. (2000)., Thermal and structural behaviour of milk fat. 1. unstable species of cream. *J. Colloid and Interface Sci.*, 229, 62–71.

36 Lumry, R., and Eyring, H (1954). Conformation changes of proteins, *J. Phys. Chem.,* 58, 110–120.

37 Marshal, R. T. and Arbuckle, W. S. (1996) Ice cream, International Thomson Publishing, New York

38 Matsui, N., Material design for hard butter from vegetable fats (1988), in Crystallization and polymorphism of fats and fatty acids, Garti N., and Sato, K. (eds) Marcel Dekker, Inc. New York, pp. 395–421.

39 McGann, T. C. A., Donelly, W. J., Kearney, R. D., and Buchleim, W. (1980) Composition and size distribution of bovine casein micelles, *Biochem., Biophys. Acta*, 630, 261–270.

40 McKenzie, H. A., and Sawyer, W. H. (1967) Effect of pH on β-lactoglobulin, *Nature* (London) 214, 1101–1104.

41 McClements, D. J., Duncan., S. R, German., J. B., Simoneau, C. and Kinsella, J. E (1993) Droplet size and emulsifier type affect crystallization and melting of hydrocarbon-in-water emulsions, *J. Food Sci.,* 58, 1148–1151.

42 Morr, C. V., E. Y. W. Ha. (1993) Whey protein concentrates and isolates: processing and functional properties, *CRC Crit. Rev. Food Sci. Nutr.*, 33, 431–476.
43 Pace, C. N., Hirley, B. A., McNutt, M., Gajiwala, K. (1996) Forces contributing to the conformational stability of proteins, *FASEB J.*, 10, 75–83.
44 Park, K. H. and Lund, D. B. (1984) Calorimetric study of thermal denaturation of β-lactoglobulin. *J. Dairy Sci.*, 67, 1699–1706.
45 Paulsson, M., Djemek, P. (1990) Thermal denaturation of whey proteins in mixture with caseins studied by differential scanning calorimetry, *J. Dairy Sci.*, 73, 590–600.
46 Pelan, B. M. C., Watts, K. M., Campbell, I. J., Lips, A. (1997) The stability of aerated milk protein emulsions in the presence of small molecule surfactants, *J. Dairy Sci.*, 80, 2631–2638.
47 Philipps, L. W. (1964) Heterogeneous and homogeneous nucleation in supercooled triglycerides and n-paraffins, *Trans Faraday Soc.*, 60, 1873–1883.
48 Privalov, P. L. (1979) Stability of proteins. Small globular proteins, *Adv. Protein Chem.*, 33, 167–241.
49 Privalov, P. L., and Potekin, S. A (1996) Scanning calorimetry in studying temperature-induced changes in proteins, *Methods Enzymol.*, 131, 4–51.
50 Raemi, A., Lambelet, Pierre, and Garti, N. (2001) Thermal behaviour of foods and food constituents, in Thermal behaviour of dispersed systems, Garti N. (ed), Marcel Dekker, Inc., New York. pp. 477–505.
51 Relkin, P., Sourdet, S., and Fosseux, P-Y. (2003) Fat crystallization in complex food emulsions. Effects of adsorbed milk proteins and of a whipping process, *J. Therm. Anal. Cal.*, 71, 187–195.
52 Relkin, P. (2002) Potentialité de l'analyse calorimétrique différentielle pour la caractérisation de l'état de dénaturation de proteines du lactosérum, *Ann Fals. Exp. Chim.*, 559, 189–197.
53 Relkin, P., (1996) Thermal unfolding of β-lactoglobulin, α-lactalbumin and bovine serum albumin. A thermodynamical approach, *Crit. Rev. Food Sci. Nutr.*, 36, 565–601.
54 Relkin, P, Meylheuc, T., Launay, B. and Raynal, K. (1998), Heat-induced gelation of globular protein mixtures. A DSC and a SEM study, *J. Therm. Anal.*, 51, 747–755.
55 Relkin, P. and Launay, B. (1991), On the partial reversibility of β-lactoglobulin heat denaturation, *J. Therm. Anal.*, 37, 1887–1895.
56 Ruegg, M. P., Morr, U., Blanc, B.(1997) A calorimetry study of the thermal denaturation of whey proteins in simulated milk ultrafiltrate, *J. Dairy Res.*, 44, 509–520.
57 Rowland, S. J. (1933) The heat denaturation of albumin and globulin in milk, *J. Dairy Res.*, 5, 46–53.
58 Roos, Y. H. (1995) Phase transitions in foods, Academin Press, Inc., London
59 Sanchez-Ruiz J. M (1992) Theoretical analysis of Lumry-Eyring models in differential scanning calorimetry, *Biophys. J.*, 61, 921–935.
60 Schäffer, B., Lőrinczy, D., and Szakály (1996), DSC and EPR investigation o the effect of fat crystallization on the consistency of butter, *J. Therm. Anal.*, 47, 515–524.
61 Schiraldi, A., Piazza L., Fessas, D., Riva, M. (1999) Thermal Analysis in foods and Food Processes, in Handbook of thermal analysis and calorimetry, Vol 4, Patrick Gallagher (Ed) Elsevier Sci., Amsterdam, p. 829–921.
62 Skoda, W. and van den Tempel, M., (1963). Crystallization of emulsified triglycerides, *J. Colloid Sci.*, 18, 568–584.

63 Sourdet S., Relkin., P., Aubry, V. and Fosseux, P-Y. (2002). Composition of fat protein layer in complex food emulsions at various weight ratios of casein-to-whey proteins, *Lait,* 82, 567–578.
64 Sourdet, S., Relkin, P., and Cesar, B. (2003) Effects of milk protein type and pre-heating on physical stability of whipped and frozen emulsions, *Colloids and Surfaces B*, (in press).
65 Sturtevant, J. M. (1987) Biochemical applications of differential scanning calorimetry, *Ann. Rev. Phys. Chem.*, 38, 463–488.
66 Walstra, P (1988).,The role of proteins in the stabilization of emulsions, in: Phillips G.O., Wedlock D. J., Williams P. A., [eds] Gums and stabilisers for the food industry 4, IRL Press, Oxford. pp. 323–336.
67 Walstra, P., Emulsion formation (1983) in Encyclopedia of emulsion technology, 1. Basic Theory, Becher P., [ed], Marcel Dekker Inc., New York. USA, pp. 57–127.
68 Walstra, P. and van Beresteyn, E. C. H. (1975). Crystallization of milk fat in the emulsified state, *Neth. Milk Dairy J.*, 29, 3565.
69 Wright, D. J., (1984) Thermoanalatycal methods in food research, in Biophysical methods in food research, H. W.-S. Chan (ed), Blackwell Scientific Publications, Oxford, p. 1–36.
70 Zhao, J. and. Reid, D. S. (1994) Thermal studies on the crystallization kinetics of triglycerides and milk fat by DSC, *Thermochim. Acta*, 246, 405–416.

Chapter 6

Structural and functional studies of muscle proteins by using differential scanning calorimetry

D. I. Levitsky*

A. N. Bach Institute of Biochemistry, Russian Academy of Sciences, Leninsky prosp. 33, Moscow 119071 and A. N. Belozersky Institute of Physico-Chemical Biology, M. V. Lomonosov Moscow State University, Moscow 119992, Russia

Introduction

DSC is the most effective and commonly employed method to study the thermal unfolding of proteins [1–4]. Last 15 years we successfully use the DSC approach for structural and functional studies of myosin, actin, and tropomyosin – the main proteins of various systems of biological motility – including all types of muscles. The cyclic interaction of myosin heads with actin filaments, which is accompanied by ATP hydrolysis in the heads and regulated by tropomyosin, is the basis of the molecular mechanism of a number of events in biological motility, from intracellular transport to muscle contraction.

The purpose of this review is to demonstrate the opportunities offered by application of the DSC to studies on myosin, actin, and tropomyosin, namely:

1) studies of nucleotide-induced structural changes in the myosin head and in actin as an example of registration and study of structural changes taking place in proteins due to their interaction with low-molecular-weight ligands;
2) interaction of F-actin with myosin heads, tropomyosin, and with other actin-binding proteins as an example of studies on protein-protein interactions;
3) combination of the DSC approach with other methods (including modifications and mutations in the myosin head and tropomyosin) to investigate the molecular mechanism and regulation of muscle contraction.

Myosin

Myosin is a representative of extensively studied last years 'molecular motors', i.e. the proteins which realize in living cells the transduction of chemical energy of ATP hydrolysis into mechanical work. Different myosins are very diverse

* levitsky@inbi.ras.ru

D. Lörinczy (ed.), The Nature of Biological Systems as Revealed by Thermal Methods, 127–158.

both in their structure and in the functions in the cell. On the basis of phylogenetic analysis all myosins are subdivided into 17 classes [5]. The characteristic feature of all myosins is the presence of highly conserved globular part (the so-called motor domain) responsible for binding and hydrolysis of ATP, interaction with actin, and transduction of chemical energy into mechanical work. All myosins studied so far are oligomers comprising a few polypeptide chains. The motor domain is formed by N-terminal part of so-called 'heavy' chain. The region of the motor domain is followed by the region of 'regulatory' domain which is α-helix containing from one to six binding sites for 'light' chains (calmodulin-like proteins taking part in the regulation of myosin functions) [6]. Following this domain is the C-terminal region ('tail') which is diverse both in size and sequence among different myosins [7, 8]. The specificity of individual myosin function in the cell is dictated by this variable part of the molecule.

The best known and well studied representatives of the myosin family are myosins of class II (myosins II) including all muscle myosins and many non-muscle myosins. The molecule of myosin II is a hexamer comprising two heavy chains (molecular mass ~ 200 kDa) and four light chains (molecular mass ~ 20 kDa). The N-terminal parts of the heavy chains form two globular heads. Each head consists of highly conserved motor domain and regulatory domain formed by the 'neck' region of the head with two associated light chains. C-Terminal parts of the myosin II heavy chains interact with each other and form long and rigid rod part of the molecule (tail) which is a double α-helix. This double-coiled coil helix is formed due to periodic repeats of hydrophobic and charged residues. By limited proteolysis of the myosin II molecule with trypsin, chymotrypsin, or papain, various fragments can be obtained in the isolated state: isolated myosin head or myosin subfragment 1 (S1); myosin rod; N- and C-terminal fragments of the myosin rod – subfragment 2 (S2) and light meromyosin (LMM), respectively; heavy meromyosin (HMM) consisting of two heads attached to S2. These fragments, retaining the properties of some parts of the myosin II molecule, are often used in experiments instead of intact myosin.

MYOSIN ROD

DSC has been used for structural studies on myosin from the late 1970's. However, until the late 1980's this method was used almost exclusively for studies of thermal unfolding of the helix rod part of muscle myosin II and its isolated fragments for the following two reasons. First, the hypothesis proposed by Harrington [9] was very popular that time; according to this hypothesis, a rapid helix-to-coil transition of the 'hinge' region between S2 and LMM in the myosin rod plays a key role in the force generation process during muscle contraction. Therefore, that time many investigators studied the thermal unfolding of the myosin rod by various methods including DSC. The second reason was that the reversible thermal denaturation of the myosin rod is much more available for calorimetric analysis rather than irreversible denaturation of the myosin head, which is accompanied by aggregation and precipitation of the protein, thus making very difficult the measurements of calorimetric parameters.

In 1979 Potekhin *et al.* [10] described for the first time the thermal denaturation of the isolated rod part of rabbit skeletal muscle myosin and some its fragments, studied by DSC. The myosin rod unfolded over a wide temperature range, from 35 to 65°C, and four maxima were clearly revealed on the calorimetric curve. After deconvolution of the calorimetric curve for the myosin rod into separate thermal transitions, six such transitions have been revealed, with maxima at 43, 46, 49, 51, 54, and 60°C [10]. The complex character of the heat sorption curve of the myosin rod has been confirmed in further studies by many other authors [11–14]; from three to six separate thermal transitions were revealed depending on the deconvolution approaches used. For the identification of these transitions, i.e. for revealing their correspondence to definite parts of the primary structure of the myosin rod, different isolated fragments of the myosin rod were studied, such as light meromyosin (LMM) or subfragment 2 (S2), containing or not containing the 'hinge' region connecting the regions of S2 and LMM in the myosin rod. The melting of such isolated fragments correlated well with definite thermal transitions of the myosin rod [10, 12, 13]. The most thermolabile transition was present in the preparations of 'long' LMM and 'long' S2 containing the parts of the 'hinge' region, but it was absent in the preparations of 'short' LMM and S2 which did not contain these parts of the primary structure [10, 12, 13, 15]. It has been suggested that this transition reflects the melting of the 'hinge' region between S2 and LMM, including also a short C-terminal segment of S2 and a short N-terminal segment of LMM [12, 13]. Thus, due to the use of DSC, the independently melting separate regions (calorimetric domains) have been revealed in the skeletal myosin rod.

In the rod part of skeletal muscle myosin the LMM region is responsible for an assembly of ordered bipolar myosin filaments at low ionic strength, whereas the S2 region together with the 'hinge' region promote movements of myosin heads between myosin filament and actin filament. It was shown by DSC that the thermal denaturation of LMM is very sensitive to the changes of ionic strength. Lowering the KCl concentration from 0.6 to 0.12 M (i.e., the conditions for an assembly of myosin filaments) shifted the maximum of the thermal transition by 10°C to higher temperature with a significant increase of the cooperativity of the transition (the peak became narrower) [16]. On the other hand, among the fragments of the myosin rod, the S2 demonstrated the most pronounced and unusually high sensitivity to pH [13, 17]. At low ionic strength lowering the pH from 7.3 to 5.9 shifted the maximum of the S2 thermal transition by 10°C, from 39.5 to 49.6°C [17]. These data correlated well with the results of the experiments on myosin filaments and muscle fibers demonstrating pH-dependent structural changes of the S2 region which led to the movement of this region relative to the surface of myosin filament [18, 19]. Thus, DSC allowed to study the functionally important structural changes in different regions of the myosin rod.

It should be noted that all the data cited above were obtained on the myosin rod from skeletal muscles of vertebrates. However, the thermal denaturation of the myosin rod from other sources can be different from that of skeletal muscle myosin rod. For example, the isolated rods of myosin II from chicken gizzard smooth muscle [16] and from *Acanthamoeba castellanii* [20, 21] unfold over a narrow temperature range, producing a single sharp endotherm. Zolkiewski *et al.* [20, 21] found no evidence for independent unfolding of the domains in the *Acanthamoeba* myosin rod and concluded that the mechanism of unfolding of the *Acanthamoeba* myosin II rod is different from that of skeletal muscle myosin rod. These data indicate that the structure of the myosin tail may be varied among different myosins depending on its functions in the myosin II molecule.

MYOSIN HEAD

By the end of the 1980's the interest of most investigators studying myosin structure and functions switched to studies on the myosin head. It was shown by that time that isolated myosin heads (myosin subfragment 1, S1) are capable of moving actin filaments in an *in vitro* motility assay [22]. Moreover, by that time a great number of 'unconventional' myosins were found, i. e. proteins devoid of the rod part characteristic for muscle myosins, but always having one or two heads with highly conserved structure [6–8]. It became clear that just the myosin head, possessing the ATPase and actin-binding sites, is the genuine 'molecular motor', i. e. that the force generation process during actin-myosin interaction occurs within the head, but not in some other parts of the myosin molecule. This can explain an enhanced interest to the structure of the myosin head and to its changes occurring during ATP hydrolysis and due to interaction with actin filaments.

Structural changes occurring in the myosin head during ATPase reaction

It is well established that the molecular mechanism of muscle contraction and many other events of biological motility is based on significant conformational changes occurring in the myosin head during ATP binding and hydrolysis. Just these changes in the myosin heads bound to F-actin alter the character of actin–myosin interaction and cause directed movement of the actin filaments. Elucidation of the mechanism of these changes in the myosin head is necessary for understanding the process of the transduction of ATP hydrolysis energy into mechanical work, which occurs in muscles and other systems of biological motility based on actin–myosin interaction.

In the course of the S1-catalysed Mg^{2+}-ATPase reaction a number of discrete intermediate states differing in the intensity of protein tryptophan fluorescence are formed. The greatest changes in fluorescence occur on formation of the intermediates $S1^{*}$-ATP and $S1^{**}$-ADP-P_i (each asterisk reflects the increase in the S1 intrinsic fluorescence [23]. However, these intermediates exist during the ATPase reaction for a very short time, which is not enough for detailed structural studies. For this purpose stable analogues of these intermediates are successfully used, i. e. the ternary complexes of S1 with ADP and P_i analogues such as orthovanadate (V_i), beryllium fluoride (BeF_x), aluminum fluoride (AlF_4^-), or scandium fluoride (ScF_x) anions [24–27]. It has been shown that the S1-ADP-BeF_x complex is different from all other complexes of S1 with ADP and P_i analogues: it resembles the $S1^*$-ATP intermediate state, whereas the complexes S1-ADP-V_i, S1-ADP-AlF_4^-, and S1-ADP-ScF_x resemble the $S1^{**}$-ADP-P_i intermediate state [28, 29].

Since 1990 the DSC method has successfully used for studying structural changes occurring in the myosin head due to formation of stable ternary complexes with ADP and P_i analogues. The first experiments demonstrated that the formation of the S1-ADP-V_i complex leads to a significant shift, by 10°C, of the heat sorption curve of S1 to higher temperature [30, 31]. The changes in the thermal unfolding of the myosin head in the presence of ADP and V_i were also observed by Lőrinczy and Belagyi in DSC experiments on myofibrils and muscle fibers [32, 33]. More detailed DSC studies showed that formation of the ternary complex S1-ADP-V_i causes a global change of the S1 conformation which is reflected not only in significant increase of the denaturation temperature (from 49 to 58°C), but also in pronounced increase of the enthalpy of the thermal transition, in considerable increase in cooperativity of the transition, and in significant change of S1 domain structure [34]. Similar although slightly less pronounced structural changes were found for the formation of the S1-ADP-BeF_x complex [35].

Figure 1 presents the heat sorption curves of skeletal S1 in the absence of nucleotides, in the presence of ADP, and in the ternary complexes S1-ADP-V_i, S1-ADP-AlF_4^-, and S1-ADP-BeF_x [36, 37]. The binding of ADP to S1 does not significantly affect the temperature of the S1 thermal transition, but it increases the cooperativity of the transition (the peak becomes much narrower). On the other hand, the formation of the S1-ADP-V_i complex leads to considerable structural changes of the entire S1 molecule [34]: the maximum of the thermal transition shifts by almost 10°C to a higher temperature, and the peak becomes very narrow, indicating a considerable increase in the cooperativity of the thermal transition. The same effects were observed on formation of the S1-ADP-AlF_4^- and S1-ADP-ScF_x complexes. A similar though somewhat less pronounced effect was also observed on formation of the S1-ADP-BeF_x complex (Fig. 1).

Thus, the DSC method is very useful for probing the structural changes that occur in the myosin head due to formation of stable ternary complexes with ADP and P_i analogues. The use of various naturally occurring nucleoside diphosphates [38] and their synthetic non-nucleoside analogues [39, 40] allowed us to conclude that these changes revealed by DSC adequately reflect those changes which occur in the S1 molecule in the course of the ATPase reaction. It has also been concluded from DSC experiments on recombinant fragments of *Dictyostelium discoideum* myosin II that the changes in the thermal unfolding, that are due to formation of stable ternary complexes with ADP and P_i analogues, occur mainly in the globular motor portion of the head [41]. Similar changes were also observed

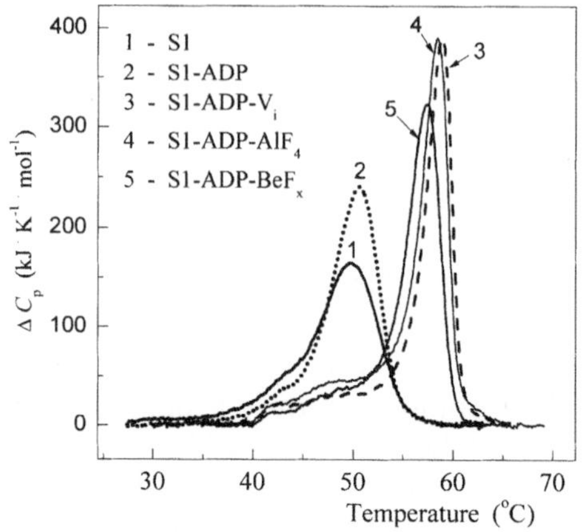

Fig. 1 Temperature dependencies of excess heat capacity (ΔC_p) of rabbit skeletal S1 in the absence of nucleotides (*1*), in the presence of ADP (*2*), and in the ternary complexes S1-ADP-V_i (*3*), S1-ADP-AlF_4^- (*4*), and S1-ADP-BeF_x (*5*). Conditions: 30 mM Hepes, pH 7.3, 1 mM $MgCl_2$. Heating rate 1 K/min

with heavy meromyosin (HMM) from smooth muscles of turkey gizzard containing dephosphorylated or fully phosphorylated regulatory light chains [42]. Thus, the presence of two heads in the HMM molecule and the extent of phosphorylation of the regulatory light chains have no significant influence on the structural changes induced in the globular motor portion of the head by the formation of stable ternary complexes with ADP and P_I analogues.

In the case of skeletal S1 the changes in the thermal unfolding induced by formation of the S1-ADP-BeF_x complex were similar to those observed for the complexes S1-ADP-V_i and S1-ADP-AlF_4^- although slightly less pronounced (Fig. 1). However, a more pronounced difference between the complexes has been observed by DSC studies with smooth muscle S1 and HMM [42], as well as with recombinant fragments of *Dictyostelium discoideum* myosin II corresponding to the motor domain of the myosin head [41]. In these cases the effects of V_i and AlF_4^- were almost the same, whereas the effect of BeF_x was much less pronounced: the shift of the protein thermal transition to higher temperature induced by formation of the complex with ADP-BeF_x was much less, by 2.5–4°C, than that observed for the ternary complexes with ADP-V_i or ADP-AlF_4^-. The most remarkable difference between the complexes has been revealed by DSC studies when skeletal S1 was specifically modified at residues Lys-83, Cys-707, or Cys-697 [37, 43, 44]. It has been shown that these modifications prevent to a great extent the conformational changes of the S1 molecule resulted from the formation of the ternary complexes S1-ADP-V_i and S1-ADP-AlF_4^- (i.e., they prevent the shift of the S1 thermal transition to higher temperature), but they have almost no influence on the changes induced by the formation of the

S1-ADP-BeF_x complex. These results suggest that during formation of the S1-ADP-V_i and S1-ADP-AlF_4^- complexes the region containing Cys-707, Cys-697, and Lys-83 (these residues are spatially located rather close to each other in the S1 molecule) plays an important role in transmission of structural changes from the active site of S1 ATPase to the entire motor portion of the myosin head, but this region does not take part in the transmission process on formation of the S1-ADP-BeF_x complex. Thus, this approach reveals a significant difference of the S1-ADP-BeF_x complex from other stable ternary complexes of S1 with ADP and P_i analogues. These DSC data support the viewpoint that the S1-ADP-BeF_x complex corresponds to a pre-hydrolysis state ($S1^*$-ATP) while the complexes S1-ADP-V_i and S1-ADP-AlF_4^- mimic the main transition state $S1^{**}$-ADP-P_i of the S1 ATPase cycle [28, 29].

Domain structure of the myosin head and its dramatic changes during ATPase cycle

The problem of elucidating the separate structural domains in the myosin head is of great importance because interdomain interactions might induce the internal motions in the head and play a key role in the transduction of energy of ATP hydrolysis into mechanical work.

The DSC method is one of the best approaches for revealing structural domains in multidomain proteins as distinct thermal transitions on the heat sorption curve. The most distinct and general feature of a domain in a globular protein is that its structure folds and unfolds cooperatively in an 'all-or-none' way with significant changes in enthalpy and entropy [3]. In accordance with this definition, domains can be detected in DSC studies as the regions in the protein molecule which unfold cooperatively and independently from each other. To reveal such domains in the myosin head, in early works we applied the 'successive annealing' method for DSC studies on the thermal unfolding of skeletal S1 [34, 35, 45–47]. This method based on the repetition of the cycle of 'heating–cooling–heating to a higher temperature' is applied to fully or partially irreversible thermal transitions. The method permits experimental decomposition of the total heat sorption curve of a protein into elementary thermal transitions corresponding to the melting of separate structural domains in the protein molecule. By means of this approach three such transitions (calorimetric domains) were revealed in the S1 molecule [34, 45, 46]. For identification of these domains, i.e. for revealing their correspondence to certain parts of the primary structure, many special approaches were applied, such as DSC studies of different preparations of S1, HMM, and their fragments, selective denaturation of some parts of the S1 molecule, the thermal gel analysis method, the measurements of temperature dependence for thermally induced inactivation of the S1 ATPase and for the changes in the position of the spectrum of either intrinsic tryptophan fluorescence or fluorescence of labels specifically attached to certain sites in the S1 molecule, etc. [41, 45–48]. On the basis of results obtained and of their comparison with literature data, I have proposed a domain model of the myosin head

[48]. According to this model proposed for the first time in 1991 [49], the least thermostable and the most thermostable calorimetric domains reflect the melting of the regulatory domain of the myosin head with associated light chains, whereas the middle domain corresponds to the globular motor portion of the head [37, 48]. Two years later, in 1993, the three-dimensional structure of the myosin head has been determined by X-ray analysis of S1 crystals [50]. An important feature of this structure is the existence, in good agreement with the DSC results, of clearly pronounced separate morphological domains in the myosin head, i. e. the motor domain and the regulatory domain.

Recently we have proposed a new approach for analyzing the domain structure of irreversibly denaturing proteins from their DSC curves, which is suitable even in the case of overlapping peaks of separate calorimetric domains [51]. An important feature of this approach is that the parameters of thermal denaturation of the most thermostable calorimetric domain are determined first. The DSC experiments are preceded by preliminary incubation of a protein at definite temperature for a definite time. Temperature for this incubation and its duration are determined from analysis of the DSC curves of a protein measured at different scanning rates. The parameters of thermal denaturation of separate calorimetric domains are verified by DSC experiments with such type treated protein at different scanning rates.

We have applied this approach to study the domain structure of the myosin head. Four separate calorimetric domains were revealed in rabbit skeletal S1 (Fig. 2*A*), in good agreement with earlier DSC results obtained by the 'successive annealing' method [34, 45–48]. On the other hand, only one calorimetric domain was revealed in the recombinant fragment M765 of the head of *Dictyostelium discoideum* myosin II corresponding to the globular motor portion of the head that lacks the 'neck' region and the light chains (Fig. 2B). By comparing these results we can conclude that the main calorimetric domain 3 of S1 (Fig. 2A) whose enthalpy is about half of the total enthalpy of the thermal unfolding of S1, corresponds to thermal denaturation of the motor domain of the S1 molecule. Other three calorimetric domains of S1 (domains 1, 2, and 4 on Fig. 2A), that are absent in the M765, can be assigned to thermal denaturation of the regulatory domain of the myosin head (i. e. the 'neck' region with associated alkali light chains, also called as 'lever arm').

Surprisingly, only one calorimetric domain was revealed in the S1 molecule in the ternary complexes S1-ADP-V_i (Fig. 2C) and S1-ADP-BeF_x [51]. The results suggest that in these complexes a tight coupling occurs between the motor and regulatory domains of the myosin head, and due to this interaction both these parts of the head denature together as a single calorimetric domain. It has been shown for scallop S1 that formation of the S1-ADP-V_i complex leads to significant movement of the lever arm, and as a result the lever arm is located rather close to the motor domain surface in this complex [52]. The DSC results presented above (Fig. 2) indicate that in the complexes S1-ADP-V_i and

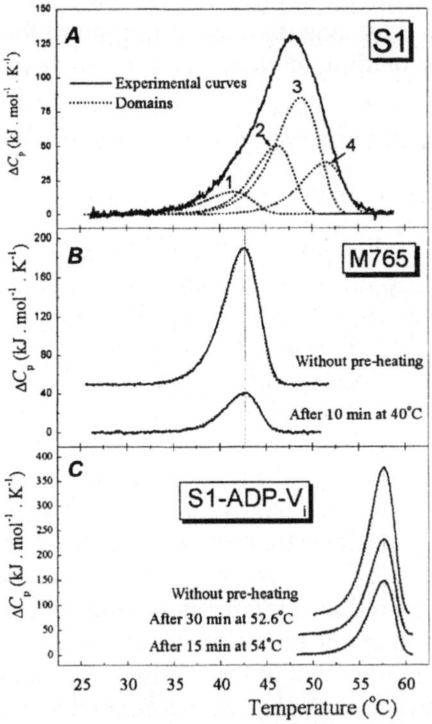

Fig. 2 Separate calorimetric domains in rabbit skeletal S1 (nucleotide-free) (***A***), M765 (the isolated motor part of the myosin head of *D. discoideum* myosin II) (***B***), and S1 in the ternary complex S1-ADP-V_i (***C***). Calorimetric domains 1, 2, 3, and 4 in the S1 molecule (***A***) were revealed by the procedure described in the text. After pre-heating the maxima of the thermal transitions for M765 and S1-ADP-V_i remain unchanged (***B, C***), and this indicates that both M765 and S1-ADP-V_i are represented by only one calorimetric domain. The heating rate was 1 K/min

S1-ADP-BeF_x the regulatory domain (lever arm) and the motor domain not only locate close to each other, but tight interaction occurs between both these domains of the myosin head. These results suggest that in the course of ATPase reaction myosin head undergoes to global changes of its domain structure, which are expressed in the tight coupling between the two main parts of the head, the motor domain and the regulatory domain.

Actin-induced structural changes in the myosin head

DSC is also used for structural studies of protein–protein interactions as a direct method to measure the thermal unfolding of proteins interacting with each other [53]. This approach reveals structural changes which occur in proteins due to their interaction.

DSC was successfully used for probing the structural changes that occur in the myosin head due to its strong binding to F-actin in the presence of ADP. It has been shown that the binding of skeletal S1 to F-actin significantly increases the S1 thermal stability by shifting the thermal transition of S1 by about 5°C to higher temperature [54]. For better separation of thermal transitions of F-actin and actin-bound myosin head on the DSC curve, we use in recent DSC experiments F-actin stabilized by phalloidin. In this case F-actin melts at very high temperature, at about 80°C, which allows more detailed study of the parameters of the thermal unfolding of the myosin head bound to F-actin (Fig. 3). While the thermal unfolding of phalloidin-stabilized F-actin is not significantly affected by the interaction with S1 (Fig. 1A), interaction with F-actin increases the thermal stability of S1. This actin-induced stabilization of the myosin head is expressed in a pronounced shift of its thermal transition to higher temperature (Fig. 3B). We use this shift, ΔT_m, as a relative measure for actin-induced structural changes in the myosin head. The size of ΔT_m depends on the source of myosin: it varies from 5–6°C for rabbit skeletal S1 (Fig. 3B) and *D. discoideum*

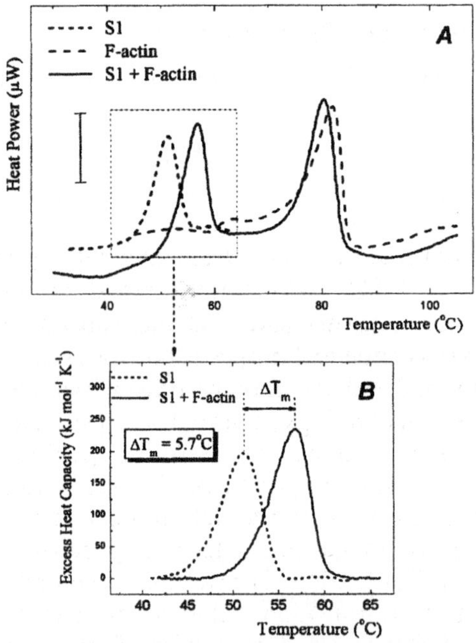

Fig. 3 (*A*) The experimental DSC curves of rabbit skeletal S1, F-actin stabilized by phalloidin, and their complex obtained in the presence of ADP. Conditions: 13 μM S1, 24 μM F-actin, 40 μM phalloidin, 20 mM Hepes, pH 7.3, 2 mM $MgCl_2$, 1 mM ADP. The vertical bar corresponds to 15 μW. Heating rate 1 K/min. (***B***) Excess heat capacity functions of S1 in the presence and in the absence of F-actin. The parameter ΔT_m (5.7°C) is defined by the difference between denaturation temperatures (T_m) of the actin-free and actin-bound S1

myosin II head fragment M765 [55] to 10–11°C for myosin I and smooth muscle HMM. It has been concluded from DSC experiments on recombinant fragments of the head of *D. discoideum* myosin II [55] that actin-induced changes in the thermal unfolding of the myosin head occur mainly in the globular motor portion of the head.

It has been shown that charge changes in the actin-binding surface loop 2 of myosin strongly affect the thermal unfolding of the myosin motor domain bound to F-actin [55]. Introducing of many additional negative changes into loop 2 of the isolated motor domain of the head of *D. discoideum* myosin II (M765) strongly decreased (from 6°C to 1.2°C) the parameter ΔT_m, whereas addition of positively charged residues to the loop produced a drastic increase, up to 9.1°C, of this parameter. All these mutant constructs did not significantly differ from each other in their ability to undergo global structural changes due to the formation of stable ternary complexes with ADP and P_i analogues [55]. Thus, the alterations in loop 2 do not affect the nucleotide-induced structural changes in the myosin head, but they affect the changes that occur in the motor domain of the head due to its strong binding to F-actin in the presence of ADP.

Direct electrostatic interaction between loop 2, a lysine-rich surface segment of the myosin head, and the negatively charged N-terminal part of actin is believed to be mainly responsible for the 'weak' binding to F-actin of the myosin head complexed with ATP or ADP-P_i. The following transition to the strongly bound state (when myosin head loses P_i and contains only ADP) is accompanied by formation of many additional contacts between actin and myosin. This transition from the weakly bound state to the strongly bound state plays a crucial role in force generation during muscle contraction as it produces a movement of actin filaments along myosin filaments.

After detailed studying the thermal unfolding of myosin head strongly bound to F-actin, we applied the DSC approach described above to study the weak binding of S1 to F-actin.

'Weak' binding of pPDM-S1 to F-actin

The weakly bound states are short-lived intermediates of actomyosin ATPase cycle, and therefore stable analogues of these states are required for structural studies. One of these analogues is rabbit skeletal S1 with SH-groups of Cys-707 and Cys-697 cross-linked by the bifunctional thiol reagent, N,N'-*p*-phenylenedimaleimide (*p*PDM). It was found that pPDM-modified S1 (*p*PDM-S1) binds weakly to F-actin even in the absence of nuscleotides, with an affinity similar to that of unmodified S1 in the presence of ATP [56, 57]. Therefore *p*PDM-S1 is often used for studies of the weak binding of the myosin head to actin.

We have applied the DSC approach described above (Fig. 3) to examine the weak binding of *p*PDM-S1 to F-actin stabilized by phalloidin [58]. It has been found that F-actin affects the thermal unfolding of *p*PDM-S1 only at very low ionic strength, when about 40% of *p*PDM-S1 binds weakly to F-actin, but not at higher ionic strength (200 mM KCl) preventing the interaction of *p*PDM-S1 with F-actin. The weak binding of *p*PDM-S1 to F-actin shifted the thermal transition of *p*PDM-S1 by about 5°C to a higher temperature [58]. This actin-induced increase in thermal stability of *p*PDM-S1 was similar to that observed with 'strong' binding of unmodified S1 to F-actin (Fig. 3).

These results show that actin-induced structural changes that are revealed by DSC in the myosin head occur not only upon strong binding but also on weak binding of the head to F-actin. This suggests that these actin-induced changes may occur before the power-stroke and play an important role in the motor function of the head. This assumption is corroborated by the DSC data demonstrating that these structural changes are strongly affected by charge changes in loop 2 [55], i. e. in the site mainly responsible for the weak binding of the myosin head to F-actin. We may speculate that since these structural changes occur in the myosin head on forming the weak binding state they must occur during the initial steps of actin–myosin interaction, and therefore they may play an important role for the transition of actin-bound myosin head from the weakly bound state to the strongly bound state.

In conclusion, these DSC results allow to propose that the actin-induced structural changes revealed by DSC in the myosin head are mainly determined by electrostatic interaction between positively charged loop 2 of myosin and negatively charged N-terminal part of actin.

Effects of specific modifications of the myosin head

Hence the use of the DSC method represents a powerful experimental approach for probing global nucleotide- and actin-induced structural changes in the myosin head (see pages 130 and 137). The main goal of these studies was to understand the mechanism of these changes, i. e. the mechanism of transmission of structural changes from the nucleotide- and actin-binding sites to the entire motor portion of the head. For this purpose, specially modified preparations of the myosin head were studied by DSC to reveal their ability to undergo global conformational changes due to interaction with F-actin and nucleotides. The most interesting modifications were those which did not directly affect the actin- and nucleotide-binding sites, but impaired the spread of conformational changes from these sites to the entire motor domain of the myosin head. Among these modifications we selected those which prevented either only nucleotide-induced or only actin-induced structural changes in the myosin head.

Modifications selectively preventing the DSC-revealed global structural changes induced in the myosin head by ADP and P_i analogues

First of all, one of these modifications is the above-described *p*PDM cross-linking between SH-groups of Cys-707 and Cys-697 in the S1 molecule (see page 137–138). The *p*PDM-S1 demonstrated the actin-induced structural changes [58], but its thermal unfolding was not affected by the addition of ADP and P_i analogues [58, 59]. However, the *p*PDM-S1 was unable to form the ternary complexes with ADP and P_i analogues [59].

Another case, which is much more interesting in this respect, is modification of both SH1 (Cys-707) and SH2 (Cys-697) groups in rabbit skeletal S1 by various thiol reagents [60]. This modification (without cross-linking between the SH-groups) had no effect on the actin-induced changes in the thermal unfolding of S1, but it almost fully prevented the changes induced by the formation of the ternary complexes S1-ADP-V_i, S1-ADP-AlF_4^-, and, to some extent, S1-ADP-BeF_x (Fig. 4). On the other hand, EPR studies on S1 spin-labeled at SH1 group showed that modification of both SH1 and SH2 groups has no effect on the local conformational changes, which occur around SH1 group due to formation of the S1 ternary complexes with ADP and P_i analogues [60]. Thus, the combined use of the DSC and EPR methods has shown that modification of both SH-groups, SH1 and SH2, does not prevent the local conformational changes induced by nucleotides around the SH1 group, but this modification strongly prevents the global nucleotide-induced structural changes of the entire S1 mole-

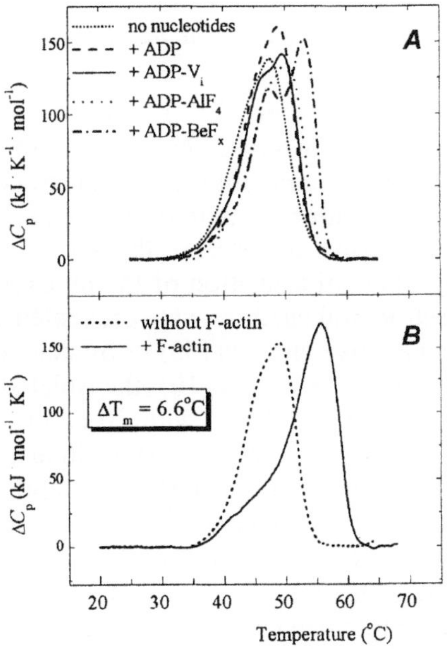

Fig. 4 Thermal unfolding of rabbit skeletal S1 modified at both reactive SH-groups, at SH1 group of Cys-707 with 4-iodoacetamido–2,2,6,6-tetramethylpiperidinoxyl (IASL), and at SH2 group of Cys-697 with N-[[(iodoacetyl) amino]ethyl]1-sulfo-5-naphtylamine (1,5-IAEDANS). (***A***) – in the presence of ADP and P_i analogues (V_i, AlF_4^-, BeF_x); (***B***) – in the presence of a two-fold molar excess of F-actin stabilized by phalloidin. In this figure and in Fig. 5B a temperature region above 70°C corresponding to the thermal denaturation of phalloidin-stabilized F-actin is not shown. Heating rate 1 K/min

cule. These results suggest that modification of SH1 (Cys-707) and SH2 (Cys-697) impair the spread of nucleotide-induced conformational changes from the ATPase site throughout the structure of the entire S1 molecule, thus disturbing a coupling between functionally important sites in the myosin head.

Similar effects were induced in the motor domain of the head of *D. discoideum* myosin II by mutation F506G, i. e. by replacing of Phe-506 by glycine in the 'relay loop' region of the head [61]. DSC data have shown that this mutation completely prevents conformational changes that normally occur upon formation of the ternary complexes with ADP and P_i analogues, but it has almost no influence on the actin-induced changes (the actin-induced shift of the thermal transition to higher temperature, ΔT_m, was equal to 4.5°C for this mutant construct). The mutant construct displayed no motor activity *in vitro* [61]. It has been proposed that the mutation F506G disrupts the communication between the nucleotide-binding site, the actin-binding site, and the 'converter' region in the motor domain of the myosin head, thus preventing transmission of structural changes from the nucleotide-binding site to the converter which is responsible for the movement of the lever arm.

Modifications selectively preventing only the actin-induced structural changes in the myosin head

It was mentioned above (see pages 136–137) that charge changes in the actin-binding loop 2 do not significantly affect the nucleotide-induced structural changes in the myosin motor domain, but they affect structural changes that occur when the motor domain is strongly bound to F-actin. Insertions with multiple positive charges were shown to increase the actin-induced shift to higher temperature, ΔT_m, of the thermal transition of the myosin motor. On the other hand, the actin-induced structural changes were prevented almost completely by insertions with multiple negative charges [55] or by deletion of the loop 2 [62]. For example, mutant constructs M765(20/-10) (the isolated motor domain of the head of *D. discoideum* myosin II, M765, with 20 additional residues inserted into loop 2, including 10 negatively charged residues) [55] and M765-NL (M765 with loop 2 deleted) [62] fully retained the ability to undergo global structural changes due to ADP binding and the formation of stable ternary complexes with ADP and P_i analogues. In contrast, actin binding to these mutant constructs had no or negligible effect on their thermal unfolding (Fig. 5).

These effects can be explained easily, as loop 2 is part of the actin-binding site and, therefore, alterations in this loop affect the actin-myosin interaction and those structural changes which occur in the myosin head due to this interaction. The DSC experiments with *p*PDM-S1 cited above (see page 137–138) have suggested that the interaction of loop 2 with actin is mainly responsible for actin-induced structural changes in the myosin head that are expressed in a pronounced shift of the thermal transition to higher temperature. In this respect, an intriguing result has recently been obtained by DSC study on S1 cleaved by

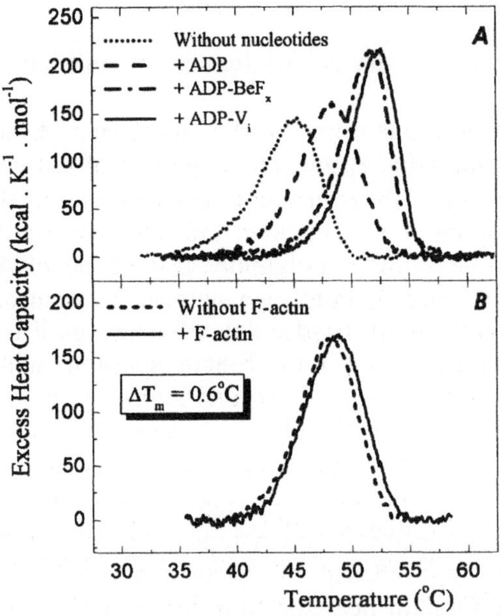

Fig. 5 Thermal unfolding of *D. discoideum* myosin head fragment M765 with deleted loop 2 (M765-NL) in the presence of ADP and P_i analogues (V_i, BeF_x) (***A***) and in the presence of a two-fold molar excess of F-actin stabilized by phalloidin (***B***). Heating rate 1 K/min

trypsin in the N-terminal region of the heavy chain, between Arg-23 and Ile-24 [63]. It has been shown that this cleavage has no effect on the nucleotide-induced structural changes in S1, but it prevents the changes that occur when S1 is bound to F-actin.

The effect of the N-terminal cleavage, i. e. the absence of the shift to higher temperature of the thermal transition of modified S1 bound to F-actin [63], cannot be explained by direct interaction between N-terminal region and loop 2 in S1 as these sites are spatially located rather far from each other in the atomic structure of S1 [50]. It seems more likely that a long-distance communication pathway exists between these sites. The cleavage between Arg-23 and Ile-24 probably disrupts this communication pathway, thus preventing the global conformational changes in the myosin head induced by actin binding to loop 2.

It can be concluded that the DSC approach makes it possible to reveal the global nucleotide-induced and actin-induced structural changes in the myosin head. This approach, in combination with other methods, allows us to investigate long-distance communication pathways between functionally important but spatially far regions in the myosin head.

Actin

Actin is found in virtually every eukariotic cell. Actin filaments have a crucial role in biological motility as the main partners of the myosin-based 'motor' systems and as the major constituent of the cytoskeleton. Filamentous actin (F-actin) is a double-stranded spiral polymer of actin monomers. Monomeric actin (G-actin) is a globular protein with molecular mass of 42 kDa. It consists of a single polypeptide chain with a bound ATP and a divalent cation. An important feature of actin is its ability to polymerization upon addition of neutral salts with formation of long polar filaments of F-actin. The polymerization of G-actin to F-actin is accompanied by the hydrolysis of bound ATP followed by slower release of P_i; as a result, each subunit of F-actin contains tightly bound ADP.

G-ACTIN

In 1984 Tatunashvili and Privalov [64] investigated by DSC the thermal denaturation of monomeric G-actin and suggested the presence of at least two interacting domains in its molecule. The existence of two domains in the G-actin molecule has also been proposed by Bertazzon *et al.* [65] after computer deconvolution of the G-actin heat sorption curve into two individual thermal transitions. It is necessary to note that thermal denaturation of actin is irreversible, and the use of such approaches for analysis of irreversible thermal transitions is rather controversial [66]. However, the existence of the domain structure in the G-actin molecule proposed from the DSC data [64, 65] has been fully confirmed by the three dimensional atomic structure of the G-actin published in 1990 [67]. It has been established that the G-actin molecule consists of two well distinguishable domains separated by a deep cleft, each domain being subdivided into two subdomains.

F-ACTIN

Polymerization of G-actin to F-actin

The use of DSC allows very clear probing of the changes caused by actin polymerization, i. e. the transformation of monomeric G-actin into F-actin filaments. These changes observed by many authors [17, 37, 65, 68] are expressed in a significant increase of the denaturation temperature, in a considerable increase of calorimetric enthalpy (the area under the peak), and in a sharp change of the peak shape (the peak becomes much narrower thus indicating a significant increase in the cooperativity of thermal denaturation) (Fig. 6).

G-actin is kept in the monomeric state in a low ionic strength buffer containing 0.2 mM $CaCl_2$ and 0.2 mM ATP (G-buffer). The polymerization is usually carried out by addition to G-buffer of either KCl to 50-150 mM, or $MgCl_2$ up to 2 mM or higher, or KCl and $MgCl_2$ simultaneously. Calorimetric curves of F-actin polymerized by these various ways were very similar to each other [69]. A similar curve was

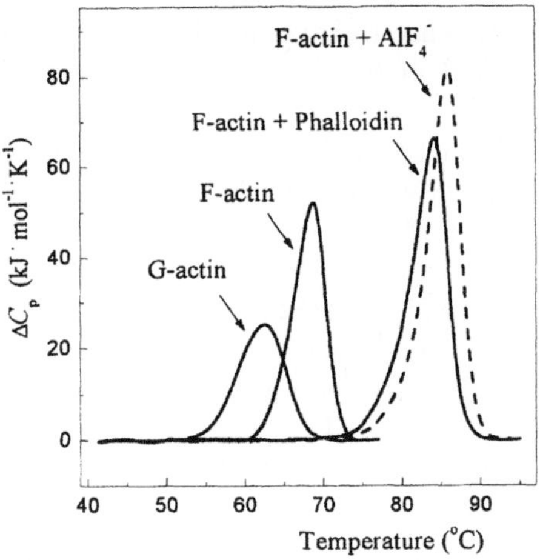

Fig. 6 Temperature dependencies of excess heat capacity (ΔC_p) of G-actin, F-actin, and F-actin stabilized by phalloidin or by aluminum fluoride (AlF_4^-). Heating rate 1.82 K/min

also obtained for F-actin polymerized by the addition of 2 mM $CaCl_2$ (so-called Ca-F-actin) [37]. However, a quite different calorimetric curve was obtained for F-actin polymerized by addition of 2 mM $MgCl_2$ in the presence of EGTA, i.e. upon replacement of tightly bound Ca^{2+} by Mg^{2+} (so-called Mg-F-actin). In this case both the temperature and the cooperativity of the thermal transition were much lower than in the case of Ca-F-actin [37]. This means that tightly bound Ca^{2+} is very important for stabilization of actin filaments.

Stabilization of F-actin by phalloidin and P_i analogues

It is well known that F-actin interacts specifically with a cyclic heptapeptide phalloidin – one of the principal toxins of a toxic mushroom *Amanita phalloides*. Phalloidin binds to F-actin with very high affinity and causes a substantial stabilization of the actin filaments, preventing depolymerization of F-actin and protecting it from proteolytic cleavage. Therefore actin filaments stabilized by phalloidin are often used for many experiments, e. g. for *in vitro* motility assays. It was shown by Le Bihan and Gicquaud [70] that the binding of phalloidin to F-actin significantly increases the temperature of F-actin thermal denaturation shifting the thermal transition by 14°C to a higher temperature. We also observed a similar effect of phalloidin in our DSC experiments [37]. This effect of phalloidin (Fig. 6) is very useful for DSC studies on the F-actin complexes with other proteins (see, for example, pages 136–141, 149–152), and therefore we often use the phalloidin-stabilized F-actin to obtain a better separation of the thermal transitions of F-actin and actin-bound proteins (e.g. myosin head, tropomyosin) on DSC curves.

The polymerization of G-actin to F-actin is accompanied by the hydrolysis of bound ATP. Like to the case of myosin, during ATP hydrolysis by actin an intermediate state is formed in actin subunits, that is a complex of actin with ADP and P_i. The structure of F-actin in this intermediate state can be studied using stable complexes of F-actin with ADP and P_i analogues – BeF_x and AlF_4^- (but not V_i) [71–73]. We have studied by DSC the thermal denaturation of F-actin in these complexes and shown that the structure of F-actin is drastically altered by interaction with BeF_x or AlF_4^- [69]. Structural changes caused by this interaction are expressed in a significant shift, by more than 16°C, of the thermal transition to higher temperature (Fig. 6). This increase in the F-actin thermal stability was similar (although more pronounced) to that observed by Bombardier *et al.* in the presence of ATP [74]. The effects of BeF_x and AlF_4^- were similar, while the effects of V_i and P_i were negligible [69].

It is interesting to note that phalloidin and AlF_4^- (or BeF_x) cause similar effects on the thermal denaturation of F-actin, which are expressed in a significant increase of the transition temperature (Fig. 6). However, when we simultaneously added to F-actin both these stabilizing agents, both phalloidin and BeF_x (or AlF_4^-), we observed an additional stabilization of F-actin expressed in the shift of the thermal transition by more than 25°C to a higher temperature [37]. This means that phalloidin and BeF_x (or AlF_4^-) stabilize F-actin independently from each other and affect, most likely, different sites of the actin molecule. This conclusion is consistent with the literature data that phalloidin and BeF_x bind to different sites on actin [72, 73].

Proposed mechanism for the thermal denaturation of F-Actin

In recent years we have undertaken a more detailed DSC study of the thermal unfolding of F-actin. It has been shown that the thermal stability of F-actin strongly depends on ADP concentration. The transition temperature, T_m, increases with increasing ADP concentration up to 1 mM and reaches a plateau at higher concentrations of ADP (Fig. 7A). This suggests that irreversible thermal denaturation of F-actin is preceded by reversible thermally induced dissociation of nucleotide (ADP) from actin subunits. The T_m value also depends on the concentration of F-actin: it increases by ~3°C as the protein concentration rises from 0.5 to 2.0 mg/ml (Fig. 7C). Similar dependence of the T_m value on protein concentration was demonstrated for F-actin stabilized by phalloidin, whereas it was much less pronounced in the presence of AlF_4^- (Fig. 7B). However, T_m was independent of protein concentration in the case of monomeric G-actin (Fig. 7C). Such dependence of T_m on protein concentration is believed to be characteristic for thermal denaturation of oligomeric proteins and denotes the presence of reversible stage of dissociation of subunits before their irreversible thermal denaturation [75, 76].

Thus, our results suggest that at least two reversible stages precede the irreversible thermal denaturation of F-actin. One of them is dissociation of nucleotide

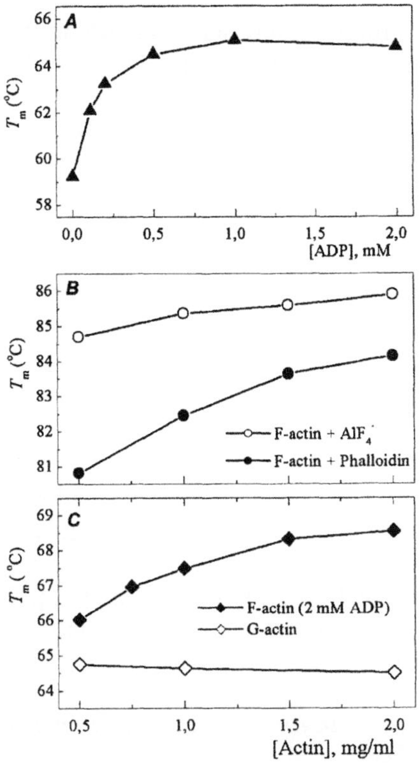

Fig. 7 (*A*) – Dependence of the maximum temperature (T_m) of the thermal transition of F-actin (1.5 mg/ml) on concentration of ADP. (***B, C***) – T_m dependence on protein concentration for G-actin (in G-buffer) and F-actin (in 30 mM Hepes, pH 7.3, containing 2 mM $MgCl_2$ and 2 mM ADP) (***C***), and for F-actin stabilized by phalloidin or AlF_4^- (***B***). Heating rate was 1 K/min

(ADP), and another is dissociation of subunits from actin filaments. Therefore we propose the following mechanism for the thermal unfolding of F-actin filaments. First, actin monomers (or short oligomers) dissociate from the pointed end of the polar actin filament. These monomers either lose bound ADP and then immediately denature, or the ADP-containing monomers may bind to the barbed end of the filament. This model explains why unfolding of F-actin depends on both ADP and protein concentration.

It is important to note that two F-actin stabilizers, phalloidin and AlF_4^-, differ from one another in their influence on the T_m dependence on protein concentration (Fig. 7B). Both these stabilizers increase the thermal stability of F-actin substantially (Fig. 6) although independently from each other [37]. Phalloidin increases the F-actin thermal stability by strengthening the bonds between adjacent subunits in the actin filament, whereas AlF_4^- demonstrates a similar effect by trapping of ADP in the nucleotide-binding site of actin. Thus, AlF_4^- (which mimics the ADP-P_i-actin intermediate state of actin polymerization) should prevent dissociation of ADP from actin subunits and, on the other hand, it should stimulate the binding of actin monomers to the barbed end of the filament. This means, according to the model proposed above, that AlF_4^- may affect both reversible stages preceding the irreversible thermal denaturation of F-actin, thus making the T_m dependence on protein concentration for F-actin similar to those characteristic for monomeric proteins.

Interaction of F-actin with other proteins and structures

Actin filaments can interact with many other proteins. First of all, interaction of F-actin with myosin heads and muscle regulatory proteins (tropomyosin and troponin in striated muscle or caldesmon and calponin in smooth muscle) plays a key role in the molecular mechanism and regulation of muscle contraction. In nonmuscle cells actin filaments are highly dynamic and participate in a range of processes such as cell polarization and movement, endocytosis, and cell division. The structure and dynamics of actin filaments are regulated by a large number of actin-binding proteins [77]. Actin filaments can also interact with the plasma membrane, directly or indirectly through actin-binding proteins, and this interaction is essential to many cellular activities such as intracellular organelle movement, plasma membrane dynamics, exocytosis, phagocytosis, and cytokinesis.

Thermal unfolding of F-actin in its complexes with myosin, tropomyosin, and muscle regulatory proteins

Although the binding of myosin heads to F-actin significantly increased the myosin thermal stability, it had no appreciable influence on the thermal unfolding of F-actin both in the absence [54] and in the presence of phalloidin [37, 55, 63] (Fig. 3A). Also, we did not observe any changes in the thermal unfolding of F-actin in its complexes with smooth muscle tropomyosin [78], caldesmon, and calponin, as well as in the complexes with skeletal tropomyosin [79]. However, we observed a pronounced change in the thermal denaturation of F-actin upon addition of troponin I, one of the components of the troponin complex. These changes were expressed in a significant decrease in the enthalpy and cooperativity of the melting of F-actin [79]. These data indicated that there is a direct interaction between troponin I and F-actin. The effect was much weaker if troponin I was added to F-actin stabilized by AlF_4^- [79]. Apparently, stabilization of F-actin by AlF_4^- prevents the effect of troponin I on the structure of actin filaments.

Effects of cofilin on the thermal unfolding of F-actin

Very little is known how the actin-binding proteins affect the thermal unfolding of F-actin. One of the first studies in this direction was our attempt to apply the DSC method to investigate the thermal unfolding of F-actin in its complexes with cofilin, a small actin-binding protein which is known to bind both G-actin

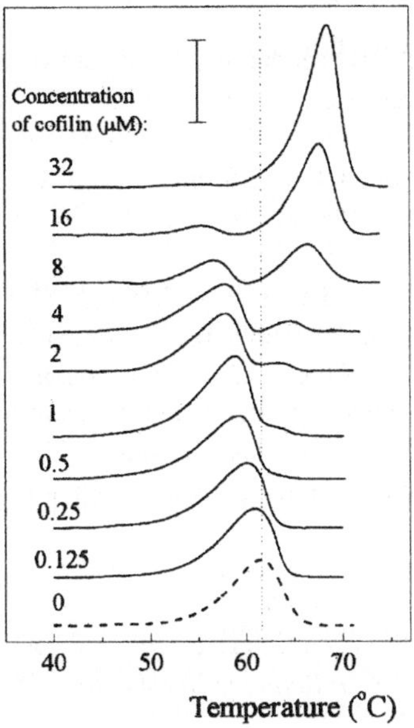

Fig. 8 DSC curves of the complexes of F-actin with cofilin obtained at various cofilin/actin molar ratios. F-actin concentration (25 μM) was held constant. Concentrations of cofilin are indicated for each curve. Dotted vertical line indicates the maximum of the thermal transition for cofilin-free F-actin (T_m= 61.5°C). The vertical bar corresponds to 25 μW. Heating rate 1 K/min

and F-actin and to promote rapid turnover actin filaments by depolymerizing/fragmenting the filaments [77]. It has been shown that cofilin binding increases the thermal stability of G-actin but two different effects are observed on F-actin, depending on a degree of saturation with cofilin. At saturating concentrations, cofilin stabilizes F-actin, but at sub-saturating concentrations it causes a strong decrease in F-actin thermal stability (Fig. 8). This destabilizing effect of cofilin is highly cooperative as it is observed even at very low cofilin/actin molar ratios, as small as 1 cofilin per 100–200 actin monomers. It has been suggested from these DSC results that cofilin, when binds to F-actin, stabilizes only those actin subunits to which it directly binds, but it destabilizes with a very high cooperativity neighboring regions of the actin filament that are free of cofilin [80]. These two opposite effects of cofilin demonstrated by DSC, stabilization and destabilization of actin filaments, may play an important role in actin dynamics in living cells.

Interaction of actin with membrane lipids

A very interesting effect was observed by Gicquaud [81] who used the DSC method for studies on direct interaction of F-actin with membrane lipids. It has been shown that due to the interaction with liposomes, actin undergoes to a major conformational change which results in the complete disappearance of its thermal transition on the DSC curves.

Proteins involved in the regulation of muscle contraction

In this section the DSC studies on the proteins responsible for regulation of actin–myosin interaction during muscle contraction will be reviewed. Only the proteins associated with actin filaments will be considered here. These will include tropomyosin and troponin in skeletal and cardiac muscles or caldesmon and calponin in smooth muscles.

TROPOMYOSIN

Tropomyosin (Tm) is a coiled-coil actin-binding protein bound along the length of actin filament in both muscle and non-muscle cells. Tm molecules bind to themselves in a end-to-end manner, and form a continuous structure along the actin filament. The presence of Tm on actin filaments confers cooperativity on the interaction between myosin heads and actin. Together with the other regulatory proteins, troponin in striated skeletal and cardiac muscles or caldesmon in smooth muscle, it takes part in the Ca^{2+} regulation of muscle contraction. The Tm subunits consist of nearly 100% α-helix, and two polypeptide chains assemble into parallel and in-register coiled-coil dimers. Two isoforms (α and β) each containing 284 residues are expressed in smooth and skeletal muscles. Smooth muscle Tm consists of a 1:1 mixture of α and β chains which are predominantly assembled into the αβ heterodimers. In skeletal muscle the α and β chains are present in the ratio of (3–4):1, and they can form all three possible dimers: αα, ββ, and αβ. Under physiological conditions, skeletal Tm is a mixture of αα-homodimers and αβ-heterodimers.

Thermal unfolding of tropomyosin and effects of an interchain disulfide bond

Thermal denaturation of Tm homodimers from skeletal and smooth muscles has been investigated by DSC in detail. As shown by Potekhin and Privalov in 1982 [82] and then confirmed by other authors [83–85], the Tm molecule consists of cooperative blocks with different thermal stabilities. For example, both αα and ββ homodimers of muscle Tm exhibited broad multistate thermal transitions composed of at least three individual transitions (or calorimetric domains) [84, 85]. In contrast, αβ heterodimers showed a sharp transition composed of only two calorimetric domains [85]. Moreover, DSC studies on smooth muscle Tm clearly demonstrated formation of αβ heterodimers during heating of a 1:1 mixture of αα and ββ homodimers [85]. Each chain of skeletal Tm dimer (both α and β) contains a cysteine residue at position 190, and therefore SH-groups of Cys-190 can form a disulfide bond between two Cys-190 belonging to two chains of the Tm dimer. Formation of this interchain disulfide bond was shown to increase substantially the thermal stability of skeletal Tm [86].

The DSC profile of skeletal α-αTm is represented by two well distinguished peaks with maxima at ~43°C and ~54°C (Fig. 9A). When the experiment is performed in the absence of β-mercaptoethanol, the low-temperature peak completely disappears during the second heating of the sample. On the other hand, this peak increases during the second heating of the sample containing β-mercaptoethanol (Fig. 9A) [79]. These results are in good agreement with the DSC data of Williams and Swenson [87]. These authors showed that the low-temperature peak of Tm corresponds to the melting of C-terminal half of the Tm molecule with reduced SH-groups of Cys-190, whereas the high-temperature peak is composed of two overlapping thermal transitions. One of them corresponds to the N-terminal half of Tm, and another – to the oxidized C-terminal half of Tm with Cys-190 cross-linked by disulfide bond between two chains of the Tm dimer. We suppose that during the first heating of Tm in the presence of β-mercaptoethanol the SH-groups of Cys-190

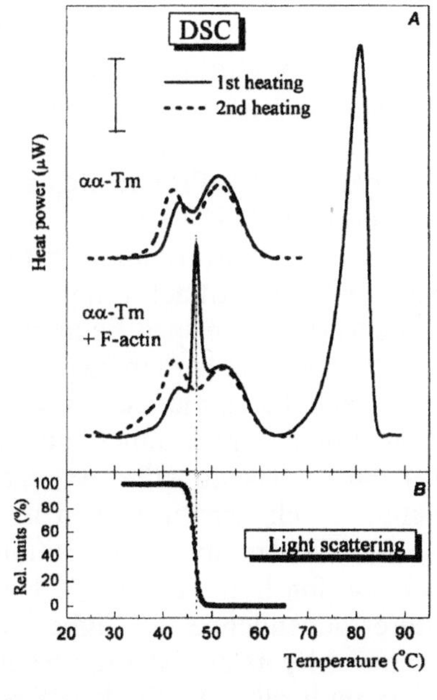

Fig. 9 (*A*) – DSC profiles of the thermal denaturation of the complex of rabbit skeletal αα-tropomyosin (αα-Tm) with F-actin stabilized by phalloidin. For comparison, the thermal unfolding of αα-Tm in the absence of F-actin is also shown. Dashed line curves were obtained by a re-heating the corresponding samples represented by solid line curves. Conditions: 30 μM αα-Tm, 46 μM F-actin, 70 μM phalloidin in 20 mM Hepes, pH 7.3, 2 mM $MgCl_2$, 1 mM β-mercaptoethanol. The vertical bar corresponds to 10 μW. Heating rate was 1 K/ min. (***B***) – Temperature dependence of dissociation of the complex of αα-Tm with F-actin. Conditions were the same as for DSC experiment in (*A*). 100% corresponds to the difference between light scattering of the αα-Tm–F-actin complex measured at 25°C and that of pure F-actin stabilized by phalloidin which was temperature independent within the temperature range used. A decrease in the light scattering intensity reflects dissociation of the αα-Tm–F-actin complex

become reduced, and therefore we observe the increase of the low-temperature peak and decrease of the high-temperature peak during the second heating of Tm [79] (Fig. 9A). Thus, the state of the SH-group of Cys-190 (reduced or oxidized with formation of disulfide bond between two chains of Tm dimer) plays a very important role in the thermal unfolding of Tm, as it determines the thermal stability of the C-terminal half of Tm molecule.

Complexes of tropomyosin with F-actin

Current views of the regulation of actomyosin interaction in striated muscle suggest that Tm can occupy three different positions or states on actin (B-state – 'blocked' or Ca^{2+}-free, C-state – 'closed' or Ca^{2+}-induced, and M-state – 'open' or myosin-induced), depending on the presence or absence of troponin, myosin, and Ca^{2+} [88]. Caldesmom-induced movements of Tm between two different positions along F-actin were shown also for smooth muscle [89]. Thus, the movements of actin-bound Tm are believed to play a crucial role in the regulation of muscle contraction. According to recent model of Tm functions, actin-bound Tm can be considered as a continuous flexible filament whose dynamic properties are modulated by external influences such as the presence of troponin and actin-bound myosin [90]. Protein flexibility may correlate with thermal instability and therefore DSC studies of thermal unfolding of actin-bound Tm may provide valuable information on the dynamic properties of Tm in its different states on the surface of actin filament.

Recently DSC was used to study the specific interaction of smooth muscle Tm with actin filaments [78]. The results demonstrated that the interaction of Tm with F-actin produces a pronounced increase in the thermal stability of Tm (shift of the Tm thermal transition to higher temperature by 2–6°C depending on the Tm/actin molar ratio), but it has no effect on the thermal unfolding of phalloidin-stabilized F-actin, which denatures at a much higher temperature than Tm. A pronounced shift of the Tm thermal transition was observed only for heterodimers of Tm, and not for homodimers. It has also been found, by measuring the temperature dependence of light scattering, that thermal unfolding of Tm is accompanied by its dissociation from F-actin [78]. Thus, the use of DSC in combination with other methods offers a new and promising approach for structural characterization of actin-bound Tm.

Very recently we have applied this approach to studies on skeletal Tm. The character of the thermal unfolding of Tm is noticeably changed if it is bound to F-actin. This is expressed in an appearance of a new highly cooperative thermal transition with maximum at ~47°C (Fig. 9A). After heating of the complex Tm–F-actin to 90°C (i. e. after complete irreversible denaturation of actin) and following cooling, only peaks at 42.5 and 53.5 °C corresponding to the thermal unfolding of free Tm were observed on heat sorption curve during reheating (dotted line curves on Fig. 9A). Thus, we conclude that the appearance of the peak at ~47°C in the presence of F-actin reflects the thermal unfolding of actin-bound Tm. A very good correlation is found between the maximum temperature of this actin-induced thermal transition revealed by DSC (Fig. 9A) and the temperature of dissociation of the Tm–F-actin complex, i. e. the temperature at which a 50% decrease in light scattering occurs (Fig. 9B). This leads to the conclusion that actin-induced changes in the thermal unfolding of Tm are associated with dissociation of Tm from F-actin. It is also important to note that the appearance of new actin-induced transition at ~47°C is accompanied by disappearance of the low-temperature peak of Tm, with no changes in the high-temperature peak (Fig. 9A).

These results allow us to propose the following mechanism of the thermal unfolding of actin-bound Tm. F-actin protects the actin-bound Tm from thermal denaturation, which only occurs upon dissociation of Tm from F-actin. Therefore the new cooperative transition at ~47°C, that appears only in the presence of actin, reflects thermal denaturation of those parts of Tm which have to denature, in the absence of actin, at temperatures lower than the temperature of dissociation (mainly the low-temperature peak of free Tm, i. e. the C-terminal half of Tm with reduced SH-groups of Cys-190). These parts of Tm denature, in the presence of F-actin, within a very narrow temperature range. All other parts of Tm (the N-terminal half and the C-terminal half with Cys-190 cross-linked by disulfide bonds between two chains of Tm), which melt at higher temperature than that of dissociation, denature independently of the presence of actin after dissociation of Tm from F-actin.

We have also applied the DSC approach to characterize the thermal unfolding of actin-bound Tm in its different states on the surface of actin filaments. The results of our experiments indicate that the transition of smooth muscle Tm to the M-state induced by the binding of myosin to actin is accompanied by significant increase in the thermal stability of actin-bound Tm: the maximum of its thermal transition shifts by 5–6°C to higher temperature without appreciable changes in the cooperativity of the transition. On the other hand, transition of skeletal Tm to the B-state induced by addition of troponin in the absence of Ca^{2+} leads to significant increase in the cooperativity of the thermal unfolding of actin-bound Tm [79].

TROPONIN

Troponin (Tn) is a protein complex composed of three subunits interacting with one another: the Ca^{2+}-binding troponin C (TnC), the inhibitory troponin I (TnI) which inhibits the ATPase activity of actomyosin, and the troponin T (TnT) responsible for Tn binding to Tm. In the absence of Ca^{2+}, TnC interacts weakly with the other two Tn components, while TnI and TnT bind strongly to Tm and fix it in the 'blocked' B-state on the actin filament. Binding of Ca^{2+} to TnC strengthens the interaction between components of the Tn complex and weakens their interaction with Tm, leading to transition of actin-bound Tm from B-state into C-state. Due to this ability, Tn plays a key role in Ca^{2+}-regulation of striated (skeletal and cardiac) muscle contraction.

DSC studies on isolated components of the Tn complex have shown that both TnI and TnT demonstrate no cooperative thermal transitions during heating up to

100°C [79, 91]. Calorimetric effects of these two proteins were observed only indirectly, when they affected the thermal unfolding of actin-bound Tm [79] (see pages 150–151). TnC is the only component of the Tn complex that denatures with a cooperative transition [92, 93]. In 1980 Tsalkova and Privalov described the thermal unfolding of TnC [92]. They have shown that in the presence of divalent cations the DSC curve of TnC is represented by two well-separated thermal transitions corresponding to two domains in the TnC molecule, onle of them containing Ca^{2+}-specific binding sites, and another – Ca^{2+}-Mg^{2+}-binding sites. Binding of Ca^{2+} to Ca^{2+}-specific sites was shown to increase dramatically the thermal stability of the domain containing these sites [92]. Thus, DSC studies reveled global structural changes which occur in TnC due to specific binding of Ca^{2+} and play an important role in the functioning of TnC as a regulatory protein.

CALDESMON AND CALPONIN

Smooth muscle contraction is regulated primarily by the phosphorylation of myosin regulatory light chains by a Ca^{2+}–calmodulin-dependent myosin light chain kinase (MLCK). Two smooth muscle actin binding proteins, caldesmon and calponin, may also be involved in regulation of smooth muscle contraction and act as a secondary control of the contraction. Both proteins inhibit the actin-activated myosin ATPase activity and the movement of actin filaments in *in vitro* motility assays. Caldesmon alters the position of Tm on actin [89] but in a manner that is different from the effect of Tn in skeletal muscle.

Our DSC experiments showed that both caldesmon and calponin do not demonstrate cooperative thermal transitions on heating up to 100°C. Furthermore, we have not found any effects of these proteins on the thermal unfolding of F-actin and actin-bound Tm. The only thermal effect of calponin observed till now by DSC was its influence on the phase transitions of phospholipid vesicles [94]. This fact indicated that calponin may interact with phospholipids and by this means anchor actin filaments to the cellular membrane.

Conclusion

The DSC approach offers new unique opportunities for structural and functional studies on muscle proteins. It allows very clear revealing and following detailed investigation of global structural changes that occur in the myosin head during ATPase reaction and due to interaction with actin. This approach, in combination with other methods, allows to investigate long-distance communication pathways between functionally important but spatially far regions in the myosin head. DSC also provides a new and promising approach for studying of highly cooperative conformational changes in actin filaments induced by actin-binding proteins. Moreover, DSC in combination with other methods is very useful for structural characterization of tropomyosin on the surface of actin filaments. Thus, DSC studies on muscle proteins offer a promising approach to probe and investigate structural changes which play an important role in functioning of these proteins.

Acknowledgements

I thank my colleagues and friends participated in DSC experiments on muscle proteins from 1989 up to now: Valery Shnyrov, Nina Golitsina, Olga Nikolaeva, Nikolai Khvorov, Andrey Bobkov, Irina Dedova, Elena Rostkova, Dmitrii Pavlov, Victor Orlov, Mikhael Ponomarev, Olga Kaspieva, Elena Kremneva, Valeriya Mikhailova, Lubov' Shakirova, Elena Siletskaya, and Eugene Zubov.
This work was supported in part by the Russian Foundation for Basic Research (Grant 03-04-48237), by the Program for the Support of the Leading Scientific Schools in Russia (Grant NSH-813.2003.04), by the Program on Physico-Chemical Biology of Russian Academy of Sciences, and by the Welcome Trust (Grant 066115/Z/01/Z).

References

1 Privalov, P. L. and Potekhin, S. A.: Scanning microcalorimetry in studying temperature-induced changes in proteins, *Methods. Enzymol.,* 131 (1986) 4–51.
2 Shnyrov, V. L., Sanchez-Ruiz, J. M., Boiko, B. N., Zhadan, G. G. and Permyakov, E. A.: Application of scanning microcalorimetry in biophysics and biochemistry, *Thermochim. Acta,* 302 (1997) 165–180.
3 Privalov, P. L.: Stability of proteins. Proteins which do not present a single cooperative system, *Adv. Protein. Chem.,* 35 (1982) 1–104.
4 Sturtevant, J. M.: Biochemical applications of differential scanning calorimetry, *Annu. Rev. Phys. Chem.,* 38 (1987) 463–488.
5 Hodge, T. and Cope, M. J. T. V.: A myosin family tree, *J. Cell Sci.,* 113 (2000) 3353–3354.
6 Cope, M. J. T. V., Whisstock, J., Rayment, I. and Kendrick-Jones, J. K.: Conservation within the myosin motor domain: implications for structure and function, *Structure,* 4 (1996) 969–987.
7 Cheney, R. E. and Mooseker, M. S.: Unconventional myosins, *Curr. Opin. Cell Biol.,* 4 (1992) 27–35.
8 Sellers, J. S., Goodson, H. V. and Wang, F.: A myosin family reunion, *J. Muscle Res. Cell Motil.,* 17 (1996) 7–22.
9 Harrington, W. F.: On the origin of the contractile force in skeletal muscle, *Proc. Natl. Acad. Sci. USA,* 76 (1979) 5066–5070.
10 Potekhin, S. A., Trapkov, V. A. and Privalov, P. L.: Stages in the thermal denaturation of spiral fragments of myosin, *Biofizika,* 24 (1979) 46–50.
11 Shnyrov, V. L., Vedenkina, N. S., Ostrovsky, A. V., Permyakov, E. A., Golitsina, N. L. and Levitsky, D. I.: Study of thermal denaturation of the rod part of myosin molecule by microcalorimetry and intrinsic fluorescence methods, *Biofizika,* 35 (1990) 415–420.
12 Lopez-Lacomba, J. L., Guzman, M., Cortijo, M., Mateo, P. L., Aguirre, R., Harvey, S. C. and Cheung, H. C.: Differential scanning calorimetric study of the thermal unfolding of myosin rod, light meromyosin, and subfragment 2, *Biopolymers,* 28 (1989) 2143–2159.

13 Bertazzon, A. and Tsong, T. Y.: Study of effects of pH on the stability of domains in myosin rod by high-resolution differential scanning calorimetry, *Biochemistry,* 29 (1990) 6453–6459.
14 Nakaya, M., Watabe, S. and Ooi, T.: Differences in the thermal stability of acclimation temperature-associated types of carp myosin and its rods on differential scanning calorimetry, *Biochemistry,* 34 (1995) 3114–3120.
15 Swenson, C. A. and Ritchie, P. A.: Conformational transitions in the subfragment-2 region of myosin, *Biochemistry,* 19 (1980) 5371–5375.
16 Cross, R. A., Bardsley, R. G., Ledward, D. A., Small, J. V. and Sobieszek, A.: Conformational stability of the myosin rod, *Eur. J. Biochem.,* 145 (1984) 305–310.
17 Bertazzon, A. and Tsong, T. Y.: Effects of ions and pH on the thermal stability of thin and thick filaments of skeletal muscle: high-sensitivity differential scanning calorimetric study, *Biochemistry,* 29 (1990) 6447–6452.
18 Ueno, H. and Harrington, W. F.: Cross-bridge movement and the conformational state of the myosin hinge in skeletal muscle, *J. Mol. Biol.,* 149 (1981) 619–640.
19 Reisler, E. and Liu, J.: Conformational changes in the myosin subfragment-2. Effect of pH on synthetic rod filaments, *J. Mol. Biol.,* 157 (1982) 659–669.
20 Zolkiewski, M., Redowicz, M. J., Korn, E. D. and Ginsburg, A.: Thermal unfolding of *Acanthamoeba* myosin II and skeletal muscle myosin, *Biophys. Chem.,* 59 (1996) 365–371.
21 Zolkiewski, M., Redowicz, M. J., Korn, E. D., Hammer J. A. III, and Ginsburg, A.: Two-state thermal unfolding of a long dimeric coiled-coil: the *Acanthamoeba* myosin II rod, *Biochemistry,* 36 (1997) 7876–7883.
22 Toyoshima, Y. Y., Kron, S. J., McNully, E. M., Niebling, K. R., Toyoshima, C. and Spudich, J. A.: Myosin subfragment-1 is sufficient to move actin filaments *in vitro, Nature,* 328 (1987) 536–539.
23 Johnson, K. A. and Taylor, E. W.: Intermediate states of subfragment 1 and actosubfragment 1 ATPase: reevaluation of the mechanism, *Biochemistry,* 17 (1979) 3432–3442.
24 Goodno, C. C.: Myosin active site trapping with vanadate ion, *Meth. Enzymol.,* 85 (1982) 116–123.
25 Phan, B. C. and Reisler, E.: Inhibition of myosin ATPase by berillium fluoride, *Biochemistry,* 31 (1992) 4787–4793.
26 Werber, M. M., Peyser, Y. M. and Muhlrad, A.: Characterization of stable beryllium fluoride, aluminum fluoride, and vanadate containing myosin subfragment 1–nucleotide complexes, *Biochemistry,* 31 (1992) 7190–7197.
27 Gopal, D. and Burke, M.: Formation of stable inhibitory complexes of myosin subfragment 1 using fluoroscandium anion, *J. Biol. Chem.,* 270 (1995) 19282–19286.
28 Fisher, A. J., Smith, C. A., Thoden, J., Smith, R., Sutoh, K., Holden, H. M. and Rayment, I.: Structural studies of myosin:nucleotide complexes: a revised model for the molecular basis of muscle contraction, *Biophys. J.,* 68 (1995) 19s–28s.
29 Ponomarev, M. A., Timofeev, V. P. and Levitsky, D. I.: The difference between ADP-beryllium fluoride and ADP-aluminum fluoride complexes of the spin-labeled myosin subfragment 1, *FEBS Lett.,* 371 (1995) 261–263.
30 Shriver, J. M. and Kamath, U.: Differential scanning calorimetry of the unfolding of myosin subfragment 1, subfragment 2, and heavy meromyosin, *Biochemistry,* 29 (1990) 2556–2564.

31 Khvorov, N. V., Levitsky, D. I., Bukatina, A. E., Shnyrov, V. L. and Poglazov, B. F.: Calorimetric evidence for two conformational states of the complex of myosin subfragment 1 with nucleotides, *Doklady Akad. Nauk SSSR,* 315 (1990) 745–748.

32 Lörinczy, D. and Belagyi, J.: Effects of nucleotide on skeletal muscle myosin unfolding in myofibrils by DSC, *Biophys. Biochem. Res. Commun.*, 217 (1995) 592–598.

33 Lörinczy, D. and Belagyi, J.: Nucleotide binding induces global and local structural changes of myosin head in muscle fibers, *Eur. J. Biochem.,* 268 (2001) 5970–5976.

34 Levitsky, D. I., Shnyrov, V. L., Khvorov, N. V., Bukatina, A. E., Vedenkina, N. S., Permyakov, E. A., Nikolaeva, O. P. and Poglazov, B. F.: Effects of nucleotide binding on thermal transitions and domain structure of myosin subfragment 1, *Eur. J. Biochem.,* 209 (1992) 829–835.

35 Bobkov, A. A., Khvorov, N. V., Golitsina, N. L. and Levitsky, D. I.: Calorimetric characterization of the stable complex of myosin subfragment 1 with ADP and beryllium fluoride, *FEBS Lett.,* 332 (1993) 64–66.

36 Levitsky, D. I., Bobkov, A. A., Golitsina, N. L., Nikolaeva, O. P., Pavlov, D. A. and Poglazov, B. F.: Calorimetric studies on the stable complexes of myosin ubfragment 1 with ADP and phosphate analogues, *Biofizika,* 41 (1996) 64–72.

37 Levitsky, D. I., Nikolaeva, O. P., Orlov, V. N., Pavlov, D. A., Ponomarev, M. A. and Rostkova, E. V.: Differential scanning calorimetric studies on myosin and actin, *Biochemistry (Moscow),* 63 (1998) 322–333.

38 Bobkov, A. A. and Levitsky, D. I.: Differential scanning calorimetric study of the complexes of myosin subfragment 1 with nucleoside diphosphates and vanadate or beryllium fluoride, *Biochemistry,* 34 (1995) 9708–9713.

39 Gopal, D., Bobkov, A. A., Schwonek, J. P., Sanders, C. R., Ikebe, M., Levitsky, D. I. and Burke, M.: Structural basis of actomyosin chemo-mechanical transduction by non-nucleoside triphosphate analogues, *Biochemistry,* 34 (1995) 12178–12184.

40 Gopal, D., Pavlov, D. A., Levitsky, D. I., Ikebe, M. and Burke, M.: Chemomechanical transducion in the actomyosin molecular motor by 2',3'-dideoxydidehydro-ATP and characterization of its interaction with myosin subfragment 1 in the presence and absence of actin, *Biochemistry,* 35 (1996) 10149–10157.

41 Levitsky, D. I., Ponomarev, M. A., Geeves, M. A., Shnyrov, V. L. and Manstein, D. J.: Differential scanning calorimetric study of the thermal unfolding of the motor domain fragments of *Dictyostelium discoideum* myosin II, *Eur. J. Biochem.,* 251 (1998) 275–280.

42 Pavlov, D. A., Sobieszek, A. and Levitsky, D. I. Calorimetric studies of the thermal unfolding of smooth muscle myosin fragments and their complexes with ADP and phosphate analogs, *Biochemistry (Moscow),* 63 (1998) 952–962.

43 Pavlov, D. A., Bobkov, A. A., Nikolaeva, O. P., Magretova, N. N., Dedova, I. V. and Levitsky, D. I.: Thermal denaturation of myosin subfragment 1 modified at residue Lys-83 and its changes induced by nucleotide binding, *Biochemistry (Moscow),* 60 (1995) 835–842.

44 Golitsina, N. L., Bobkov, A. A., Dedova, I. V., Pavlov, D. A., Nikolaeva, O. P., Orlov, V. N. and Levitsky, D. I.: Differential scanning calorimetric study of the complexes of modified myosin subfragment 1 with ADP and vanadate or beryllium fluoride, *J. Muscle Res. Cell Motil.,* 17 (1996) 475–485.

45 Levitsky, D. I., Khvorov, N. V., Shnyrov, V. L., Vedenkina, N. S., Permyakov, E. A. and Poglazov, B. F.: Domain structure of myosin subframent-1. Selective denaturation of the 50 kDa segment, *FEBS Lett.,* 264 (1990) 176–178.
46 Levitsky, D. I., Nikolaeva, O. P., Vedenkina, N. S., Shnyrov, V. L., Golitsina, N. L., Khvorov, N. V., Permyakov, E. A. and Poglazov, B. F.: The effect of alkali light chains on the thermal stability of myosin subfragment 1, *Biomedical Science,* 2 (1991) 140–146.
47 Golitsina, N. L., Shnyrov, V. L. and Levitsky, D. I.: Thermal denaturation of the alkali light chain–20 kDa fragment complex obtained from myosin subfragment 1, *FEBS Lett.,* 303 (1992) 255–257.
48 Levitsky, D. I.: Domain Structure of the Myosin Head. In *Soviet Sci. Rev. - Physico-Chem. Biol.,* (Skulachev, V. P., ed), v. 12, Pt. 1, Harwood Acad. Publ. GmbH (1994) pp. 1–53.
49 Levitsky, D. I.: The structure of the myosin head, *Biokhimiya,* 56 (1991) 1539–1566.
50 Rayment, I., Rypniewski, W. R., Schmidt-Base, K., Smith, R., Tomchick, D. R., Benning, M. M., Winkelmann, D. A., Wesenberg, G. and Holden, H. M.: Three-dimansional structure of myosin subfragment 1: a molecular motor, *Science,* 261 (1993) 50–58.
51 Zubov, E O. and Levitsky, D. I.: Tight coupling between the motor domain and the regulatory domain of the myosin head complexed with ADP and phospahate analogues, *J. Muscle Res.Cell Motil.,* 23 (2002) 15.
52 Houdusse, A., Szent-Györgyi, A. G. and Cohen, C.: Three conformational states of scallop myosin S1, *Proc. Natl. Acad. Sci. USA,* 97 (2000) 11238–11243.
53 Brandts, J. F. and Lin, L.-N.: Study of strong to ultratight protein interactions using differential scanning calorimetry, *Biochemistry,* 29 (1990) 6927–6940.
54 Nikolaeva, O. P., Orlov, V. N., Dedova, I. V., Drachev, V. A. and Levitsky, D. I.: Interaction of myosin subfragment 1 with F-actin studied by differential scanning calorimetry, *Biochem. Mol. Biol. Internat.,* 40 (1996) 653–661.
55 Ponomarev, M. A., Furch, M., Levitsky, D. I. and Manstein, D. J.: Charge changes in loop 2 affect the thermal unfolding of the myosin motor domain bound to F-actin, *Biochemistry,* 39 (2000) 4527–4532.
56 Chalovich, J. M., Greene, L. E. and Eisenberg, E.: Crosslinked myosin subfragment 1: a stable analogue of the subfragment-1 ATP complex, *Proc. Natl. Acad. Sci. USA,* 80 (1983) 4909–4913.
57 Xie, L. and Schoenberg, M.: Binding of SH–SH2-modified myosin subfragment-1 to actin, *Biochemistry,* 37 (1998) 8048–8053.
58 Kaspieva, O. V., Nikolaeva, O. P., Orlov, V. N., Ponomarev, M. A. and Levitsky, D. I.: Changes in the thermal unfolding of *p*-phenylenedimaleimide-modified myosin subfragment 1 induced by its "weak" binding to F-actin, *FEBS Lett.,* 489 (2001) 144–148.
59 Bobkov, A. A. and Reisler, E.: Is SH1–SH2-cross-linked myosin subfragment 1 a structural analog of the weakly-bound state of myosin? *Biophys. J.,* 79 (2000) 460–467.
60 Levitsky, D. I., Shakirova, L. I., Mikhailova, V. V., Siletskaya, E. I. and Timofeev, V. P.: Nucleotide-induced and actin-induced structural changes in myosin subfragment 1 modified at both SH1 and SH2 groups, *Abstracts of XXX European Muscle Conference,* (Pavia, Italy, 2001) 85.
61 Tsiavaliaris, G., Fujita-Becker, S., Batra, R., Levitsky, D. I., Kull, F. J., Geeves, M. A. and Manstein, D. J.: Mutations in the relay loop region result in dominant-negative inhibition of myosin II function in *Dictyostelium, EMBO Reports,* 3 (2002) 1099–1105.

62 Ponomarev, M., Furch, M., Knetsch, M., Manstein, D. and Levitsky, D.: Changes in loop 2 affect the thermal unfolding of myosin head fragments while complexed to F-actin, *J. Muscle Res.Cell Motil.,* 20 (1999) 72.
63 Nikolaeva, O. P., Orlov, V. N., Bobkov, A. A. and Levitsky, D. I.: Differential scanning calorimetric study of myosin subfragment 1 with tryptic cleavage at the N-terminal region of the heavy chain, *Eur. J. Biochem.,* 269 (2002) 5678–5688.
64 Tatunashvili, L. V. and Privalov, P. L.: Calorimetric investigation of G-actin denaturation, *Biofizika,* 29 (1984) 583–585.
65 Bertazzon, A., Tian, G. H., Lamblin, A. and Tsong, T. Y.: Enthalpic and entropic contributions to actin stability: calorimetry, circular dichroism, and fluorescence study and effects of calcium, *Biochemistry,* 29 (1990) 291–298.
66 Le Bihan, T. and Gicquaud, C.: Kinetic study of the thermal denaturation of G-actin using differential scanning calorimetry and intrinsic fluorescence spectroscopy, *Biochem. Biophys. Res. Commun.,* 194 (1993) 1065–1073.
67 Kabsch, W., Mannherz, H. G., Suck, D., Pai, E. F. and Holmes, K. C.: Atomic structure of the actin–Dnase I complex, *Nature,* 347 (1990) 37–44.
68 Lörinczy, D., Könczol, F., Gaszner, B. and Belagyi, J.: Structural stability of actin filaments as studied by DSC and EPR, *Thermochim. Acta,* 322 (1998) 95–100.
69 Nikolaeva, O. P., Dedova, I. V., Khvorova, I. S. and Levitsky, D. I.: Interaction of F-actin with phosphate analogues studied by differential scanning calorimetry, *FEBS Lett.,* 351 (1994) 15–18.
70 Le Bihan, T. and Gicquaud, C.: Stabilization of actin by phalloidin: a differential scanning calorimetric study, *Biochem. Biophys. Res. Commun.,* 181 (1991) 542–547.
71 Combeau, C. and Carlier, M.-F.: Probing the mechanism of ATP hydrolysis on F-actin using vanadate and the structural analogs of phosphate BeF_3^- and AlF_4^-, *J. Biol. Chem.,* 263 (1988) 17429–17436.
72 Orlova, A. and Egelman, E. H.: Structural basis for the destabilization of F-actin by phosphate release following ATP hydrolysis, *J. Mol. Biol.,* 232 (1993) 334–341.
73 Muhlrad, A., Cheung, P., Phan, B. C., Miller, C. and Reisler, E.: Dynamic properties of actin. Structural changes induced by beryllium fluoride, *J. Biol. Chem.,* 269 (1994) 11852–11858.
74 Bombardier, H., Wong, P. and Gicquaud, C.: Effects of nucleotides on the denaturation of F-actin: a differential scanning calorimetry and FTIR spectroscopy study, *Biochem. Biophys. Res. Commun.,* 236 (1997) 798–803.
75 Sanchez-Ruiz, J. M.: Theretical analysis of Lumry-Eyring models in differential scanning calorimetry, *Biophys. J.,* 61 (1992) 921–935.
76 Kurganov, B. I., Kornilaev, B. A., Chebotareva, N. A., Malikov, V. Ph., Orlov, V. N., Lyubarev, A. E. and Livanova, N. B.: Dissociative mechanism of thermal denaturation of rabbit skeletal muscle glycogen phosphorylase *b*, *Biochemistry,* 39 (2000) 13144–13152.
77 dos Remedios, C. G., Chhabra, D., Kekic, M., Dedova, I. V., Tsubakihara, M., Berry, D. A. and Nosworthy, N. J.: Actin binding proteins and regulation of cytoskeleton microfilaments, *Physiol. Rev.,* 83 (2003) 433–473.
78 Levitsky, D. I., Rostkova, E. V., Orlov, V. N., Nikolaeva, O. P., Moiseeva, L. N., Teplova, M. V. and Gusev, N. B.: Complexes of smooth muscle tropomyosin with F-actin studied by differential scanning calorimetry, *Eur. J. Biochem.,* 267 (2000) 1869–1877.

79 Kremneva, E. V., Nikolaeva, O. P., Gusev, N. B. and Levitsky, D. I.: Effects of troponin on the thermal unfolding of actin-bound tropomyosin, *Biochemistry (Moscow),* 68 (2003) 802–809.
80 Nikolaeva, O. P., Dedova, I. V., Mikhailova, V. V. and Levitsky, D. I.: Effects of cofilin on the thermal unfolding of actin, *J. Muscle Res. Cell Motil.,* 23 (2002) 24–25.
81 Gicquaud, C.: Actin conformation is drastically altered by direct interaction with membrane lipids: a differential scanning calorimetry study, *Biochemistry,* 32 (1993) 11873–11877.
82 Potekhin, S. A. and Privalov, P. L.: Cooperative blocks in tropomyosin, *J. Mol. Biol.,* 159 (1982) 519–535.
83 Sturtevant, J. M., Holtzer, M. E. and Holtzer, A.: A scanning calorimetric study of the thermally induced unfolding of various forms of tropomyosin, *Biopolymers,* 31 (1991) 489–495.
84 O'Brien, R., Sturtevant, J. M., Wrabl, J., Holtzer, M. E. and Holtzer, A.: A scanning calorimetric study of unfolding equilibria in homodimeric chicken gizzard tropomyosin, *Biophys. J.,* 70 (1996) 2403–2407.
85 Orlov, V. N., Rostkova, E. V., Nikolaeva, O. P., Drachev, V. A. Gusev, N. B. and Levitsky, D. I.: Thermally induced chain exchange of smooth muscle tropomyosin dimers studied by differential scanning calorimetry, *FEBS Lett.,* 433 (1998) 241–244.
86 Krishnan, K. S., Brandts, J. F. and Lehrer, S. S.: Effects of an interchain disulfide bond on tropomyosin structure, *FEBS Lett.,* 91 (1978) 206–208.
87 Williams, D. L. Jr. and Swenson, C. A.: Tropomyosin stability: assignment of thermally induced conformational transitions to separate regions of the molecule, *Biochemistry,* 20 (1981) 3856–3864.
88 Lehman, W., Hatch, V., Korman, V., Rosol, M., Thomas, L., Maytum, R., Geeves, M. A., Van Eyk, J. E., Tobacman, L. S. and Craig, R.: Tropomyosin and actin isoforms modulate the localization of tropomyosin strands on actin filaments, *J. Mol. Biol.,* 302 (2000) 593–606.
89 Lehman, W., Vibert, P. and Craig, R.: Visualization of caldesmon on smooth muscle thin filaments, *J. Mol. Biol.,* 274 (1997) 310–317.
90 Smith, D. A., Maytum, R. and Geeves, M. A.: Cooperative regulation of myosin-actin interactions by a continuous flexible chain. I: Actin-tropomyosin systems, *Biophys. J.,* 84 (2003) 3155–3167.
91 Morozova, L. A., Gusev, N. B., Shnyrov, V. L. and Permyakov, E. A.: Study of the physico-chemical properties of troponins I and T from the heart and skeletal muscles using protein fluorescence and calorimetry methods, *Biokhimiya,* 53 (1988) 531–540.
92 Tsalkova, T. N. and Provalov, P. L.: Stability of troponin C, *Biochim. Biophys. Acta,* 624 (1980) 196–204.
93 Ingraham, R. H. and Swenson, C. A.: Stability of the Ca^{2+}-specific and Ca^{2+}–Mg^{2+} domains of troponin C. Effect of pH, *Eur. J. Biochem.,* 132 (1983) 85–88.
94 Bogatcheva, N. V. and Gusev, N. B.: Interaction of smooth muscle calponin with phospholipids, *FEBS Lett.,* 371 (1995) 123–126.

Chapter 7

Effect of nucleotides and environmental factors on the intermediate states of ATP hydrolysis cycle in skeletal muscle fibres

D. Lőrinczy*

Biophysical Department of University Pécs, Faculty of Medicine
H-7624 Pécs, Szigeti str. 12, Hungary

Introduction

It is widely accepted that during muscle function the contractile force is generated through the actin-myosin interaction while ATP is consumed. The head fragment of myosin (subfragment 1 [S1]) is rigidly attached to actin in the absence of nucleotides and they form together the actomyosin complex. The ATP hydrolysis starts after ATP binding to myosin in the presence of Mg^{2+}. The chemical energy arising from this process is temporaly stored by the myosin molecule – the hydrolitic process and the internal conformational changes are separated in time – the molecule transfers into a conformation of higher free enthalpy. The products of hydrolysis (inorganic phosphate [P_i], ADP) are released during the transition into the ground state or just before it. After the release of products a stuctural change is going on which is followed by the internal motion and rotation of head portion while the molecule remains attached to actin. The power-stroke produces about 10 pN force development resulting a displacement of actin filament along the longitudinal axis of muscle [1–4].

Determination of the structure of S1 and actin - S1 as well as structural changes in the vicinity of active place in the presence of ATP and its analogues became possible mainly by X-ray diffraction investigations. These data gave a remarkable contribution to a detailed molecular structural picture and validated the molecular mechanism of contraction [5, 6]. It was observed by crystallographic investigations that the greater globular region of S1 (catalytic domain) is attached to a long -helical region (regulatory domain). The link between the catalytic and regulatory domains involves the converter domain. Two light chains (ELC and RLC) are connected to the regulatory domain. The converter region plays a crucial role in the force transmission. ATP and actin binding

* denes.lorinczy@aok.pte.hu

D. Lőrinczy (ed.), The Nature of Biological Systems as Revealed by Thermal Methods, 159–186.

places are located in the catalytic domain. It is supposed that the narrow slit between the upper 50 kD and N-terminal subdomains represents the nucleotide binding place. The adenine and ribose rings have greater motional freedom, whereas the triphosphate part fits tightly to a structural unit defined as phosphate tube (P_i tube). This structural conformation assists in the formation of protein – ligand complex [3, 6, 7].

The long-helical region of S1 behaves as a lever arm as a consequence of the ATP – myosin – actin interaction, making the sliding motion of filaments possible. The Bagshaw-Trendham scheme describes the interaction of myosin with actin and nucleotides [8, 9] (Scheme 1). The sceheme was only partly modified in its details by the latest investigations [10]:

$$AM + ATP \leftrightarrow A + M{\cdot}ATP \leftrightarrow A + M^{*}{\cdot}ATP \leftrightarrow AM^{**}{\cdot}ADP{\cdot}P_i$$

$$AM^{*}{\cdot}ADP + P_i \leftrightarrow AM + ADP + P_i$$

Scheme 1 The main steps of ATP hydrolysis cycle based on protein solution experiments

where A and M stand for actin and myosin, * refers for intermediate conformation. The X-ray diffraction investigations made the conformational change possible in the loop named switch-II and in its vicinity at 'relay helix' during the binding of nucleotide or just after it. The interaction of relay helix with SH helix-containing SH1 and SH2 groups – and converter domain resulted in the rotation of region of rod shape. Nucleotide dependent rotation of rod-like part of S1 is supported by a lot of experiments. It was shown partly by X-ray diffraction [7, 11] and fluorescence method partly by PC simulation that the P_i tube undergoes to an 'open-closed' transition during ATP hydrolysis [12, 13] states and characterised them to a good extent. Present models suggest that the conformational change of switch-II region makes possible the loosening of the pocket, that binds the γ-phosphate group resulting the dissociation of γ-phosphate from myosin. The release of phosphate product is activated by the binding of actin to myosin. The detailed investigation of the fluorescence of conservative W501 tryptophane side chain made possible the separation of quick open-closed conformational change from the slower ATP hydrolysis [13, 14]. The nucleotide induced internal motion could be significant, it was proved by cross-linking that it could be a potential interaction between the 27 kDa N-terminal and C-terminal fragments [15]. It could be supposed too that an internal rotation in the free S1 state (dissociated actin-myosin complex in the absence of nucleotides) leads to a conformation in which the structure of myosin stores the energy of ATP hydrolysis, and this energy appears in the sliding motion of filaments after the actin binding. These experimental findings came mainly from investigations performed on protein solutions, therefore these could be an approximation of the real processes going on in the muscle fibres. To understand the working principle of muscle machine the main goal should be to perform experiments at least on the muscle fibre level. Muscle fibres have complex supramolecular structure. A further technical problem to follow the structural consequences of ATP hydrolysis is the different time scale of the kinetics of process and the measuring methods of structural changes. To eliminate this technical problem stable intermediate states of ATP hydrolysis cycle were developed for rabbit psoas muscle fibres with long lifetime [16].

Intermediate states and their monitoring

During the ATP hydrolysis cycle we distinguish as main steps the *strong binding*: *rigor* (AM) and *ADP binding* (AM.ADP) as well as *weak binding* (AM.ADP.P_i) states which are the most important intermediates of muscle contraction. Different spectroscopic methods (EPR, NMR, fluorescence), electron microscope (ELMI) as well as X-ray scattering provided evidence for the existence of a flexible link between the catalytic and regulatory domains [17–20] and were used to detect the rotational motion of myosin head in the distinct intermediate states of ATP digestion [21–27]. These techniques give different information; ELMI and steady-state X-ray scattering are unable to describe precisely the molecular dynamic phenomena while the motion sensitive molecular probes can not reproduce the static structure and are unable for image reconstruction.

In the vicinity of the active region of myosin head there is an SH helix containing two sulfhydryl groups which could be labelled by paramagnetic and fluorescence markers, this way the local conformational changes of the intermediate states of ATP hydrolysis could be observed. Although these labels could affect the enzymatic properties of myosin [28], their use provided important details in previous studies. Maleimide probes attached to reactive –SH group (Cys-707) of myosin in skeletal muscle are able to distinguish the myosin state specially bounded to actin, the relaxed state, and the orientational changes in the myosin head could be verified. The outcome of these experiments was the 'order-disorder' hypothesis. According to the hypothesis – there are two basic S1 conformations – which are in close correlation with the affinity of actin for myosin [29–31]. The S1 is in the specially oriented state compared to actin or it undergoes to a restricted motion in its rotation. Force generation is realised in oriented state. Using a proper muscle model the local and global consequences of these structural changes could be monitored.

Glycerol-extracted muscle fibre bundles prepared from rabbit psoas muscle could also be used to characterise the muscle function. The advantage of the use of muscle fibres is that the force development and ATPase activity is only slightly affected compared to the native one [32]. During the preparation of fibres fibre bundles from glycerinated muscle are washed for 60 minutes in rigor buffer and then transferred to fresh buffer. This state models the rigor state of the muscle. When MgADP is added to the rigor solution the myosin binds ADP and the complex simulates the strong binding ADP-state of acto-myosin. The other analogues of intermediates in the ATPase pathway can be generated by the addition of ATP (5 mM) and orthovanadate (5 mM), which together bind stoichiometrically at the active site of myosin to form a stable complex $AM^{+}ADP.V_{i}$ after the hydrolysis. This complex is believed to mimick the steady-state intermediate $AM^{**}ADP.P_{i}$ (M^{+} and M^{**} denote different conformations of myosin [16] (Scheme I). Another analogues of intermediates in the ATPase pathway (believed to mimick the weak binding state) are formed by AMP.PNP or by ATP (5 mM) and beryllium/aluminium fluoride which stoichiometrically bind to myosin to form a stable complex. These states have long enough lifetime to perform DSC or spectroscopic (e.g. EPR) measurements. The 4-maleimido-2,2,6,6-tetramethylpiperidinooxyl (4-maleimido-TEMPO, MSL) or the 4-isothiocianate-2,2,6,6-tetrametil-piperidine-1-nitroxide (4-isothiacianate-TEMPO, TCSL) probes are the most frequently used labels for EPR measurements. These labels are attached to the SH1 group (Cys-707) of myosin. Glycerinated muscle fibres are pretreated with DTNB in a buffer of low ionic strength to assure the specificity of labelling [6, 31]. The spectroscopic data can be obtained by using the standard protocol of Fajer and Marsh [33].

RIGOR AND ADP BINDING STATES

Rigor and AM.ADP states are frequently investigated intermediate states [7]. Domain movements in the myosin head play a decisive role in the energy transduction process which produces several conformational states of myosin. It is known that the nucleotide-binding pocket did not experience large conformational changes during the hydrolysis cycle and rigor as well as strong-binding states are believed to be identical from molecular dynamical point of view [21, 34–38]. However, the small nucleotide-induced conformational changes in the motor domain should be converted into larger movements of the elongated lever arm. Data agree that while the structure of motor domain remains similar to rigor, the regulatory domain swings around a point in the distal end of the motor domain [1]. The changes in the 50 kDa domain might affect the segment of the 20 kDa domain that contains the essential sulfhydryl groups.

Spectroscopic probes widely used to get information about orientations and rotational motion of myosin head. In muscle fibre studies mainly maleimide--based nitroxides (MSL) attached to the reactive sulfhydryl sites are used [34]. In EPR experiments the main problem that limits the interpretation of the measurements is the determination of the orientation of bound spin labels relative to the protein matrix. Generally two orientations can be used to obtain information regarding the probe orientation: the longer axis of fibres is oriented either parallel or perpendicular to the magnetic field. The labels have different chemical structure thus a different orientation with respect to the longer axis of S-1. This way the labels report different manner the change in the orientation of the entire S-1 or an alteration in its internal structure. It was observed that an isothiocyanate-based (TCSL) spin label is more sensitive to the domain orientation in myosin head than the widely used maleimide spin label. Very likely, TCSL has a little more flexible linkage to the protein, and does sense the internal rearrangement of the segments [39, 40]. To clarify this problem it is reasonable to use different labels to understand the molecular motion of S-1 in the presence of nucleotides and to combine the spectroscopic investigations with DSC (differential scanning calorimetry) measurements that reports domain stability and interactions to find correlation between local and global structural changes in the intermediate states of the ATPase cycle.

The unfolding of proteins in muscle fibres by thermal excitation is a complex process and depends on the state of the actomyosin complex. The experimental data suggest at least four transitions in the main melting temperature range (45–70°C). A minor transition at T_m=18°C might be attributed to the interaction of LC-2 light chain with the long γ-helical part of the myosin head [41]. This transition could be evaluated as the sign of an interdomain communication. In the 0–100°C temperature range the muscle fibres are irreversibly denaturated. In recent studies [42–44] it was shown that the irreversible DSC signals could be treated by using the principals of equilibrium thermodynamics. According to the Lumry-Eyring theory [45] the irreversible denaturation contains at least two steps: (a) reversible (D) denaturation of native (N) protein, (b) irreversible transition of denaturated protein into a final conformation (F):

$$N \leftrightarrow D \rightarrow F$$

The degree of irreversibility depends on denaturating temperature and time constant of phenomenon that is the second step is determined by the velocity constant of kinetic process instead of by the temperature dependent equilibrium constant. This is proved by the scanning rate dependence of melting temperature of denaturation. The melting temperature-scanning rate function has a saturation characteristic at higher rates and the shape of DSC signals does not depend on the scanning rate which means that the methods of equilibrium thermodynamics can be applied to interpret the thermal denaturation [46].

Accepting these results the starting point of the evaluation for rigor and ADP binding states can be that the main transitions between 45 and 70°C are the superposition of endotherms that are believed to correlate mainly with T_ms (the peak temperature of melting curves) of the larger domains of myosin and a smaller contribution from thin filaments. On isolated myosin three endotherms were reported [47] at 39, 47 and 51°C, therefore in the deconvolution procedure three transitions with Gaussian peak function were assumed [29]. In the absence of nucleotides in muscle fibres three transitions could be also decomposed from the DSC scan with 52, 58.3 and 67.5°C meltings, and in the presence of MgADP (Fig. 1.) there is an alteration in the first two melting temperatures (53.5 and 57°C) and enthalpies while the last transition temperature is unchanged with lower calorimetric enthalpy change, which could be the energetic consequence of the internal rearrangement of myosin structure in ADP binding state [35, 37].

The complexity of melting profiles is increasing with the size of fragments and the investigation of thermal stability of proteins gives global information about the actually existing domain level. The dynamic and energetic characteristics of proteins is changed by the supramolecular organisation (it is a stabilising factor) and it is manifested by the shift of melting temperatures to the higher range. According to experiments performed on S-1 [37, 47, 48] the endotherm at 52°C could be the motor domain of myosin head while the remaining part of myosin melts at around 58°C. The transition at 67.5°C in muscle fibre could be the actin binding domain and the contribution of F-actin. In the presence of ADP only this last melting temperature is unchanged (Fig. 1). Very probably the binding of ADP can cause local perturbation in the association between S-1 domains which is manifested in the increased transition temperature (53.5°C, motor domain becomes more stabile) and the widening of the half value of T_m which means the decrease of cooperativity between the structural domains.

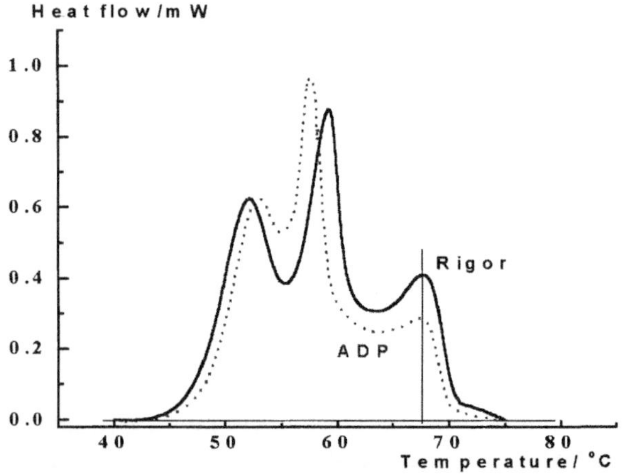

Fig. 1 Global conformational differences between the two strongly binding states in muscle fibres as revealed by thermal denaturation (only the last melting is uneffected by the nucleotide binding). With permission of JTAC

The ST (saturation transfer) EPR measurements on TCSL-fibres in rigor and also in ADP state showed that the labels were strongly immobilized on the millisecond time domain (Fig. 2., upper two panels only for rigor). The characteristic spectral parameter was greater than 0.8 which evidences that the rotational correlation time of the label is about 60 – 80 μs. We could not find significant difference between rigor and ADP state. This suggests - in agreement with previous data - that there is only a small difference in the rotational mobility of the powerstroke state and rigor (AM.ADP and AM states) of the muscle mashine [49].

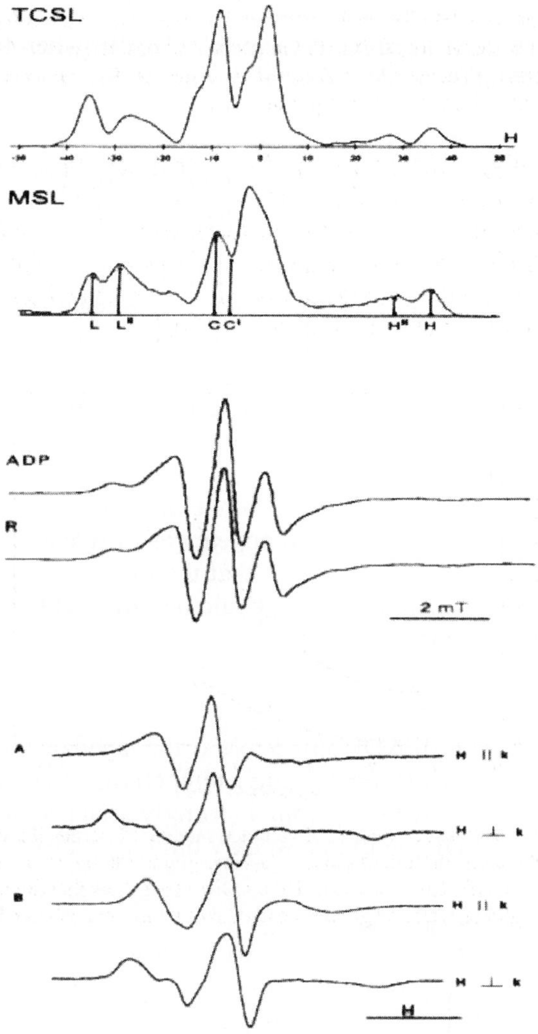

Fig. 2 ST EPR measurements on muscle fibres show a perfect immobilization for both labels (upper two curves) in rigor state. Conventional EPR using MSL label reports about the same orientational order in muscle fibres being in rigor as well as ADP state (two middle spectra). TCSL fibres in rigor (A) as well as in ADP state (B) exhibit different orientation dependence oriented either parallel or perpendicular compared to the external magnetic field (four EPR scans at the bottom)

This statement is supported by onventional EPR spectroscopic data as well. Experiments provided direct information about the orientation of myosin heads; in rigor the myosin heads had only one mode of binding. The fixed position of the catalytic domain was also observed in FRET experiments. In oriented fibre system, the mean angle of the Gaussian distribution of attached label was found to be ϑ=82° and the angular spread was σ=6° [34, 50] when the fibres were labelled with a maleimide-based spin label. Using TCSL probes, the EPR spectra also reported a high dependence on orientation, but in comparison to MSL-fibres with different mean angle and angular spread (ϑ=75°, σ=16°, [39]). Addition of MgADP to buffer solution did not affect the orientational order of spin labels in MSL-fibres (Fig. 2., two middle spectra), that is, myosin heads in rigor and with bound ADP exhibit the same orientational order. However, the incubation of TCSL-fibres in rigor buffer with MgADP resulted in significant changes of orientation dependence (Fig. 2., four EPR scans). It is probably due to the fact that TCSL molecules posses more flexible attaching linkage than MSL probes, as can be derived from the ST-EPR spectra of the labelled fibres [51]. It supports the view that TCSL probes can reflect internal structural changes in the catalytic domain of myosin induced by nucleotides. DSC scans give further evidences on the overall consequences of this process (Fig. 1.), that is in rigor and ADP state the attached heads have the same orientation, but the internal structure undergoes local conformational change. This alteration appears as different DSC fingerprint for rigor and ADP state which should mean that the myosin head exhibits different free enthalpy landscape, therefore there should be a different global conformation in these two intermediate states.

WEAKLY BINDING STATES

In the presence of 5 mM MgADP plus 5 mM orthovanadate the separation in the first two peaks is less while for the high temperature transition is more pronounced (Fig. 3). The conversion from a strongly attached state of myosin for actin (rigor state and ADP-state) to a weakly binding state (ADP.V_i state) is ac-

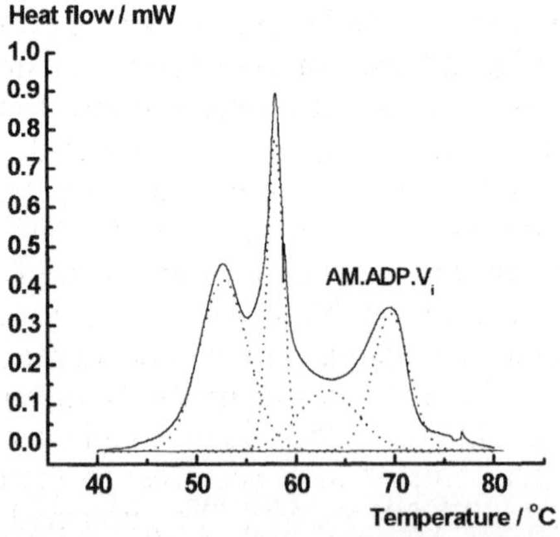

Fig. 3 DSC scan of a muscle fibre in AM.ADP.V_i state. Four possible melting domains could be distinguished with a pronounced contribution of the denaturation of F-actin (third component)

companied with increase of the last transition temperature ($T_m \sim 70°C$) and enthalpy which are due to the change of the increased affinity of nucleotide binding to myosin. In weakly binding state of myosin to actin the myosin heads are loosely attached to actin (in the deconvolution around 62°C appears the thermal contribution of F-actin, which is more pronounced in ATP.V_i state), and the fibres do not produce tension. The difference in the melting temperatures clearly indicates differences in the conformation of myosin head region [49]. This latter finding is in good agreement with the model from Rayment *et al.* [52], because in rigor the narrow left between the upper and lower domains of the 50 kDa segment is in a closed conformation, while in AM.ADP.V_i state it opens.

The conventional EPR spectra showed changes in the presence of nucleotides (ADP or ATP plus orthovanadate) in the order of the probe molecules in fibres (Fig. 4., left panel). Addition of 5 mM MgADP resulted in a change in the mean angle of the distribution of spin labels, it decreased from 75° to 56° and the angular spread increased by four degree. However, the orientation order remained preserved. In contrast, 5 mM MgATP plus 5 mM orthovanadate to rigor buffer produced an orientation disorder of myosin heads, only one spectral component could be detected which was charateristic to random population of spin labels fibres (Fig. 4., right panel). It shows the dissociation of myosin heads from actin or a weak interaction between M^{+}ADP.V_i and actin [49].

The comparison of the melting curves in the presence of nucleotides, ADP.V_i or AMP.PNP showed that the first two transitions were slightly affected by the binding of nucleotides (Fig. 5). Earlier experiments using tryptic diges-

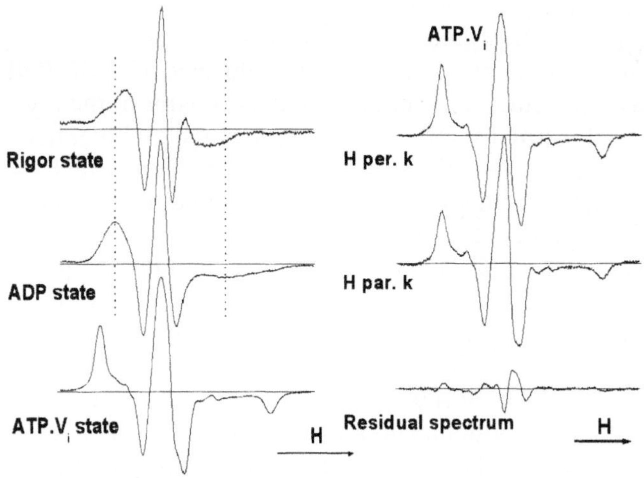

Fig. 4 The change of the orientational order of probe molecules in muscle fibres in the presence of different nucleotides (left panel). Using ATP.V_i only a random population of labels can be detected (right panel, the residual spectrum is the difference of parallel and perpendicular setting compared to the external magnetic field)

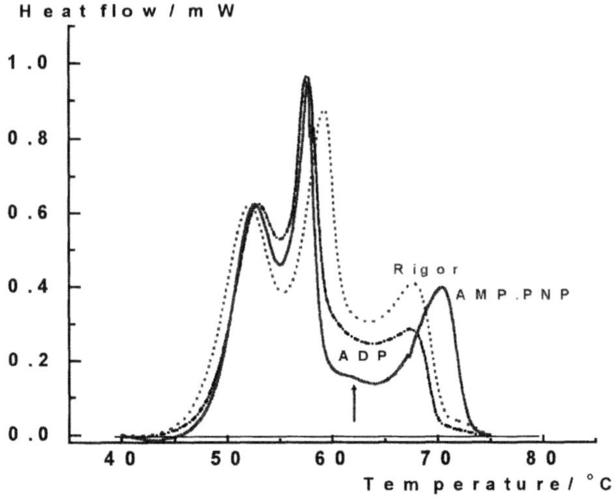

Fig. 5 Effect of nucleotides (ADP and AMP.PNP) on the thermal denaturation of muscle fibres. The weaker actin-myosin interaction in AMP.PNP state appears as more separable actin contribution at around 62°C. With permission of JTAC

tion gave evidence that the most labile part of myosin was the 50 kDa segment (actin-binding domain) or a part of it, therefore it is believed that the first thermal transition can be assigned to the 50 kDa domain even in muscle fibres [53]. The binding of ADP to myosin induced only little change in the DSC pattern, indicating that ADP alone produced small local conformational change in the catalytic domain of myosin heads [49].

The experiments on isolated myosin heads (S-1) showed the appearance of a new transition at higher temperature in the presence of AMP.PNP, which was attributed to the nucleotide interaction with the domains of the myosin head [35, 54]. The increased thermal stability of the dissociated globular heads induced by this interaction appears very likely in the last transition, which is shifted from 67.8°C to 70.1°C. The third weak transition at around 62.0°C can be assigned to the contribution of actin [55–57].

In the presence of ATP plus BeF_x or AlF_4^-, the high temperature transition was shifted from 67.7°C to T_m=74.6°C and T_m=77.4°C, respectively (Fig. 6). Biochemical experiments reported dissociation of heads from actin in ADP.AlF_4^- and ADP.BeF_x states [58]. From these experiments on muscle fibres we could confirm that the transition at 77.4°C (last transition) characterized the interaction of the nucleotide binding domain with nucleotides or nucleotide analogues. X-ray diffraction studies reported that ADP.BeF_x in complex with myosin mimicked the ATP bound state, whereas ADP.AlF_4^- complex was the analogue of the metastabil ADP.P_i state after ATP hydrolysis [48, 59]. According to biochemical experiments the complexes of ADP with phosphate analogues are long-lived complexes, and this can appear as a larger shift of temperature for the complex of myosin with

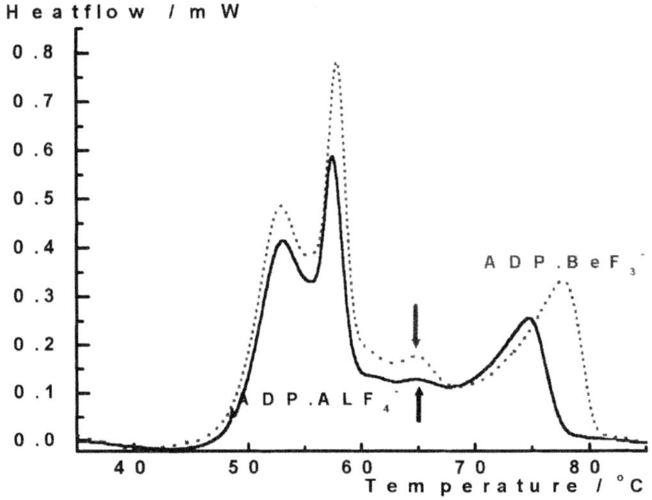

Fig. 6 Aluminium and beryllium fluoride shift the last DSC peak towards to the higher temperature. These two weakly binding states represent different intermediate steps in ATP hydrolysis cycle, where in ADP.BeF_x the actin contribution is more definite. With permission of JTAC

ADP and BeF_3 or AlF_4^-, in comparison with AMP.PNP. DSC measurements on subfragment-1 with ADP and beryllium fluoride also revealed significant shift of T_m in comparison with nucleotide free subfragment-1 [48].

In the presence of AMP.PNP the EPR spectrum deconvolution indicated the presence of two populations; about 50 % of labels belonged to the ordered fraction, and 50 % of labels was randomly oriented [60, 61]. AMP.PNP increased the orientation disorder of myosin heads, a random population of spin labels was superimposed on the ordered fraction evidencing motional changes in the internal structure of myosin heads. ST EPR measurements reported increased rotational mobility of spin labels in the presence of AMP.PNP, the population of heads belonging to the disordered fraction either dissociated from actin filaments or exhibited binding property differing from rigor. The myosin heads that exhibited high degree of order were in the strongly binding ADP-state, the heads being attached to actin differ from those of heads in rigor. Subtracting an ADP.V_i spectrum from the AMP.PNP spectrum, the difference spectrum was characteristic of ordered population of spin labels (ADP-like spectrum, third spectrum in (Fig. 7). Residual spectrum = difference spectrum – AM.ADP [high degree of order] state).

Almost no orientation dependence was detected in the presence of ATP and aluminium or beryllium fluoride (Fig. 8). The hyperfine splitting constants of the conventional EPR spectra were different; 6.6610±0.04 mT (ADP.BeF_x complex) and 6.712±0.03 mT (ADP.AlF_4^- complex), respectively. The myosin heads represented disordered populations with reduced rate of rotational motion, char-

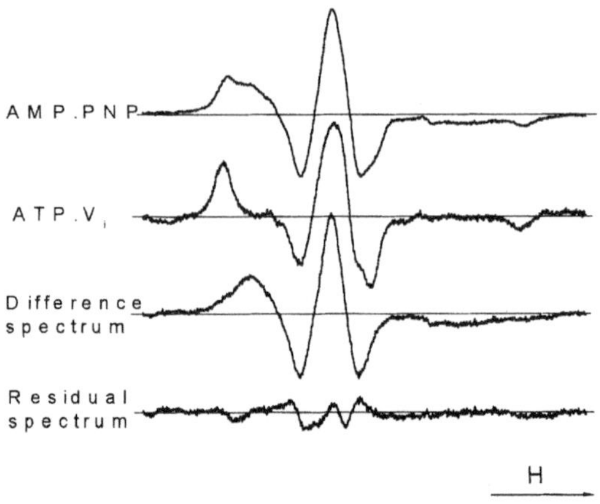

Fig. 7 The complexity of an AMP.PNP conventional EPR spectrum: subtracting the randomly oriented scan (second one) from AMP.PNP (first spectrum) resulted in an ADP-like (difference) spectrum which is proved by the residual spectrum (difference spectrum – ADP spectrum). With permission of JTAC

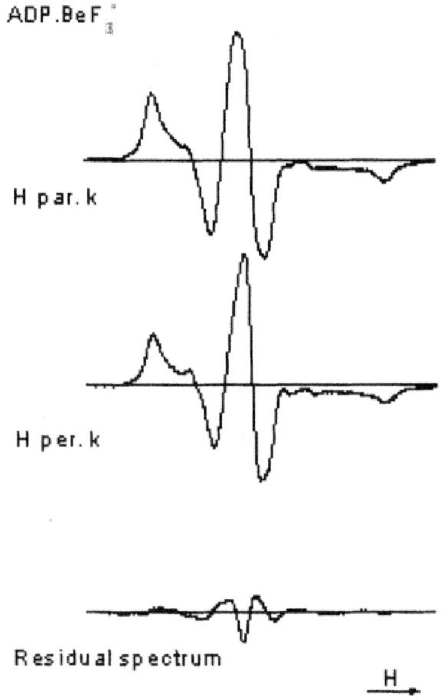

Fig. 8 There is no orientation dependence in AM.ADP.BeF$_x$ state: the difference of the parallel and perpendicular oriented fibre spectra (residual spectrum) is practically zero

acterising of dissociated myosin heads or non-specific binding of myosin heads to actin. The spectrum of ADP.BeF$_x$ complex seemed to be the superposition of an ADP.V$_i$-like spectrum and the spectrum of a protein moiety which rotated

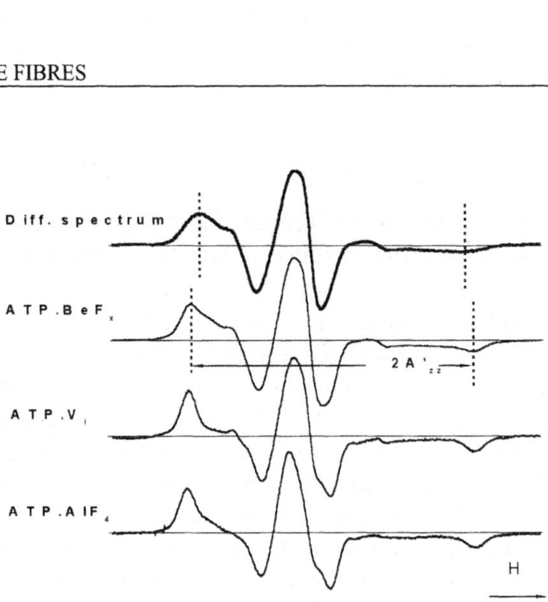

Fig. 9 The difference spectrum (first scan; ATP.BeF$_x$ – ATP.V$_i$ or ATP.AlF$_4$) indicates, that the Be spectrum is a superposition of two spin label population. With permission of JTAC

with an effective rotational correlation time of 12–15 ns [62]. Digital subtraction of either an ADP.V$_i$ spectrum or an ADP.AlF$_4^-$ spectrum from the ADP.BeF$_x$ spectrum resulted in two fractions (Fig. 9). This result suggests that under experimental conditions two conformers existed in ADP. BeF$_x$ state, mimicking the M*.ATP and M**.ADP.P$_i$ states.

Effect of free radicals on actin/myosin and muscle fibres

It was shown by physiological and biochemical studies that during the death of heart muscle the acidosis inside the cell and the activation of proteolytic enzymes is resulted in the dissociation of structurally bound myosin light chains (mainly the regulatory light chain LC-2) [63–65]. The death of muscle cells caused by the irreversible damage of sarcolemma makes possible the enter of external agents into the cell and their interaction with myosin. An exogenously administered, oxygen free radical-generating system has the capacity to cause cardiac dysfunction [66]. The fragments of contractile proteins fall into the circulation where the concentration of cardiac myosin LC-2 is proportional to the extent of heart infarct [67]. In glycerol-extracted muscle fibres the essential –SH groups are involved in the interaction of oxygen free radicals with myosin [68]. This way the molecular dynamic state and thermal stability of muscle proteins could be a good monitor of the damage caused by the free radical attack.

THERMAL STABILITY OF MYOSIN AFFECTED BY FREE RADICALS

Oxygen free radicals can be evoked by H_2O_2 treatment and UV irradiation of bovine cardiac myosin [69]. Two conformations, the rigor and the MgADP or strong-binding state of calf cardiac myosin show the most significant changes

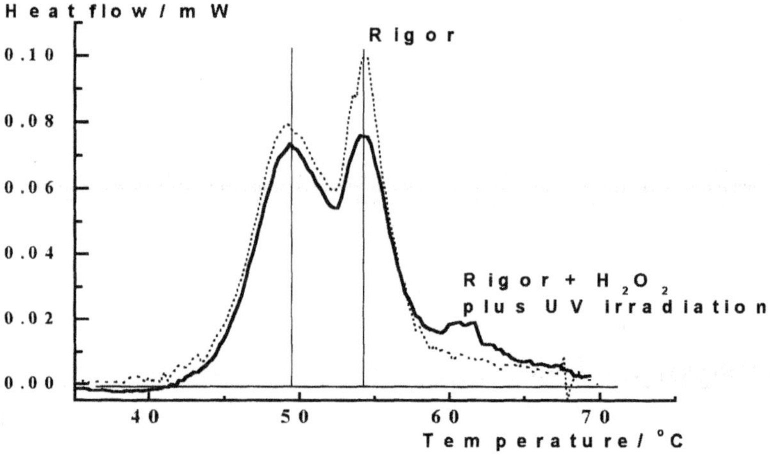

Fig. 10 Rigor state of calf myosin treated with hydrogen peroxide plus UV irradiation exhibits a significantly different thermal denaturation compared to the untreated one

[70–71]. Using a rigor ('native'), and rigor + H_2O_2 treated plus UV irradiated samples the thermal denaturation of rigor state shows a usual DSC scan (Fig. 10) with two big endotherms in the main transition. In the present of hydrogen peroxide (figure is not shown) the separation of two main transition is more pronounced including a temperature shift of catalytic domain to a lower (46.9°C) and for the rod part to the higher temperature (55.1°C).

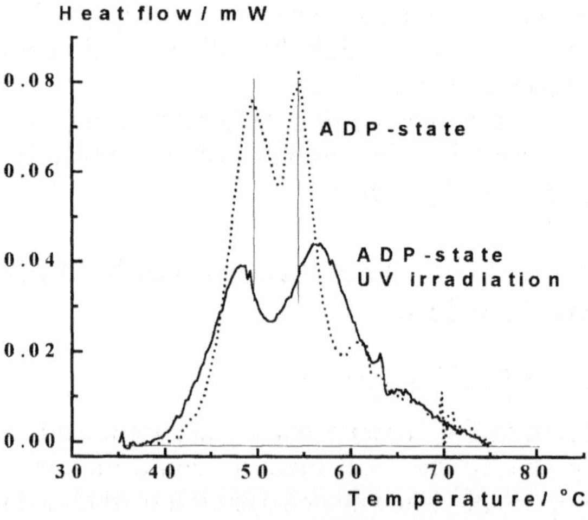

Fig. 11 The side chains of calf myosin are more effectively attacked by oxygen free radicals in ADP state than that in rigor one

In the strong-binding state, the effect of hydrogen peroxide is more effective, the catalytic domain becomes less stabile (T_{m1}=45.5°C) while the rod part shows an increased thermal stability (T_{m2}=56.3°C,)(Fig. 11). After an UV irradiation of treated samples in rigor we can observe only a broadening of the melting temperature range practically with the same transition temperatures while in MgADP state the catalytic domain will be less (47.9°C) and the rod part of myosin has greater (56.1°C) thermal stability. In both state there is significant decrease in the total enthalpy change.

After UV irradiation the main profile of the thermal transition did not change basically, no new transition was detected. It indicates that the hydroxyl free radicals interacted with the protein did not induce large structural changes in the internal structure of the cardiac myosin. However, the remarkable shifts of the transition temperatures and the changes of the form of the heat absorption curves show that the interaction of the free radicals with the side chains of the protein resulted in the altered affinity of myosin to nucleotides and in the altered flexibility of the internal structure of the protein. The modifications can contribute to the change of the ATPase activity and to the increased probability of the LC-2 light chain dissociation observed in cardiac muscle after ischaemic injury. It should be noted that after addition of 8 mM H_2O_2, shifts in the melting temperature and enthalpy changes were already observed in the heat absorption curves (Figs are not presented). UV irradiation induced further changes in the DSC patterns, but the total DSC profile approximated the nucleotide-free DSC profile. It indicates that not only the local conformation, but the global conformation of the cardiac myosin reflected the enhanced dissociation of the hydrolysis products from the nucleotide binding site.

INTERACTION OF SPIN LABELLED MYOSIN WITH OXYGEN FREE RADICALS

UV irradiation in the presence of hydrogen peroxide generates hydroxyl free radicals in buffer solution. Figure 12 shows the concentration of MSL spin labels attached to myosin as a function of time of UV irradiation. The semi-logaritmic plot of the spectral intensity gave a straight line evidencing that the interaction of the attached spin labels with the generated hydroxyl free radicals followed a pseudo first order chemical reaction. The characteristic time of the reaction was 1.3 min.

It is known that the binding of nucleotides, ADP or ADP plus orthovanadate (V_i) to myosin resulted in a significant increase in the mobility of the strongly immobilized labels. Since the reactive sulfhydryl sites are near to the nucleotide binding pocket in the crystal structure of myosin [3], changes are expected in the environment of the probe molecules. Experiments performed on cardiac myosin showed that the addition of 4 mM MgADP or 4 mM MgADP and orthovanadate to TCSL labelled myosin affected strongly the mobility of the attached spin labels [41, 72]. The change in the rotational mobility of spin labels was much larger when MgADP and orthovanadate was added to cardiac myosin (Fig. 13). The hyperfine splitting constant changed from 6.456 mT to 6.199 mT in ADP.V_i state of myosin.The spec-

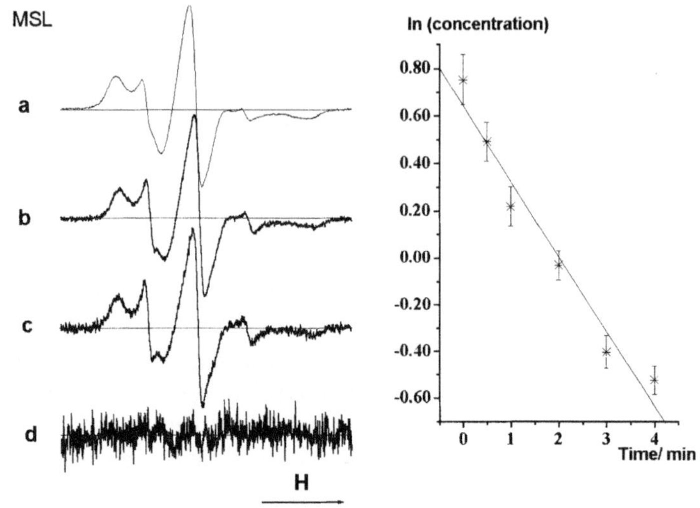

Fig. 12 Effect of time of irradiation on EPR spectra of MSL myosin (left panels: a./ non irradiated; b./ after an irradiation of 30s; c./ 60s irradiation and d./ at the and of irradiation). The semi-logarithmic plot of generated free radical concentration in the function of time of irradiation (right panel)

tral intensity of the MSL- or TCSL-myosin changed immediately after addition of 8 mM H_2O_2. It indicated that hydroxyl free radicals were already generated in the buffer solution containing nucleotide and H_2O_2. After UV irradiation the change of the spectral intensity of the samples was more pronounced. In some cases, when the time of irradiation was increased, a singlet superimposed on the spin label signal was also detected. Independent of the nucleotide, ADP or ADP.V_i, the spectral intensity decreased (Fig. 13). The decrease of the spectral intensity was accompanied

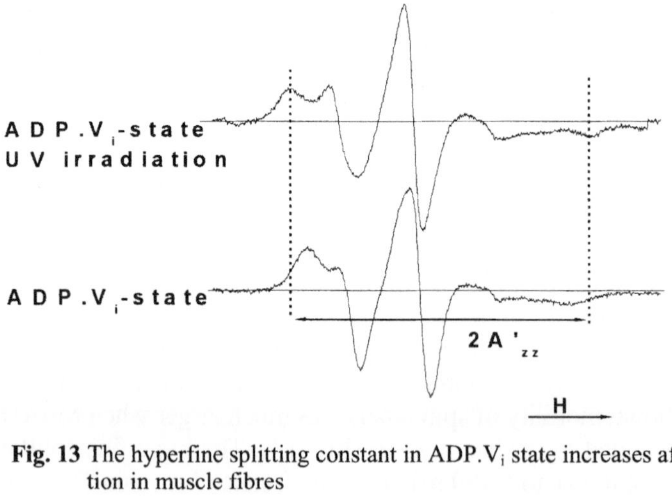

Fig. 13 The hyperfine splitting constant in ADP.V_i state increases after UV irradiation in muscle fibres

by the increase of the hyperfine splitting. The increase was about 0.1 mT in the case of MSL-myosin in ADP.V_i state. This shows that the hydroxyl free radicals interact not only with the spin labels located on the essential thiol site of the protein, but affect the intermediate states of the ATP hydrolysis cycle. It was reported earlier for skeletal myosin that irradiation by UV light modified the rate of ATP hydrolysis, and accelerated the dissociation of ADP and V_i [73]. In nucleotide-free state greater hyperfine splitting is expected. Under specific conditions in muscle fibres we estimated the portion of the –SH groups participating in the reaction with hydroxyl free radicals [68]. The comparison of the spectral intensity of the protein sample with the rate of the ATP hydrolysis showed that about 40 % of the essential thiols reacted with the hydroxyl free radicals. The experiments support the view that both local and global conformational changes play an important role in the interaction of oxygen free radicals with motor proteins that leads to protein damage. It can be suggested that suitable chemicals that suppress and/or quench the generation of free radicals in biological systems can reduce the ischemic injury.

EFFECT OF UV IRRADIATION ON MUSCLE FIBRES IN AM$^+$.ADP.V_I STATE

Short irradiation (30s) of control muscle fibres in the quartz sample cell used for EPR observations generated hydroxyl free radicals in the presence of H_2O_2 [69]. The EPR signal recorded had no hyperfine structure, a broad singlet was observed

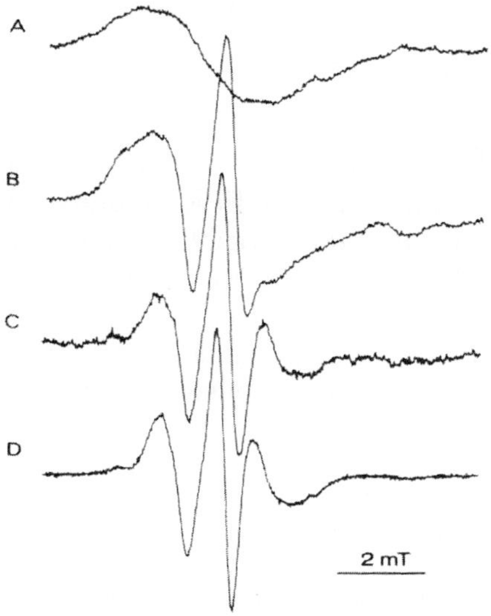

Fig. 14 Conventional EPR spectra of isothiocyanate labelled (TCSL) muscle fibres after UV irradiation. Panels **A.**: control fibres in AM.ADP.V_i state after 90s irradiation; **B**: TCSL fibres in. AM.ADP.V_i state after 90s irradiation; **C**: rigor like residual spectrum after subtraction of EPR spectrum of intact AM.ADP.V_i state from spectrum **B**; **D**: spectrum of rigor state (All the fibres were oriented parallel to the laboratory magnetic field. The scan width was 10 mT)

(Fig. 14A, top spectrum), the spectral intensity of that increased with the time of irradiation. Longer irradiation of spin-labelled fibres produced already significant decrease of the spectral intensity of the TCSL signal, therefore as a compromise 90 s was chosen for irradiation. Until now, we have no evidence that $\dot{O}H$ radicals influence differently the intermediate states of the myosin motor. The $\dot{O}H$ free radicals interacted very quickly with myosin and with the spin labels located on it, resulting in the reduction of the bound TCSL labels.

Immediately after UV irradiation a composed EPR spectrum was recorded (Fig. 14B). As a first step of the spectrum analysis, the spectrum obtained after UV irradiation of control fibres was subtracted from the complex spectrum. The contribution of the $\dot{O}H$ free radicals to the total absorption was about 15–20 %. In the second step, the spectrum recorded in the presence of MgADP and orthovanadate was subtracted from the same spectrum obtained after UV irradiation. The residual spectrum (Fig. 14C) resembles to the spectrum obtained in rigor (bottom spectrum, Fig. 14D). The concentration of this fraction varied between 5–11 % of the total absorption. This is a strong evidence that UV irradiation induced a process that enhanced the dissociation of ADP and V_i from the myosin heads, and forced the transition of cross-bridges from weakly binding state into rigor. It was shown that irradiation with UV light of the stable MgADP.V_i - myosin subfragment 1 complex modified covalently the enzyme and induced the release of trapped MgADP and V_i [74]. Experiments on fibres show that the photomodification of myosin by UV light might take place even in a more complex system as well.

OXIDATION OF MYOSIN BY CE(IV)

The mild oxidation of myosin by Ce(IV) in muscle fibres in the presence of PBN (spin trap phenyl-N-tertier-butylnitrone) resulted in a strongly immobilised nitroxide EPR spectrum (Fig. 15A). ST-EPR measurements revealed that the rotational correlation time is in the millisecond time range (Fig. 15B). The ratio of the first two diagnostic peaks L″/L is near to one; e.g. the PBN spin adduct is very rigidly attached to the oxidised thiol group. Ce(IV) complexed to nitrilotriacetic acid (NTA) oxidises sulfhydryl compounds *via* thiol free radicals, which can be trapped by PBN. In earlier papers [75] it has been suggested that the reaction

$$\text{myosin} - \text{SH} + \text{Ce(IV)} \rightarrow \text{Ce(III)} + \text{myosin} - \dot{\text{S}}$$

$$\text{myosin} - \dot{\text{S}} + \text{PBN} \rightarrow \text{myosin} - \text{S} - \text{PB}\dot{\text{N}}$$

took place after addition of Ce(IV)-nitrilotriacetic acid to myosin solution.

The spectrum of Ce(IV)-PBN treated muscle fibres exhibited no signal arising from weakly immobilised PBN spin adduct, implying that the nitroxide radicals were strongly bound to the protein structure. The average concentration of free radicals trapped by myosin was 0.28 mol PBN/mol myosin after 10 min incubation. It should be noted that the efforts to obtain spectrum that is characteristic to oriented nitroxide groups were not successful. Ordering of probe mol-

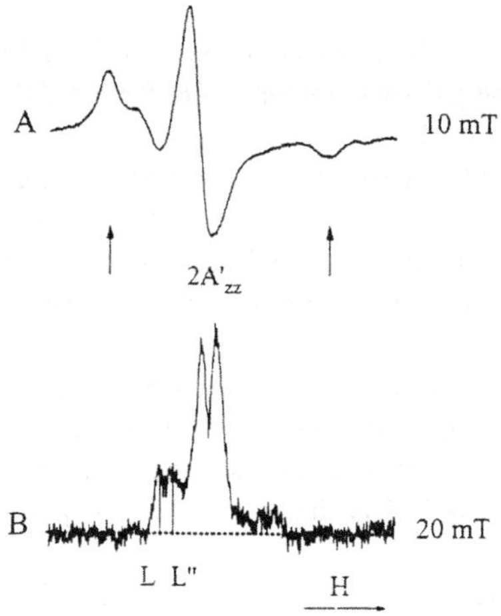

Fig. 15 Conventional (**A**, field scan was 10 mT) and ST EPR spectrum (**B**, field scan was 20 mT) of muscle fibres after Ce(IV)-nitrilotriacetic acid treatment in the presence of 30 mM PBN. Measurements were performed just after 10 min incubation (All the fibres were oriented parallel to the laboratory

ecules in MSL- and TCSL-fibres in rigor and in ADP-state was already reported [39]. The absence of orientation dependence might have two reasons: (i) not only the SH-1 groups are responsible for the observed EPR spectra, or (ii) the PBN nitroxide radical might have different orientations with respect to the longer axis of the muscle fibre after binding to the oxidised SH-1 thiol group. Graceffa [75] has shown in his experiments that the K^+-EDTA ATPase activity of myosin isolated from rabbit muscle did not change significantly after Ce(IV) oxidation and PBN spin trapping. This would mean that the thiyl radicals involved in the oxidation process by Ce(IV) partly differ from the essential thiols. However, in muscle fibres the Ce(IV) plus PBN treatment decreased the K^+-EDTA ATPase activity of myosin that accompanies the modification of the essential thiols [34]. The fractional inhibition of the ATPase activity was 0.54, larger than it would be expected on the basis of PBN free radical concentration. One explanation of that would be the short lifetime of PBN radicals located on the essential thiols. Former experiments reported that the essential thiols are highly reactive even in muscle fibres [76]. Therefore, it is possible that only the essential thiols are accessible to Ce(IV), the other thiol groups of myosin are buried in the organised structure of muscle fibres. On the other hand, it cannot be excluded that Ce(IV)-nitrilotriacetic acid perturbs the local structure of myosin heads around the segment containing the essential thiols, and this affects the ordering of probe molecules in the head region of myosin and/or the perturbation opens the structure and other SH groups are accessible to Ce(IV) as well. These observations suggest a linkage between the function of myosin motor and the particular effect of free radicals and oxidation on myosin.

EFFECT OF FREE RADICALS ON ACTIN BY DSC

It is well established that actin can also be sensitive to the changes of the environment. Cation exchange [77–80], the replacement of bound ATP with ADP [81, 82] or the shift in the pH [83] modified the conformational and dynamic properties of actin monomers and filaments. Earlier experiments on globular and filamentous actin by DSC showed that the main thermal transitions were at 47.3°C and 51°C for G-actin, and much higher transitions were detected for F-actin, they were at 59.7, 60.6 and 61.3°C [56]. The monomer actin consists of structural domains which are separated by a cleft, and the bound nucleotide (ATP for G-actin and ADP for F-actin) is localised in the cleft. The larger and smaller domains can move as two units relative to one another. The transitions obtained as a result of the deconvolution could be assigned to these structural domains. During polymerisation the monomer-monomer interaction enhances the thermal stability, and this leads to increased melting enthalpy and transition temperatures.

The free radical reactions (mild oxidation of actin by Ce(IV)-NTA performed similar manner as in the case of myosin) produce significant alterations in the DSC pattern of F-actin. The narrow peak broadens drastically and is shifted by about 5°C to lower temperature evidencing that the monomer-monomer interaction and/or the filament-filament association are affected (Fig. 16). The oxidation of the Cys-374 residues in F-actin results in conformational changes in the environment of the subdomain-1 which has consequence in the global structure of the protein. The significant broadening of the melting curve is probably due to the fact that the effect of the radical reactions modified the compact structure of F-actin, and this led to a decreased cooperativity between the protomers. The weekened interaction appears as a reduced transition enthalpy.

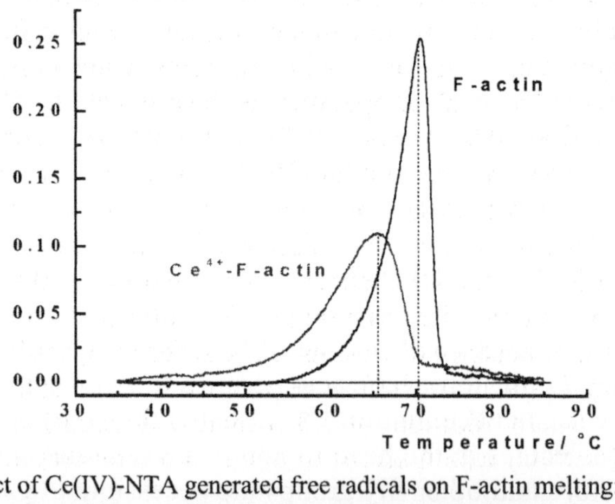

Fig. 16 Effect of Ce(IV)-NTA generated free radicals on F-actin melting: melting temperature shift towards to the lower value and the broadening of DSC scan are the signs of structure modification and decreased cooperativity of actin subdomains. With permission of JTAC

EFFECT OF FREE RADICALS ON ACTIN BY EPR

The EPR spectrum on actin was characteristic of strongly immobilized spin labels, the hyperfine splittting constants obtained were about the same for both forms of actin implying that the mild oxidation by Ce(IV)-NTA could modify the protein structure in the neighbourhood of the thiol sites. The free radical concentration after Ce(IV)-NTA treatment was 0.42±0.06 (n=5) mole of free radical/mole of protein in G-actin, whereas only 0.17±0.04 (n=4) mole of free radical/mole of actin was detected in F-form (Fig. 17). However, a very low concentration of PBN spin adduct was detected when F-actin was pretreated with N-ethyl-maleimide (NEM) before addition of Ce(IV)-NTA. The reaction with NEM results in the blocking of Cys-374 residue. This suggests that very likely the reactive Cys-257 thiol groups are involved in the reactions, and only a smaller fraction of Cys-374 thiol groups contributes to the PBN signal [84]. Polymerisation of actin buries the exposed Cys-257 groups (and the other three less reactive thiol sites), and therefore only the thiol sites of the Cys-374 residues participate in the reaction.

The amount of strongly immobilised spin labels on the thiol sites can be influenced by the non-specific binding properties of PBN spin adduct to the proteins. Both proteins, actin and myosin have hydrophobic areas which can bind small molecules. The binding of PBN radical adducts to hydrophobic areas can contribute to the apparent spin concentration, assuming similar rotational corre-

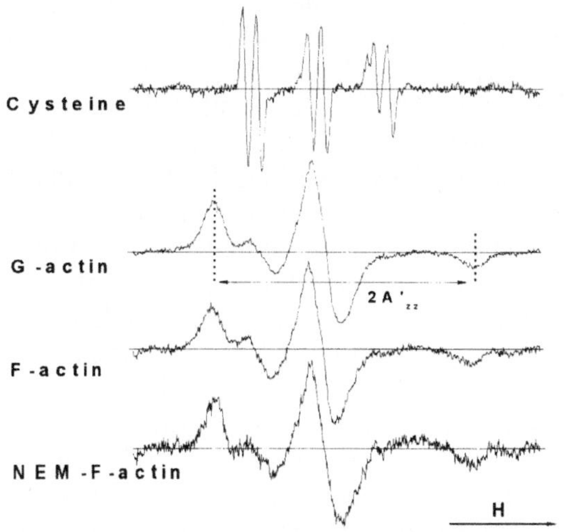

Fig. 17 EPR spectra immediately after thiyl radical generation: in contrast to free cysteine the PBN adduct molecules are strongly immobilized on monomer actin. When F-actin was pretreated with 0.1 mM NEM for 10 min before addition of Ce(IV)-NTA, only a very low concentration of PBN spin adducts was detected. The hyperfine splitting constant obtained were the same for both forms of actin therefore the mild oxidation by Ce(IV)-NTA could modify the protein structure in the neighbourhood of the thiol sites

lation time of the paramagnetic molecules. In order to test this possibility, experiments were performed on samples which contained bovine serum albumine (BSA) in hydroxyl free radical generating system. Albumin has many hydrophobic areas, and therefore binds to small molecules as steroid hormones, fatty acids that are only slightly soluble in the blood serum [85]. The EPR spectrum of BSA was superposition of two spectral components, one fraction of weakly immobilized PBN spin adducts and a second fraction of strongly immobilized PBN spin adducts, implying that the nitroxide radicals were strongly bound to the protein structure. The binding of radical products to blood plasma proteins suggests that the role of non-covalent binding in the free radical processes cannot be excluded.

Conclusions

The combination of differential scanning calorimetry (DSC) with electron paramagnetic resonance spectrometry (EPR) seems to be a good choice to monitor the global and local consequences of the molecular dynamic effects during the ATP hydrolysis cycle in muscle – mainly motor-proteins. This way it was demonstrated that the rigor and AM.ADP states represent two different myosin head conformation in local and global sense too. The starting point of the evaluation of DSC scans is that the main transitions between 45 and 75°C are the superposition of endotherms that are believed to correlate mainly with T_ms of the larger domains of myosin: the *globular heads* (the motor), which contain the nucleotide and the actin binding sites, the *long tail* with the *two light chains*, and the *rod-like parts*. On isolated myosin two large endotherms are detected – characteristic of the head and rod portion of myosin – but in the deconvolution procedure usually three or four transitions with Gaussian peak function are assumed to resolve the internal domains of the head [41, 47, 86]. It was shown that the most labile part of myosin is the 50-kDa segment or a part of it, therefore it is belived that the first thermal transition can be assigned to the 50-kDa domain [35, 53]. This transition might involve the unfolding of the subfragment-2 domain [87], and the LMM rod part of the myosin as well [88, 89]. In the presence of nucleotides the largest changes were measured at the higher temperature transitions.

Reactive oxygen species (ROS: hydrogen peroxide, t-BuOOH etc.) are randomly interacting at the side chains of proteins and other macromolecules, and are able to generate thiyl free radicals on the reactive –SH groups of myosin, serum albumin and hemoglobin, except actin. In the case of actin when we apply Ce-treatment we make a specific, directed attack on cystein side chains where thyil free radicals will be generated. The observed hyperfine splittings on the EPR spectra of different proteins agree quite well with the hyperfine splitting of the spin-labelled proteins by –SH-directed spin labels, which suggests the selective oxidation of the proteins, and a site-specific damage at particular residues.

According to DSC measurements the oxidation of thiol groups can cause significant global conformational change in the protein structure.

References

1 Holmes, K. C. (1998) A molecular model for muscle contraction. *Acta Cryst.,* A54, 789–797.

2 Holmes, K. C. (1998) A powerful stroke. *Nature Struct. Biol.,* 5, 940–942.

3 Rayment, I. Holden, H. M. Whittaker, M. Yohn, C. B. Lorenz, M. Holmes, K. C. Milligan, R. A (1993) Structure of the actin-myosin complex and its implications for muscle contraction. *Science,* 261, 58–65.

4 Baumann, B. A. J. Hambly, B. D. Hideg, K. Fajer, P. G. (2001) The regulatory domain of the myosin head behaves as a rigid lever. *Biochemistry,* 40, 7868–7873.

5 Dominguez, R. Freyzon, Y. Trybus, K. M. Cohen, C. (1998) Chystal structure of a vertebrate smooth muscle myosin motor domain and its complex with essential light chain: visualization of the pre-power stroke state. *Cell,* 94, 659–671.

6 Houdusse, A. Kalabokis, V. N. Himmel, D. Szent-Györgyi, A. G. Cohen, C. (1999) Atomic structure of scallop myosin subfragment S1 complexed with MgADP: a novel conformation of the myosin head. *Cell,* 97, 459–470.

7 Geeves, M. A. Holmes, K. C. (1999) Structural mechanisms of muscle contraction. *Annu. Rev. Biochem.,* 68, 687–728.

8 Trentham, D.R. Bardsley, R. G. Eccleston, J. P. Weeds, A.G. (1972) Elementary processes of the magnesium ion-dependent adenosine triphosphatase activity of heavy meromyosin. *Biochem. J.,* 126, 635–644.

9 Bagshaw, C. R. Trentham, D. R. (1974) The characterization of myosin-product complexes and of product release step during magnesium ion-dependent adenosine triphosphatase reaction.. *Biochem. J.,* 141, 331–349.

10 Eisenberg, E. and Greene, L.E. (1980) The relation of muscle biochemistry to muscle physiology. *Ann. Rev. Physiol.,* 42, 293–309.

11 Smith, C., A. and Rayment, I. (1996) X-ray structure of the magnesium(II).ADP.vanadate complex of the Dictyostelium discoideum myosin motor domain to 1.9 Å resolution. *Biochemistry,* 35, 5404–5417.

12 Pate, E. Naber, N. Matuska, M, Franks-Skiba, K. Cooke, R. (1997) Opening of the myosin nucleotide triphosphate binding domain during the ATPase cycle. *Biochemistry,* 36, 12155–12166.

13 Málnási-Csizmadia, A. Pearson, D. S. Kovács, M. Wooley, R. J. Geeves, M. A. Bagshaw, C. R. (2001) Kinetic resolution of a conformational transition and the ATP hydrolysis step using relaxation methods with Dictyostelium myosin II mutant containing a single tryptophan mutant. *Biochemistry,* 40, 12727–12737.

14 Málnási-Csizmadia, A. Kovács, M. Wooley, R. J. Botchway, S. W. Bagshaw, C. R. (2001) The dynamics of the relay loop tryptophan residue in the Dictyostelium myosin motor domain and the origin of spectroscopic signals. *J. Biol. Chem.,* 276, 19483–19490.

15 Pliszka, B. Karczewska, E. Wawro, B. (2000) Nucleotide-induced movement in the myosin head near the converter region. *Biochim. Biophys. Acta,* 1481, 55–62.

16 Goodno C. C. (1979) Inhibition of myosin ATPase by vanadate ion. *Proc. Natl. Acad. Sci.* USA 76, 2620–2624.

17 Barnett, V. A. Thomas, D. D. (1987) Resolution of conformational states of spin-labeled myosin during steady-state ATP hydrolysis. *Biochemistry,* 26, 314–323.

18 Ajtai, K. Peyser, M. Park, S. Burghardt, T. P. Muhlrad, A. (1999) Trinitrophenylated reactive lysine residue in myosin detects lever arm movement during consecutive steps of ATP hydrolysis. *Biochemistry,* 38, 6428–6440.

19 Nyitrai, M. G. Hild, A. Lukács, Bódis, E. and Somogyi, B. (2000) Conformational distributions and proximity relationships in the rigor complex of actin and myosin subfragment-1. *J. Biol. Chem.*, 275, 2404–9,

20 Nyitrai, M., G. Hild, E. Bódis, Lukács, A. and Somogyi, B. (2000) Flexibility of myosin-subfragment-1 in its complex with actin as revealed by fluorescence resonance energy transfer. *Eur. J. Biochem.,* 267, 4334–8.

21 Cooke, R. (1986) The mechanism of muscle contraction. *CRC Crit. Rev. Biochem,.* 21, 53–118.

22 Fisher, A. J. Smith, C. A. Thoden, J. Smith, R. Sutoh, K. Holden, H. M., Rayment, I. (1995) Structural studies of myosin:nucleotide complexes: A revised model fo the molecular basis of muscle contraction. *Biophys. J.*, 68, 19s–28s.

23 Fisher, A. J. Smith, C. A. Thoden, J. Smith R. Sutoh, K. Holden, H. M. Rayment, I. (1995) X-ray structures of the myosin motor domain of Dictyostelium discoideum complexed with MgADP. BeF_x and MgADP.AlF_4. *Biochemistry,* 34, 8960–8972.

24 Palm, T. Sale, K. Brown, L. Li, H. Hambly, B. D. Fajer, P. G. (1999) Intradomain distances in the regulatory domain of the myosin head in prepower and postpower stroke state: fluorescence energy transfer. *Biochemistry,* 38, 13026–13034.

25 Ling, N. Shrimpton, C. Sleep, J. Kendrick-Jones, J. Irving, M. (1996) Fluorescent probes on orientation of myosin regulatory light chains in relaxed, rigor and contracting muscle. *Biophys. J.,* 70, 1836–1846.

26 Fajer, P. G. Fajer, E. A. Schoenberg, M. Thomas, D. D. (1991) Orientational disorder and motion of weakly attached cross-bridges. *Biophys. J.,* 60, 642–649.

27 Hambly, B. Franks, K. Cooke, R. (1992) Paramagnetic spin probes attached to a light chain on the myosin head are highly disordered in active muscle fibres. *Biophys. J.,* 63, 1306–1313.

28 Sekine, T. Kielley, W. W. (1964) The enzymatic properties of N-ethylmaleimide modified myosin. *BBA,* 81, 336–345.

29 Zhao, L. Naber, N. Cooke, R. (1995) Muscle cross-bridges bound to actin are disordered in the presence of 2,3-butanedione monoxime. *Biophys. J.,* 68, 1980–1990.

30 Frisbie, S. M. Xu, S. Chalovich, J. M. Yu, L. C. (1998) Characterization of cross-bridges in the presence of saturating concentrations of MgAMP-PNP in rabbit permeabilized psoas muscle. *Biophys. J.,* 74, 3072–3082.

31 Zhao, L. Gollub, J. Cooke, R. (1996) Orientation of paramagnetic probes attached to gizzard regulatory light chain bound to myosin heads in rabbit skeletal muscle. *Biochemistry,* 35, 10158–10165.

32 Crowder, M. S. Cooke, R. (1984) The effect of myosin sulphydryl modification on the mechanics of fibre contraction. *J. Muscle Cell Motil.,* 5, 131–146.

33 Fajer, P. G. and Marsh, D. (1982) Microwave and modulation field inhomogenities and effect of cavity Q in saturation transfer EPR spectra. Dependence of sample size. *J. Mag. Res.,* 49, 212–224.

34 Thomas, D. D. Cooke, R. (1980) Orientation of spin-labeled myosin heads in glycerinated muscle fibres. *Biophys. J.*, 32, 891–906.
35 Levitsky, D. I. Shnyrov, V. L. Khvorov, N. V. Bukatina, A. E. Vedenkina, N. S. Permyakov, E. A. Nikolaeva, O. P. Poglazov, B. F. (1992) Effects of nucleotide binding on thermal transitions and domain structure of myosin subfragment 1. *Eur. J. Biochem.*, 209, 829–835.
36 Wakabayashi, K. Tokunaga, M. Kohno, I. Sugimoto, Y. Hamanaka, T. Takezawa, Y. Wakabayashi, T. Ameniya, Y. (1992) Small-angle synchrotron x-ray scattaring reveals distinct shape changes of the myosin head during hydrolysis of ATP. *Science*, 258, 443–447.
37 Bobkov, A. A. Levitsky, D. I. (1995) Differential scanning calorimetric study of the complexes of myosin subfragment-1 with nucleoside diphosphates and vanadate or beryllium fluoride. *Biochemistry,* 34, 9708–9713.
38 Baker, J. E. Brust-Mascher, I. Ramachandran, S. LaConte, L. E. W. Thomas, D. D. (1998) A large and distict rotation of the myosin light chain domain occurs upon muscle contraction. *Proc. Natl. Acad. Sci.*, USA 95, 2944–2949.
39 Belagyi, J. Frey I. Pótó, L. (1994) ADP- induced changes in ordering of spin-labelled myosin heads in muscle fibres. *Eur. J. Biochem.*, 224, 215–222.
40 Belagyi, J. Lőrinczy, D. (1996) Internal motion in myosin head: effect of ADP and ATP. *Biochem. Biophys. Res. Comm.*, 219, 936–940.
41 Lőrinczy, D. Hoffmann, U. Pótó, L. Belágyi, J. Laggner, P. (1990) Conformational changes in bovine heart myosin as studied by EPR and DSC techniques. *Gen. Physiol. Biophys.*, 9, 589–603.
42 Sanchez-Ruiz, J. M. Lopez-Lacomba, J. L. Cortijo, M. Mateo, P. L. (1988) Differential scanning calorimetry of the irreversible thermal denaturation of thermolysin. *Biochemistry,* 27, 1648–1652.
43 Conjero-Lara, F. Mateo, P. L. Aviles, F. X. Sanchez-Ruiz, J. M. (1991) Effect of Zn^{2+} on the thermal denaturation of carboxypepdidase B. *Biochemistry,* 30, 2067–2072.
44 Thorolfsson, M. Ibarra-Molero, B. Fojan, P. Petersen, S. B. Sanchez-Ruiz, J. M. Martinez, A. (2002) L-Phenylalanine binding and domain organization in human phenylalanine hydroxylase: a differential scanning calorimetry study. *Biochemistry,* 41, 7573–7585.
45 Lumry, R. Eyring, H. (1954) Conformation changes of proteins. *J. Phys. Chem.*, 58, 110–120.
46 Vogl, T. Jatzke, C. Hinz, H-J. Benz, J. Huber, R. (1997) Thermodynamic stability of annexin V E17G: equilibrium parameters from an irreversible unfolding reaction. *Biochemistry,* 36, 1657–1668.
47 Bertazzon, A. Tian, G. H. Tsong, T. Y. (1988) Differential scanning calorimetric (DSC) study of thermal unfolding of myosin and its subfragments in several forms of assemblies. *Biophys. J.*, 53, 236a.
48 Bobkov, A. A. Khovorov, N. K. Golitsina, N. L. and Levitsky, D. I. (1993) Calorimetric characterization of the stable complex of myosin subfragment 1 with ADP and beryllium fluoride. *FEBS Lettr.*, 332, 64–66.
49 Kiss, M. Belagyi, J. Lőrinczy, D. (2003) Vanadate (V_i) and ADP induced domain motions in myosin head by DSC and EPR. *J. Therm. Anal. Cal.*, 72, 573–580.
50 Fajer, P. G. (1994) Determination of spin-label orientation within the myosin head. *Proc. Natl. Acad. Sci.*, USA, 91, 937–941.

51 Lőrinczy, D. Hartvig, N. Farkas, N. Belagyi, J. (2001) Binding of nucleotides at the active site modulates the local and global conformation of myosin in muscle fibres. *J. Therm. Anal. Cal.,* 65, 351–358.

52 Rayment, I. Holden, H. M. Whittaker, M. Yohn, C. B. Lorenz, M. Holmes, K. C. and Milligan, R. A. (1993) Structure of the actin-myosin complex and its implications for muscle contraction. *Science,* 261, 58–65.

53 Setton, A. and Muhlrad, A. 1984. Effect of mild heat treatment on the ATPase activity and proteolytic sensitivity of myosin subfragment-1. *Arch. Biochem. Biophys.,* 235, 411–417.

54 Levitsky, D. I. Khvorov, N. V. Shnyrov, V. L. Vedenkina, N. S. Permyakov, E. A., Poglazov, B. F. (1990) Domain structure of myosin subfragment-1. Selective denaturation of the 50 kDa segment. *FEBS Letters,* 264, 176–178.

55 Lőrinczy, D. Belagyi, J. (1995) Scanning calorimetric and EPR studies on the thermal stability of actin. *Thermochim. Acta,* 259, 153–164.

56 Lőrinczy, D. Könczöl, F. Gaszner, B. Belagyi, J. (1998) Structural stability of actin filaments as studied by DSC and EPR. *Thermochim. Acta,* 322, 95–100.

57 Lőrinczy, D. Hartvig, N. Belagyi, J. (2001) Nucleotide analogue induces global and local changes in muscle fibres. *J. Therm. Anal. Cal.,* 64, 651–658.

58 Raucher, D. Fajer, P. G. (1994) Orientation and dynamics of myosin heads in aluminum fluoride induced pre-power stroke states: an EPR study. *Biochemistry,* 33, 11993–99.

59 Gulick, A. M. Bauer, C. B. Thoden, J. B. Rayment, I. (1997) X-ray structure of the MgADP, MgATPgammaS, and MgAMP.PNP complexes of the Dictyostelium discoideum myosin motor domain. *Biochemistry,* 36, 11619–11628.

60 Fajer, P. G. Fajer, E. A. Brunsvold, N. J. Thomas, D. D. (1988) Effects of AMPPNP on the orientation and rotational dynamics of spin-labeled muscle cross-bridges. *Biophys. J.,* 53, 513–524.

61 Hartvig, N. Lőrinczy, D. Farkas, N. Belagyi, J. (2002) Effect of adenosine 5'-[β,γ-imido]triphosphate on myosin head domain motions. Saturation transfer EPR measurements without low-power phase setting. *Eur. J. Biochem.,* 269, 2168–2177.

62 Belagyi, J. Hartvig, N. Lőrinczy, D. Farkas, N. (2001) EPR study of BeF_3 and AlF_4 containing myosin nucleotide complexes. *Muscle Res. Cell Motil.,* 22, 585.

63 Katus, H. A. Yasuda, T. Gold, H. K. Leinbach, R. C. Strauss, H. W. Waksmonski, C. Haber, E. Khaw, B. A. (1984) Diagnosis of acute myocardial infarction by detection of circulating cardiac myosin light chains. *Am. J. Cardiol.,* 54, 964–970.

64 Katus, H. A. Diederich, K. W. Uellner, A. Remppis, A. Schuler, G. Kubler, W. (1988) Myosin light chain release in acute myocardial infarction: non-invasive estimation of infarct size. *Cardiovasc. Res.,* 22, 456–463.

65 Apple, F. S. (1992) Acute myocardial infarction and coronary reperfusion. Serum cardiac markers for the 1990s. *Am. J. Clin. Pathol.,* 97, 217–226.

66 Gupta, M. Singal, P. K. (1987) Oxygen radical injury in the presence of desferal, a specific iron-chelating agent. *Biochem. Pharmacol.,* 36, 3774–3777.

67 Kaneko, M. Masude, H. Suzuki, H. Matsumoto, Y. Kobayashi, A. Yamazaki, N. (1993) Modification of contractile proteins by oxygen free radicals in rat heart. *Mol. Cell. Biochem.,* 125, 163–169.

68 Könczöl, F. Lőrinczy, D. Belágyi, J. (1998) Effect of oxygen free radicals on myosin in muscle fibres. *FEBS Letters,* 427, 341–344.

69 Rustgi, S. Riesz, P. (1978) E.s.r. and spin-trapping studies of the reactions of hydrated electrons with dipeptides. *Int. J. Radiat. Biol. Relat. Stud. Phys. Chem. Med.*, 34, 127–148.
70 Lőrinczy, D. Gaszner, B. Könczöl, F. Belagyi, J. (2000) Effect of oxygen free radicals on myosin in muscle fibres. DSC and EPR study. *J. Therm. Anal. Cal.*, 61, 597–605.
71 Lőrinczy, D. Könczöl, F. Farkas, L. Gaszner, B. Belagyi, J. (2000) UV generated oxygen free radicals in cardiac myosin. DSC and EPR study. *Thermochim. Acta,* 343, 35–41.
72 Lőrinczy, D. Belágyi, J. (1996) Internal flexibility of cardiac myosin. *J. Therm. Anal.*, 47, 503–514.
73 Grammer, J. C. Cremo, C. R. Yount, R. G. (1988) UV-induced vanadate-dependent modification and cleavage of skeletal myosin subfragment 1 heavy chain. 1. Evidence for active site modification. *Biochemistry,* 27, 8408–8415.
74 Cremo, C. R. Grammer, J. C. Yount, R. G. (1988) UV-induced vanadate-dependent modification and cleavage of skeletal myosin subfragment 1 heavy chain. 2. Oxidation of serine in the 23-kDa NH2-terminal tryptic peptide. *Biochemistry,* 27, 8415–8420.
75 Graceffa, P. (1983) Spin labeling of protein sulfhydryl groups by spin trapping a sulfur radical: application to bovine serum albumin and myosin. *Arch. Biochem. Biophys.*, 225, 802–808.
76 Kielley, W. W. Bradley, L. B. (1956) The relationship between sulfhydryl groups and the activation of myosin adenosinetriphosphatase. *J. Biol. Chem.*, 218, 653–659.
77 Nyitrai, M. Hild, G.,Belágyi, J. Somogyi, B. (1997) Spectroscopic study of conformational changes in subdomain 1 of G-actin: influence of divalent cations. *Biophys. J.*, 73, 2023–2032.
78 Nyitrai, M. Hild, G. Lakos, Zs. Somogyi, B. (1998) . Effect of Ca^{2+}-Mg^{2+} exchange on the flexibility and/or conformation of the small domain in monomeric actin. *Biophys. J.*, 74, 2474–2481.
79 Hild, G. Nyitrai, M. Belágyi, J. Somogyi, B. (1998) The influence of divalent cations on the dynamic properties of actin filaments: a spectroscopic study. *Biophys. J.*, 75, 3015–3022.
80 Nyitrai, M. Hild, G. Belágyi, J. Somogyi, B. (1999) The flexibility of actin filaments as revealed by fluorescence resonance energy transfer. The influence of divalent cations. *J. Biol. Chem.*, 274, 12996–13001.
81 Gaszner, B. Nyitrai, M. Hartvig, N. Kőszegi, T. Somogyi, B. Belágyi, J. (1999) Replacement of ATP with ADP affects the dynamic and conformational properties of actin monomer. *Biochemistry,* 38(39), 12885–12892.
82 Nyitrai, M. Hild, G. Hartvig, N. Belágyi, J. Somogyi, B. (2000) Conformational and dynamic differences between actin filaments polymerized from ATP- or ADP-actin monomers. J. *Biol. Chem.*, 275, 41143–41149.
83 Hild, G. Nyitrai, M. Somogyi, B. (2002) Intermonomer flexibility of Ca- and Mg-actin filaments at different pH values. *Eur. J. Biochem.*, 269, 842–849.
84 Liu, D. F. Wang, D. Stracher, A. (1990) The accessibility of the thiol groups on G- and F-actin of rabbit muscle. *Biochem. J.*, 266, 453–459.
85 Kuznetsov, A. N. Ebert, B. Lassmann, G. Shapiro, A. B. (1975) Adsorption of small molecules to bovine serum albumin studied by the spin-probe method. *Biochim. Biophys. Acta.*, 379, 139–146.
86 Zolkiewski, M. Redowicz, M. J. Korn, E. D. Ginsburg, A. (1995) Thermally induced unfolding of Acanthamoeba myosin II and skeletal muscle myosin: nucleotide effects. *Arch. Biochem. Biophys.*, 318, 207–214.

87 Shriver, J. W. Kamath, U. (1990) Differential scanning calorimetry of the unfolding of myosin subfragment 1, subfragment 2, and heavy meromyosin. *Biochemistry,* 29, 2556–2564.
88 Swenson, C. A. Ritchie, P. A. (1980) Conformational transitions in the subfragment-2 region of myosin. *Biochemistry,* 19, 5371– 5375.
89 King, L. Lehrer, S. (1989) Thermal unfolding of myosin rod and light meromyosin: circular dichroism and tryptophan fluorescence studies. *Biochemistry,* 28, 3498–3507.

Chapter 8

Thermal investigation on whole plants and plant tissues

I. Lamprecht*[1] and E. Schmolz[2]

[1]*Free University of Berlin, Institute for Biology, Ehrenbergstraße 26-28 D-14195 Berlin, Germany*
[2]*Free University of Berlin, Institute for Biology/Zoology Königin-Luise-Straße 1-3 D-14195 Berlin, Germany*

Introduction

Plant thermal analysis – although not comparable in the general attention of the scientific community with those for microorganisms, small animals and animal or human tissues and isolated cells - nevertheless offers a large variety of different aspects. They cover the range from plant tissue over organs to whole plants, from simple thermometry or thermography over combustion and differential scanning calorimetry to adiabatic and isothermal experiments, from germination of seeds to biochemical regulation of heat output and metabolic flare-up at blooming in thermogenic flowers. Moreover, wood as one of the most important plant products opens another field of thermal analysis. All of them will be touched – more or less intensively – in this short survey. But attention will be paid mainly to higher plants to avoid too much broadening.

Plant calorimetry was always assumed to be difficult since plants are poikilothermic living entities with an unfavourable surface to volume ratio. Plants need light of special wavelengths to be photosynthetically active. Metabolic rates are – with a few exceptions – low compared to that of animals or especially microorganisms. Moreover, evaporation with its high degree of energy consumption plays an essential role in the life of plants and may cover all other calorimetric signals if not carefully matched. Table 1 provides a list of heat flow rates from different plants or plant parts, divided into non-thermogenic and thermogenic objects. As a rule, one may expect at least 1 mW g^{-1} wet weight (w.w.) for the first group of plants.

If temperature sensing is taken as a genuine part of Thermal Analysis and qualitative statements are allowed, J. B. A. Lamarck was the first to contribute

* biophys@zedat.fu-berlin.de

D. Lőrinczy (ed.), The Nature of Biological Systems as Revealed by Thermal Methods, 187–214.

Table 1. Some selected plant tissues, organs and whole plants and their mass specific heat production rates (p) per g wet weight (d.w. = dry weight) at ambient temperature T_a or that of the organ (*). First part: Non-thermogenic plants: Second part: Thermogenic plants in normal state or during metabolic flare-up. In some cases heat production rates were calculated from O_2 consumption or CO_2 production

Trivial name	Botanical name	T_a °C	p mW g^{-1} w.w.	Reference
Non-thermogenic plants				
Apple: leaf segment	*Malus sp.*	23	2.8	[16]
Barley roots	*Hordeum vulgare*	24	1.4	[50]
Coast redwood, meristem tissue	*Sequoia sempervirens*	24	0.4 – 1.4	[43]
Coffee roots	*Coffea arabica*	25	2.4	[56]
Spinach leafs	*Spinacia oleracea*	25	0.9	[28]
Tomato, undifferentiated cell culture	*Lycopersicon esculentum*	23	1.7	[98]
Thermogenic plants				
Cuckoo-pint	*Arum maculatum*	-	385 d.w.	[99]
Eastern skunk cabbage	*Symplocarpus foetidus*	0	68	[100]
Philodendron	*Philodendron selloum (spadix)*	5	64	[101]
	(sterile male flowers)	37*	160	[101]
Voodoo lily; appendix	*Sauromatum guttatum*	26	95 d.w.	[102]
Voodoo lily (whole plant); quiescent	*Sauromatum guttatum*	25	1	[22]
Voodoo lily (whole plant); flare up	*Sauromatum guttatum*	25	7	[22]

to thermal investigations on plants [1]. He wrote about several members of the aroid family that the flowering catkin in a special state of ripeness or development is warm to appear burning ('il est chaud au point de paroître brûlant'), and not at all at the temperature of the rest of the plant. *Arum* belongs to the family of *thermogenic plants* that show a metabolic explosion during blooming. Many further thermometric results followed by other authors [2], but always oriented towards temperature, never to energy turnover or heat production.

Presumably the first calorimetric results were published by Rodewald at the end of the 1880th [3, 4]. He investigated plant objects of spherical form (apple, kohlrabi, onion) and by distributing a thermopile of 12 or 36 couples over the surface to determine temperature difference between the biological sample and the constant environment (Fig. 1). The corresponding heat flow was calculated with Newton's cooling law after a sophisticated calibration, and rates of evaporation, oxygen consumption and carbon dioxide release were determined simul-

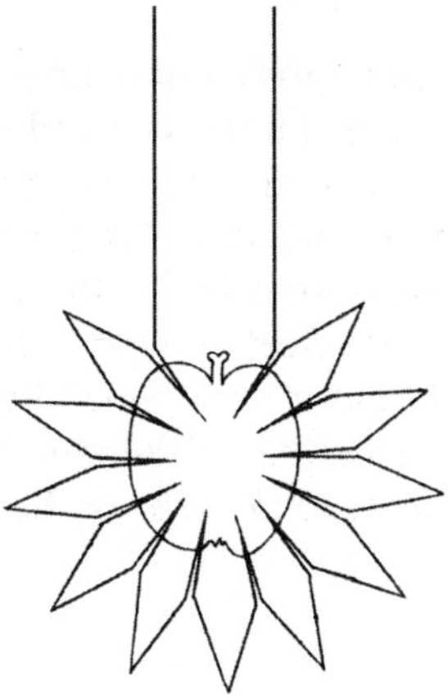

Fig. 1 Distribution of the thermopile couples along the surface of an apple. In first experiments 12 thermocouples were used, in later ones 36 [3]. Adapted from [3]

taneously. Apples and kohlrabies showed mean metabolic heat losses of 0.13 and 0.37 mW g^{-1} w.w., resp. The respiratory quotient (carbon dioxide production divided by oxygen consumption) was practically 1, the heat output per mol O_2 or CO_2 around 440 kJ, near to the value of 455 kJ mol^{-1} expected from Thornton's rule [5]. The author supposed that it was the first time to show that energy produced in respiration processes is nearly completely dissipated via heat flow and external work – in this case evaporation of water [3, 4].

Another very early paper on (true) plant calorimetry should be mentioned briefly, Pierce's Dewar-vessel experiments on germinating peas [6]. Within 8 days they heated up from 17 to 56°C, while chemically killed peas kept a temperature of around 15°C. As it became obvious that besides germination fermentation occurred also in the pea sample, peas were sterilized by chemical means before the next run. In this case, temperature rose to about 40°C within 3 days, but no effort was made to calculate the amount of heat released in these experiments.

WOOD

Wood belongs besides fruits and crops to the most important parts of plants and plays a significant role in the daily life, private as well as in engineering and industry. As the present review shall concentrate on 'living', that means actively growing or at least metabolising plant material, and since a comprehensive, thermal analytically oriented wood survey appeared recently [7], wood shall be neglected in favour of other plant parts.

Earlier reviews

Periodically, review papers on plant heat production appeared in monographs dedicated to the whole spectrum of biological calorimetry [8–11] or in special issues of scientific journals from the field of Thermal Analysis [12–14]. They inform about instrumental techniques applied for plant investigations [15], some aspects of plant calorimetry [8, 9] or the means to find 'differences between apples and oranges' [16, 17]. Special attention should be paid to the comprehensive contribution of Criddle and Hansen to the *Handbook of Thermal Analysis and Calorimetry* [11], which is a true gold mine for historical and modern aspects of plant calorimetry and facilitates the access to this fascinating and growing field.

In his 1995 survey of microcalorimetric techniques for plant material investigation Wadsö presented a number of existing instruments and vessels that could be used for this end [15]. They would allow for mere batch experiments in closed ampoules with changing gas atmosphere, for stirring of tissue cultures, for illumination in the sense of photocalorimetry, for gas perfusion and determination of carbon dioxide production. Although multifunctional instrument systems would be ideal tools for more complex analyses and appropriate physiological tissue conditions, the author nevertheless recommended the use of simple static ampoules as the most dependable process monitors [15]. His statement is still valid, but meanwhile he himself presented several sophisticated solutions for advanced tissue studies (see Instrumentation).

In the same special issue of *Thermochimica Acta* Hansen and his colleagues published part 2 of their *Plant calorimetry* [17] that reports about new technical efforts to ameliorate plant calorimetry, e.g. by coupling gas chromatographs or mass spectrometers to traditional instruments. Their work deals with aspects of thermogenic plants (see below) and presents a broad discussion about modelling the relation between growth rate and respiratory variables. Reviews to the stimulating area of thermogenic plants may be found in early surveys of Leick [2, 18], papers of Seymour [19, 20] and the present authors [21–24].

Instrumentation

It is advantageous to perform simultaneous determinations of the rates of heat and CO_2 production and oxygen consumption in order to get a more complex picture of the underlying plant metabolism. But there are limitations for the application of the corresponding electrodes when the vessels are small, the electrodes not submerged in the solution, the solutions not stirred, or the samples solid. To overcome such problems Criddle *et al.* [25] made threefold experiments: a first run with only the sample in the vessel, the second with an additional container for a CO_2 absorbing base, and the third repeating the first (Fig. 2). The upward shift in the steady heat flow corresponds to the enthalpy shift due to the absorption and neutralization of CO_2. This is an elegant solution for short term experiments of a few hours, but hardly applicable with changing atmospheres or long-term investigations.

A further step forward to simultaneous determinations were done with a differential heat conduction calorimeter housing three vessels for samples and

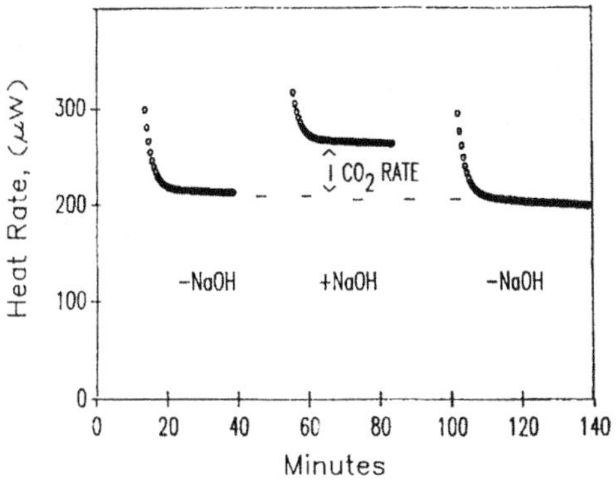

Fig. 2 Heat production rate of corn meristem tissue without (first and last trace) and with CO_2 absorber in the calorimetric vessel. With permission from [25]

one as reference. All three sealed 1 ml-reaction vessels were connected with pressure sensors [26]. The pressure signal in the absence of CO_2 absorption in the vessel renders the difference between oxygen consumption and carbon dioxide production, with CO_2 absorption, however, it provides the O_2 consumption rate. The rate of CO_2 production may then be determined as the difference between both signals and additionally from the shift in the heat flow rate as described in the last paragraph and shown in Fig. 2. The crucial point in such experiments is to know the volume of the head space in the vessels. The functioning of the new setup was tested with redwood meristem tissue [26].

In spite of the successful plant experiments with the ampoule calorimeter, the Lund school of calorimetry developed a gas perfusion instrument for plant tissue under dark conditions [27]: a combination of two twin heat flow calorimeter with 4.5 ml perfusion vessels. The first unit houses the biological sample, the second the CO_2 absorbing liquid that dissipates –97 kJ mol^{-1} CO_2 heat of solution and neutralization. The gas flow that passes through a prehumidifier and a CO_2 trap before entering the vessel may be changed during the analysis. Experiments with potato tuber slices of about 1.2 g fresh weight were run over 15 h with maximum heat outputs of 0.7 mW and CO_2 rates of 1.8 nmol s^{-1} and a ratio of –450 kJ mol^{-1} CO_2 [27].

Two years later a photo microcalorimeter system was presented with three twin elements of larger 20 ml volume: unit 1 the sample unit proper, unit 2 the photocalorimetric reference, and unit 3 the CO_2 absorber with an NaOH solution [28]. This triple instrument was checked with a 130 mg-leaf-tissue sample from spinach and a light flow of about 2500 μW (or 8 W m^{-2}). The metabolically active tissue dissipated 0.91 mW g^{-1} (Fig. 3) and rendered a positive ratio of +478 kJ mol^{-1} CO_2 during illumination, nearly identical with the Gibbs energy change of ΔG^{o} = +479 kJ mol^{-1} CO_2 for the formation of glucose from CO_2 and H_2O at standard conditions [28].

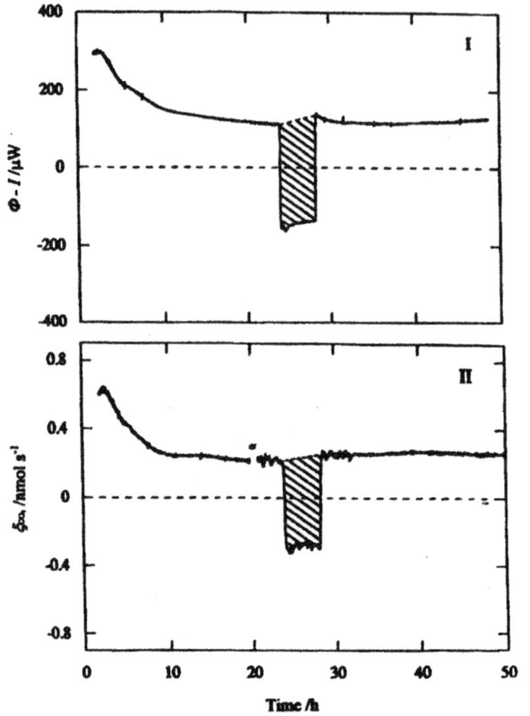

Fig. 3 Rates of energy expenditure (solid line) and energy uptake (shaded area) (upper part) and of CO_2 expenditure (solid line) and CO_2 uptake (shaded area) (lower part) for spinach leaf tissue. During the time of the shaded areas the sample was illuminated. With permission from [28]

A photocalorimetric method was developed by the Kazan/Russia group on the basis of two differential calorimeter, a Calvet-type and a LKB-type one. The additional module consisted of a 100 W water-cooled lamp, a set of filters to eliminate UV and IR light and to select special wavelengths in the visible range, and a quartz light-guide between the lamp outlet and the calorimetric vessel [29]. This equipment was successfully applied for the determination of energy storage in the unicellular alga *Chlorella* and several agricultural plants [29, 30].

An interesting new calorespirometric instrument was presented recently by L. Wadsö and Y. Li (see Conclusion) that connects the vessels of two twin isothermal calorimeters by an outside operated valve and each of them to a pressure sensor. This offers the opportunity to study simultaneously calorimetry and Warburg respirometry in a way similar to that used by Criddle, Hansen and their colleagues [25, 26].

In order to investigate whole plants and not just active organs or small samples, Lamprecht and colleagues [22] constructed a single calorimeter with an aluminium cylinder of 9.5 cm inner diameter, 39.5 cm inner height and an active volume of 2800 ml. The cylinder was connected via 12 Peltier elements (4.0 x 4.0 cm^2) to a corresponding rectangular upright cube as heat sink. Three methods of calibration rendered a sensitivity of 52.8 mV W^{-1} and a time constant of 13.2 min. The cube was enclosed in an isolating Styrofoam mantle, and the whole setup housed in an air thermostat of an LKB 10700 calorimeter that was expanded on top by a wooden box to provide the necessary height. This calorimeter was designed to house a complete voodoo lily from the bulb to the tip of the appendix (about 170 g w.w.). Due to the enormous heat output during the metabolic flare-up of more than 1 W and a calorimetric signal of 60 mV or more, it is possible to run this calorimeter in a single and not in the usual twin mode with convenient baseline stability. A perspex cover of the calorimeter proper allowed for illumination during the experiments.

A completely different way was followed by the group of Roger Seymour in Adelaide [31]. To broaden their metabolic investigations of thermogenic flowers and especially of the sacred lotus *Nelumbo nucifera* they designed a light and portable twin heat flow calorimeter that could be used outside in the lotus pond of the Adelaide Botanical Gardens. The idea was to have an inverted calorimeter open at the bottom that could be put over the blossom from the top and be closed by foam stoppers around the stalk. The calorimetric vessels proper were inverted thin-walled tinned steel cans with an inner diameter of 80 mm and a height of 145 mm. They were

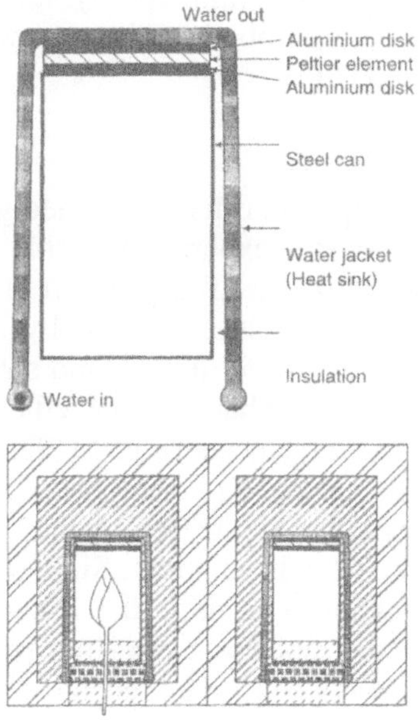

Fig. 4 Two cross-sections through one calorimetric unit (top) and the complete differential setup (bottom) of the lotus calorimeter designed in Adelaide. The Styrofoam insulation and the three foam stoppers are indicated. With permission from [31]

placed in two double-walled styrene wine coolers (100 mm diameter and 185 mm height). 3-mm aluminium plates were glued to the wine cooler inside bottom for better heat conduction. The coolers were equipped with water in- and outlets and connected to a refrigerated, thermostatic water bath and served as heat sinks for the cans. A Peltier element (40x40x4 mm^3) was placed between cans and coolers in tight contacts to both surfaces. The whole set-up is seen in Fig. 4. This calorimeter had a sensitivity of about 25 mV W^{-1} and a time constant of 476 s for a drop to 1/e of the initial signal. After changing the working temperature from 10 to 30°C baseline stability was achieved after 1 hour. The baseline fluctuated less than 0.5 mV during 24 h in the 50 mV range that was used throughout the flower investigations. Such a stability was good enough for the normal biological fluctuations. The calorimeter was constructed in such a manner that simultaneous respiratory investigations could be performed without disturbing the calorimeter and that the heat flux due to air flow could be calculated from temperature differences.

Heat production in plant tissue

Investigating plant tissues and seeds in a calorimeter, a *caveat* is necessary: except for tissue cultures, samples are not sterile and microbial growth may strongly interfere in the obtained heat dissipation. An exponentially growing heat output during experiments with plant tissues is a clear sign of microbial contamination. The danger can be minimized by surface sterilization with diluted hypochlorite, addition of microbial inhibitors and use of solutions that do not support bacterial growth [16].

EARLIER OBSERVATIONS

Early calorimetric plant investigation often concerned germination of seeds. Prat [8, 9] described Calvet-calorimeter experiments with 1 g of wheat grains of a resting metabolism very near to zero. After addition of 1 ml water three distinct phases became visible in the first hours: a mere physicochemical effect with a rapid increase of heat flow due to water uptake and swelling of the grains and a drop back under the zero line ('dead phase') followed by a steady increase of heat flow due to the biological thermogenesis (Fig. 5). These phases are typical for germination, but they can be changed by drying the seeds prior to the experiment (augmentation of the first peak), illumination of the germinating plants and by variation of the gas mixture in the head space. Similar germination experiments were already performed on various seeds in the 1880ties by G. Bonnier with a Berthelot calorimeter and a thermocalorimeter after Regnault [32] (Figs. 6a, b). These classical observations were confirmed recently when studying germination and root elongation in quinoa (*Chenopodium quinoa*) seeds and looking for the presence of water transporting channels (aquiporines) [33].

In the 60th and 70th of the last century Zholkevitch and his colleagues in Moscow made intensive investigations of plant metabolism by means of a

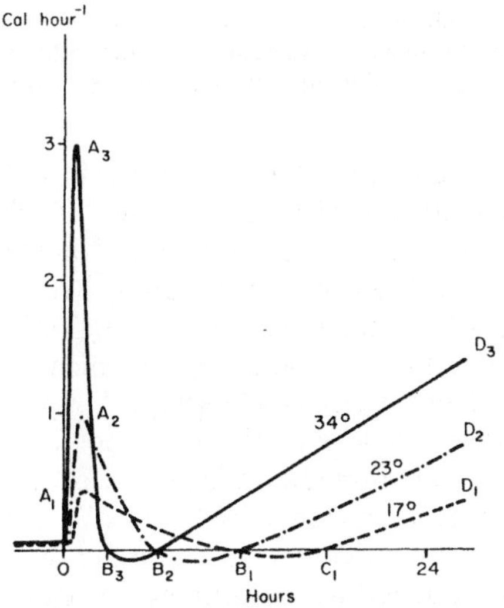

Fig. 5 Course of thermogenesis in 32 wheat grains (~ 1 g) at 24°C in a Calvet calorimeter with the three phases described in the text. Adapted from [9]

Calvet microcalorimeter and a conventional Warburg apparatus that remained unnoticed by western calorimetrists [34–38]. The aim of their experiments was to compare energy liberation as result of respiration with heat dissipation to the surrounding in various plant and plant tissue reactions: (i) pollination of isolated pistils of winter rye directly in the calorimetric vessel including swelling of pollen and rapid increase of energy turnover rate [34]; (ii) cellular redistribution of energy short time before elongation or division [35]; (iii) establishing an intracellular energy balance in plant tissue under varying conditions of water supply [36]; (iv) the relation between respiration and heat production in slightly withering plants [37]; and (v) low temperature influence on the energy metabolism in cucumber leaf tissue [38].

Kreshek and coworkers [39] analysed the heat production rate during elongation of *Avena* coleoptiles in an adiabatic solution calorimeter and determined rates around 1.2 mW g^{-1}. Comparison with respiration measurements ensured that oxidative metabolism accounted for essentially all energy changes in the cell. Moreover, it was found that the auxin indole-3-acetic acid stimulated the elongation by a factor of 10, but heat production rate by only 25 %. This is consistent with earlier findings that this auxin influences the aerobic respiration, inducing a significant increase at lower concentrations and a decrease at higher ones [39].

GENERAL OBSERVATIONS

Due to the limited space in this review, only a few examples of plant calorimetry will be presented here in a condensed form. More details can be easily found in

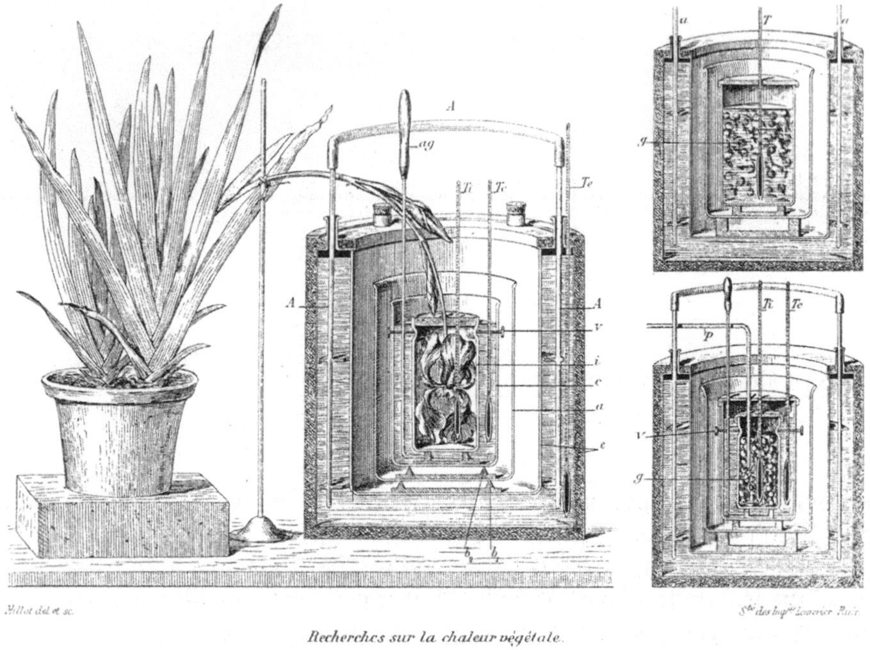

Fig. 6 a) Investigation of plant heat production by means of a modified Berthelot calorimeter in the beginning of the 1890th by G. Bonnier [32]. The heat produced by the flower in vessel (i) leads to an increase of the air temperature (T_i) and of the heat sink (T_c). A ring-like stirrer (ag) enables a homogeneous water temperature in the heat sink (c). The external water mantle (e) with another ring-like stirrer (A) and thermometer (T_e) protects the calorimeter proper against disturbances. The two smaller pictures show similar or slightly modified experiments with germinating peas in water (top, right) and in air (bottom, right). Adapted with pleasure from [32]

the cited references. A background discussion about kinetics of plant growth and metabolism [40], about the general influence of the alternative oxidative phosphorylation pathway on heat generation [41] and on four methods to evaluate the enthalpy change during anabolism [42] may serve as a stimulating introduction to the field.

It could be shown by heat-flow calorimetry that the integrated growth rate of the coast redwood (*Sequoia sempervirens*) is correlated to the dark metabolic heat rate [43]. This observation may provide a chance to discriminate clones with a high growth rate. Calorimetrically determined dark respiration of the coastal Douglas-fir (*Pseudotsuga menziesii*) was used to evaluate its drought hardiness in 3-years-old seedlings from different families. Heat and CO_2 production rates measured from 20 to 50°C together with Arrhenius plots rendered significant differences between drought sensitive and drought hardy families [44]. Increased heat production rates were observed calorespirometrically after

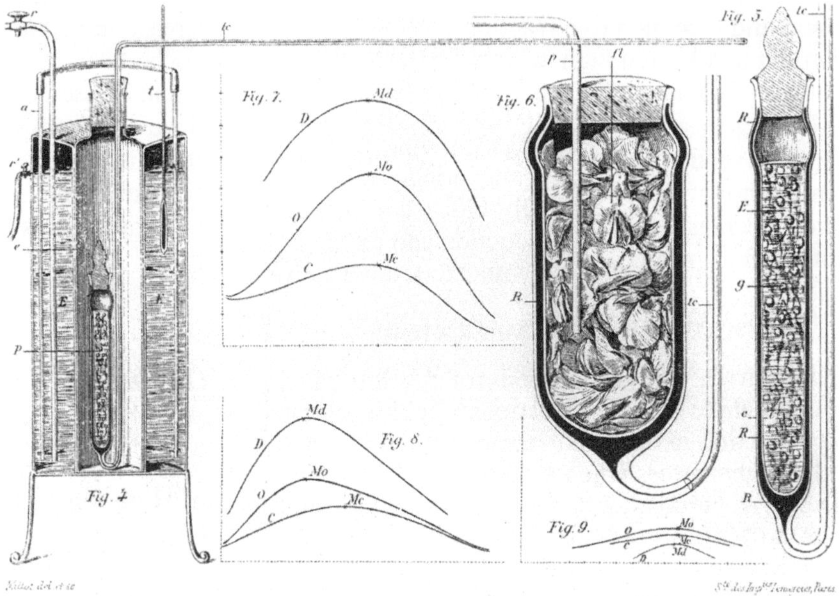

Fig. 6 b) Investigations of plant heat production by means of a thermocalorimeter Regnault for peas in water (left and inner part enlarged: right) and for blossoms (half-right). The black areas (R) correspond to the "thermometer" mercury that expands due to heat transfer and shifts its meniscus along the horizontal capillary (t_c). The graphs in the center show heat production rates as function of time determined directly (D) or calculated from oxygen consumption (O) and carbon dioxide production (C). Adapted with pleasure from [32]

wounding of carnation (*Dianthus caryophyllus* L.) shoot tips [45] as to be expected from the intensive thermodynamic discussion of regenerative processes by Zotin [46] and shown for several groups of animals from worms to lizards [47]. Corresponding results with strongly increased heat dissipation (up to six fold) were reported for sliced potato tubers (*Solanum tuberosum* L.) [48].

HEAT PRODUCTION UNDER STRESS

In conquering the surface of the Earth plants had to learn how to cope with hostile environments. Temperatures and high salt concentration in the soil belong to the most prominent stress factors. In many cases, plants became highly adapted to such conditions and only react metabolically to changes in both parameters. Here, calorimetry offers the possibility to detect subtle degrees of adaptation. In a more general paper Smith and colleagues dealt with questions of metabolism, photosynthesis and growth of different plants under stresses by temperature and salt [49]. In another paper the heat output of barley (*Hordeum vulgare* L.) root tips under salinity stress demonstrated two levels of inhibition with increasing NaCl concentration and a cooperative reaction responsible for the decrease in metabolism and nutrient uptake at a high salinity [50]. Connected with such observations are the changes in the rate of heat output on ion balance shifts seen in excised roots of wheat (*Triticum* spec.) seedlings growing in $CaCl_2$ [51]. Two ion transporters were applied rendering significant varieties in the energy metabolism of the roots. Cerulin is known to specifically block the fatty-acid synthesis and thus should provoke a decrease in heat production and oxygen consumption rates. This could be confirmed calorimetrically with excised wheat roots [52].

HEAT PRODUCTION AND TEMPERATURE

Seedlings from five populations of the big sagebrush (*Artemisia tridentata*) growing in three different locations were investigated among others for heat and CO_2 production in relation to temperature [53]. Adaptation to their site of origin and significant stress when seedlings were transplanted to other sites were confirmed, by monitoring changes in heat output and growth. High temperatures from 30 to 40°C meant high stress, low values (5 to 10°C) were without stress.

Chilling-sensitive plants can already be injured by temperatures significantly above the freezing point (5 to 15°C). When they are brought back from low to higher temperatures they show a respiratory burst, which can be easily detected by microcalorimetry [54]. Heat output – connected with the alternative pathway – of leaf discs increased up to 98 % after chilling in sensitive species, but only up to 22 % in resistant plants [55]. Coffee seedlings (*Coffea arabica* L.) belong to the sensitive group and are hindered in growth by temperatures below 15°C, closely correlated with the reduction of heat production rates. An Arrhenius plot of the heat rate revealed a break in the line at 15°C, sign of a metabolic transition at this temperature [56]. Further chilling connected experiments are described for soybean (*Glycine max*) cultivar leaves [57], for husk tomato (*Physalis ixocarpa*) leaves [58] and for dormant vegetative apple (*Malus* sp.) buds [59]. Seeds from different populations of the cold-desert subshrub winterfat (*Eurotia lanata*) germinated at temperatures between 0 and 20°C. Heat and CO_2 production rates were determined at the same temperatures rendering metabolic efficiencies and specific growth rates. It became obvious that the seedlings' responses reflected the climate at the site of their origin [60].

HEAT PRODUCTION OF THERMOGENIC PLANTS

True thermogenic plants that produce heat for their own sake and not as by-product of usual metabolic activity are found in several plant families, for instance in Araceae (arum lilies), Nymphaeaceae (water lilies) and Nelumbonaceae (true lotus) (see e.g. [19, 20, 23, 61]). The specific heat output of flowers of such plants can reach that of a hovering hummingbird or raise the inflorescence temperature

35 K above that of the environment [19]. Some of them are even able to regulate their temperature (in the sense of warm-blooded animals) for hours or days [62]. Several reasons for this astonishing phenomenon are presented in the literature: in a first place the volatilisation of odour molecules to attract insect pollinators and the provision of a heated shelter for these insects as well as the possibility to bloom at low temperature and to protect sensitive parts in the flowers. The biochemical background of the metabolic burst is (i) a shift from the normal to a cyanide-insensitive pathway rendering only one third of the ATP amount obtained by the usual phosphorylation and dissipating the residual energy as heat and (ii) even more important a strongly increased respiration rate [63].

Most of the thermogenic plants are large and soil-bound so that inflorescences as place of heat production have to be cut and investigated separately from the rest of the plant. Moreover, these organs are usually so large that only few calorimeters are large enough to house a whole inflorescence. Thus, they must be used as tissues slices. Problems accompanying slicing of plant tissue (and thus minimizing diffusion barriers for oxygen) are arrestingly described for voodoo lily (*Sauromatum guttatum*) tissue showing a heat output of 174.7 mW g^{-1} dry weight for a 10 mg sample, but 9097.1 mW g^{-1} (!) for a 0.1 mg piece [64]. The voodoo lily is the only member of the thermogenic family that is able to flower just from the corm without soil and water, predestined for a 'whole-body' calorimetry. In a special plant calorimeter described above complete voodoo lilies were investigated. Figure 7 shows the metabolic burst of a 155 g lily in the early morning hours of D-day with a maximum calorimetric signal of 60 mV or 1020 mW, rendering 6.8 mW g^{-1} w.w. for the whole plant, but about 100 mW g^{-1} for the metabolically active appendix. Integration of heat output over a 4.5-h period comes to 11.6 kJ, the oxygen consumption in the same period to 11.4 kJ. This underlines that the flare-up is carried by a strongly increased respiration intensity [22]. The pre-history of flowering in the voodoo lily was investigated calorimetrically for the five days preceding the metabolic burst [65]. The authors used tissue slices

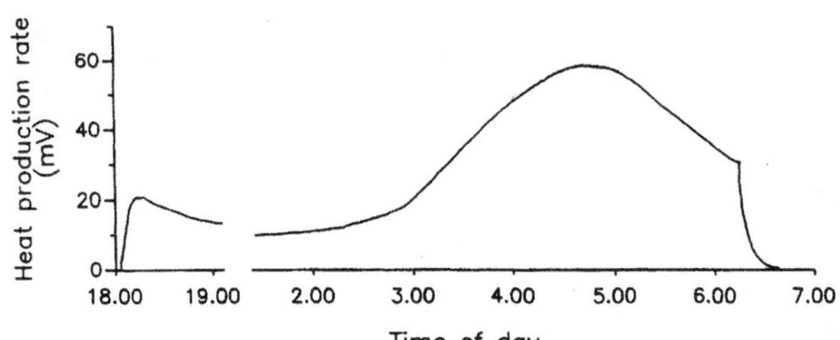

Fig. 7 Anthesis of a voodoo lily (*S. guttatum*) of 155 g wet weight in a 3-liter heat flow calorimeter. The decrease in heat production after 5:00 h is an artefact due to the decreasing oxygen concentration in the calorimeter. With permission from [22].

from three parts of the appendix and determined the influence of salicylic acid as natural inducer of heat production and of the head space atmosphere. The same authors detected an oscillatory behaviour of the heat production in the thermogenic male cones of two cycad species (palm ferns) with maximum amplitudes of 14 to 18 mW g^{-1} w.w. and confirmed the presence of the alternative respiratory pathway [66].

The first true calorespirometric investigations on plants with a combination of a gradient layer calorimeter of the Benzinger/Kitzinger type and an external gas analyser concerned the severed spadix of *Philodendron selloum*, a member of the arum family [67, 68]. Its spadix warms up to 38 to 46°C independent of the ambient temperature changing between 4 and 39°C. It indicates that this organ enforced its metabolism when the temperature decreased. While heating endured for several hours under normal conditions, it reduced to 1 to 2 hours for cut samples. Nevertheless, they obtained the same maximum respiratory heat production around 50 mW g^{-1} w.w. Additionally, bomb calorimetry was applied to determine the heat content of cut sterile male florets (the main seat of heat production with a maximum rate of 150 mW g^{-1} w.w.) before, during and after the metabolic burst.

Direct and indirect calorimetric field experiments on flowers of the sacred lotus *Nelumbo nucifera* were performed in an outdoor pond of the Adelaide Botanical Gardens [31] by means of the light, transportable differential calorimeter already described above. Its adjustable water-bath heat-sink allowed to deceive the flower and to simulate a cold environment during day and a warm at night. Thus, it could be shown that heat production in lotus flowers depends on the ambient temperature and not on the light cycle [69]. The flowers kept their temperature rather constant at 30.7 and 34.2°C at mean set calorimeter values of 18.4 and 30.4 °C, respectively. The maximum heat production dropped from 0.51 W (60 mW g^{-1} w.w.) to 0.25 W (30 mW g^{-1}) at high temperatures. Figure 8 presents the outcome of a five-day experiment in the lotus pond with a reversed daily temperature regime of the calorimeter (15°C cool during the day, 30°C warm at night). T_r and T_c represent the receptacle and the calorimeter temperatures. Neglecting the over- and undershoots, one observes plateau phases in both half-days. The receptacle was always warmer than the calorimeter, but at a decreasing difference with time. Longer periods of constant metabolic heat production rates Φ_p (upper part of the figure) and of dry heat loss Φ_l (lower part) are seen in the graph interrupted by metabolic bursts when the calorimeter temperature was lowered. During these periods the differences between T_r and T_c amounted to 12.3°C, in the other time only 3.8°C – a clear sign of the stated thermoregulation.

In addition to the field investigations, cut flowers were placed in the same gradient layer calorimeter arrangement as used before for the *Philodendron* spadix [67, 68]. An energy calculation showed that the metabolic heat produc-

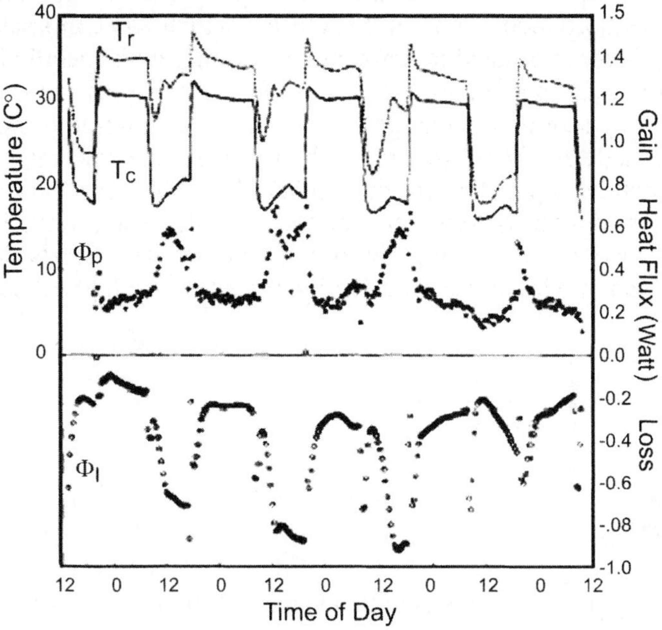

Fig. 8 Graph of a 5-days outside experiment on a flower of the sacred lotus *Nelumbo nucifera*. The right ordinate gives gains (upwards) and loses (downwards) of the metabolic heat production Φ_p calculated from the oxygen consumption rate and for the dry heat loss Φ_l. Due to heat take-up in the 6 m long hose from the water thermostat to the calorimeter and the warming by the flower the mean low calorimeter temperature was 18.4 °C at a thermostat setting of 15°C. For more information see text). With permission from [31]

tion was almost completely balanced by evaporative heat loss and that there was no conservation of energy in metabolic processes during thermogenesis.

This lotus field research was – to our knowledge – the first outdoor application of direct calorimetry to plants while there were already a number of indirect field investigations on several aroids including *Philodendron* and eastern skunk cabbage or on the tropical water lily *Victoria cruziana*.

OTHER TECHNIQUES

Thermal analysis – in the title of this monograph – means not only just isothermal microcalorimetry, but also other techniques like combustion calorimetry, differential thermal analysis/differential scanning calorimetry, thermogravimetry or infrared false color thermography. But as the daily routine and the main interest of the authors focus on microcalorimetry, these other fields will be touched more marginally saying nothing about their importance in thermal analysis. Including some papers in this review enables the deeper-interested reader to track the activities of the chosen authors and to get access to related fields by their cited references.

COMBUSTION CALORIMETRY (CC)

Combustion calorimetry is one of the oldest techniques in Thermal Analysis dating back to Black, Crawford, Lavoisier, and Count Rumford; in its modern form to Berthelot. In its earliest application energy content of wood played an essential role. Later on, other plant material like leaves, fruits and roots were enclosed and CC is nowadays an important tool in ecological energy balances. Recently, a contribution by one of the present authors (I. L.) appeared in the *Handbook of Thermal Analysis and Calorimetry* [70] covering the whole field from instrumentation (see also [71]), sample preparation and experimental approaches to applications in different areas including plant material. Therefore, this section may be kept short. Meanwhile, further papers were published from the Santiago group around Prof. Nunez Regueira dealing with caloric values and flammability of living forest species or forest waste biomass and the risk of wildfires in Galicia [72–76] from which a risk index map of this area emerged [77]. Wild fires represent an enormous problem in Spain as they devastated thousands of square kilometre forest and bush land in the last 40 years with a value of billions of Euro.

DIFFERENTIAL SCANNING CALORIMETRY (DSC)

In a review on biologic applications of DSC [12] plants played an important role with questions concerning the state of water in plant cells, of cold resistance, supercooling, freezing cryo-protection and dehydration, but also of wood, its components and fungal degradation and finally of ancient plant material like papyri or fig-tree bark for historical paper production. More information on DSC used in wood analysis may be found in a handbook surview [7].

While in the 'classical' DSC fresh or dried milligram samples of plant material are investigated in the range from room temperature to about 600°C, with scanning rates of several K min^{-1} and in air as well as in inert atmosphere, a modern alternative for plant investigation was opened following an inseminating paper by Sturtevant and coworkers [78] on heat production of murine macrophages in a strongly reduced temperature range and at low scan rates. This technique is frequently applied today in groups of plant calorimetrists [79]. Some examples were already cited in the 'chilling' paragraph [57–59] and only one shall follow concerning the temperature dependence of tissue metabolism from barley (*Hordeum vulgare* L.) roots [80]. Instead of a stepwise increase of temperature as in true isothermal experiments the sample was slowly scanned (2.4 to 3.5 K h^{-1}) from 5 to 45°C (Fig. 9). This rate was sufficiently slow so that the tissue metabolism occurred at a steady, quasi-isothermal state at each time. This is underlined by the 4 square points at the curve 'CM72 + ISOTHERMAL' that result from isothermal calorimetry at 5 K different temperatures. Smooth

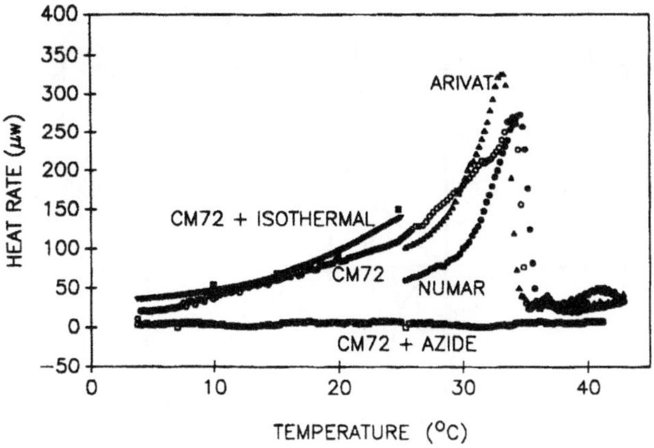

Fig. 9 DSC thermograms of 3 barley cultivars from 5 to 45°C (CM72) and 25 to 45°C (Arivat, Numar). The graph (CM72 + isothermal) compares the DSC thermogram with 4 isothermally obtained data points (■) in 5 K steps. With permission from [80]

curves result for all barley cultivars up to about 30°C where a threefold increase in activation energy becomes obvious. While this change was reversible, the next one at 34°C lead to an irreversible decrease of the heat rate to nearly zero. In the whole DSC thermogram from 5 to above 30°C no physical state changes of membranes or proteins became visible although sensitivity was high enough to show them [80].

THERMOGRAVIMETRY

Thermogravimetry alone or coupled with mass spectrometry is not so often applied to plant tissue. In thermogravimetric experiments the weight loss of the sample is determined during a heating up program with usually constant heating rates of a few K min^{-1}. The obtained signals show, as a function of time (or temperature), the increase in temperature (T), the relative weight loss in percent (TG, taking the initial weight as 100%, thus starting with a weight loss of 0 %), the first time derivative of this loss (DTG) and the differential thermal analysis signal (DTA) as the temperature difference between the sample and the reference crucible. By convention, exothermic peaks point upward, endothermic ones downward.

Commercial raw plant drugs, material consisting of leaves, flowers, roots, rhizomes and bark, were heated up to 900°C at a rate of 5 K min^{-1} and the obtained DTA, TG and DTG curves were analysed [81, 82]. Moreover, the non-metallic and metallic elements in the samples were determined. It became obvious in the majority of cases that samples taken from the same plant species render similar results within the five kinds of plant material listed above: Fig. 10 shows a typical set of curves (T: sample temperature) for three different flowers. Only one further

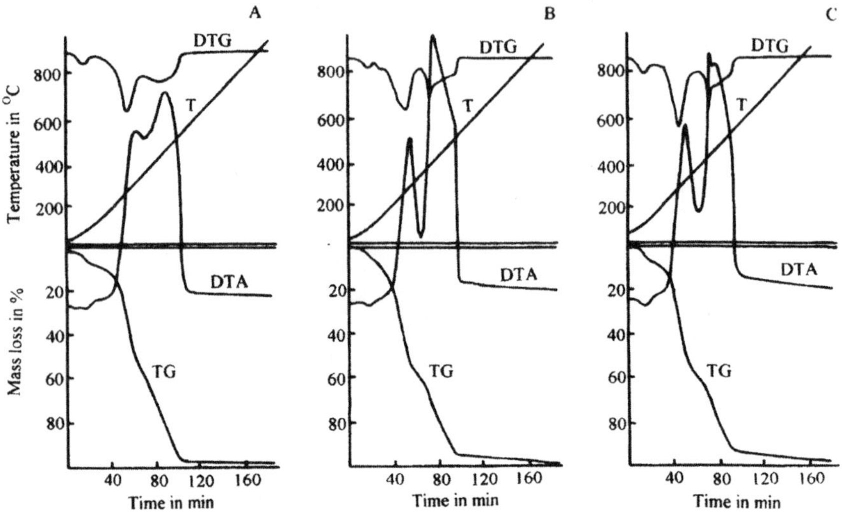

Fig. 10 DTA, T, TG and DTG curves obtained during thermal decomposition of three flowers. A: Scotch heather (*Calluna vulgaris*); B: coltsfoot (*Tussilago farfara* L.); C: yarrow (*Achilla millefolium* L.). Sample size about 100 mg, heating rate 5 K min^{-1}. The DTA curve consists of one weak endothermic and two pronounced exothermic peaks. With permission from [82]

example of plant thermogravimetry shall be presented here, a coupling of a modified thermobalance with a mass spectrometer [83] for the investigation of two herbaceous plants with high biomass production potential. Small samples of a few mg were heated from 20 to 900°C with a rate of 20 K min^{-1}. This is slow enough to resolve the different steps of temperature dependent biomass decomposition: starting with moisture evolution and followed by decomposition of hemicellulose and cellulose. The simultaneous mass spectrometry data provide information about the low molecular weight volatile products of pyrolysis as function of temperature and thus together with the DTG curves also information about the underlying degradation mechanism. They are essential for the understanding and optimisation of different biomass conversion techniques [83].

IR THERMOGRAPHY

Infrared false-colour thermography [84] is well introduced in various fields of industry, research and medicine (including veterinarian), but only recently more intensively used for surface investigations of plants. Thermographic cameras collect near infrared radiation in the wavelength range from about 5 to 15 µm, has a typical sensitivity of 0.1 K at 30°C and may discriminate between points of 2 mm distance and 0.2 K temperature difference. The detected IR radiation is proportional to the temperature distribution over the surface of the object. The camera transforms this IR picture into a false-colour picture in the visible range;

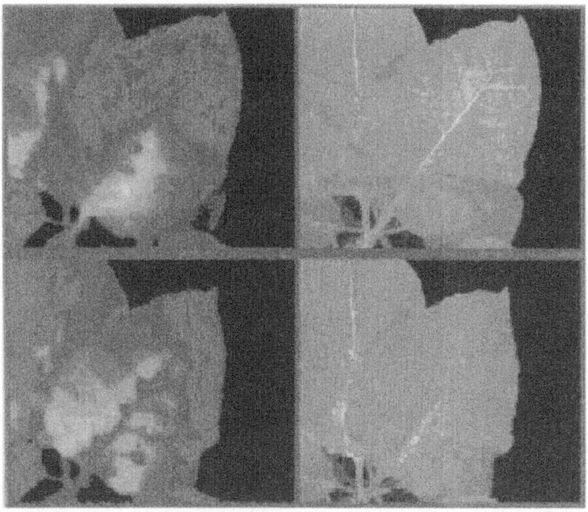

Fig. 11 Thermographic and visual imaging of cell death (yellow parts) in bacterio-opsin tobacco 32 h (upper two pictures) and 40h (lower two) after first detection of a thermal effect. The maximum temperature difference amounts to 0.6 K. With permission from [85] *(See colour section, p. 347).*

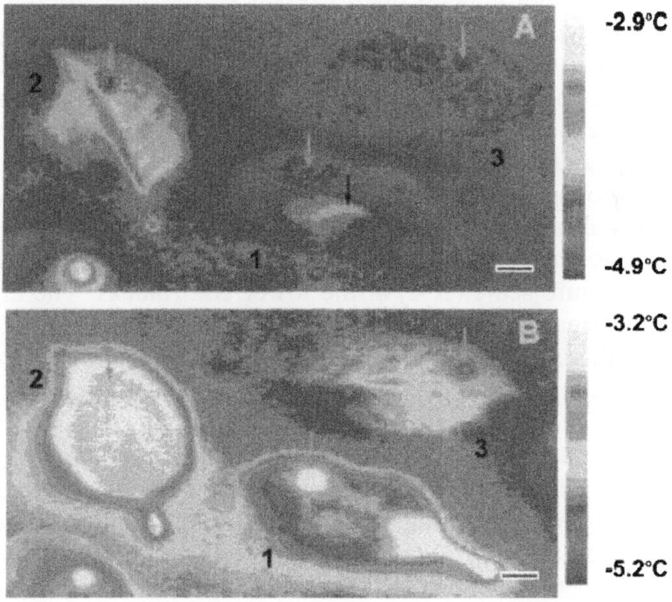

Fig. 12 Holly leaves (*Ilex* sp.) during freezing shown in false colours of a 2 K temperature range. Picture B was taken about 3 min after A. The pale blue to whitish areas (A,B) indicate an initial exothermic effect of low intensity, the yellish colours (B) a second stronger exothermic effect. Green arrows point to water droplets put on the leaves before cooling started. With permission from [90] *(See colour section, p. 347).*

colours or grey intensities can be freely chosen to render an optimal resolution for the interesting temperature range. In recent years increasing, but still small numbers of papers were published concerning plant IR thermography dedicated to freezing and ice nucleation, infections, leaf energy balances, metabolic flare-up in thermogenic flowers and seeding quality assessment. A few shall be presented in this survey.

Chaerle and Van Der Straeten recently published a paper that could render the title for this chapter on thermography: 'Seeing is believing: imaging techniques to monitor plant health' [85]. It deals with the IR observation of stress-induced changes in plants before the human eye can detect them, with screening for mutants of increased stress tolerance and with the application of such results for plant engineering [85]. In the same sense, the presymptomatic thermographic visualization of plant destruction by tobacco mosaic virus infection was demonstrated by the same authors [86, 87]. Already 8 h before the visual detection lesions were clearly visible in the IR pictures as 'hot spots', 0.3 to 0.4 K warmer than the surrounding (Fig. 11). These spots were colocalized with later on formed lesions due to an accumulation of salicylic acid, a compound that is known to stimulate heat production in thermogenic plants (see below), to induce thermogenicity in non-thermogenic leaves [88] and to serve as a signal against pathogens [89].

One of the major general plant stresses is freezing, an effect of enormous importance for wild plants as well as for crops and thus for agriculture also. Freezing in plants is accompanied by an (exothermic) heat release and temperature increase of up to 1 K that can easily be detected by thermography. A number of papers (e.g. see [90–93]) appeared dedicated to nucleation of ice and its propagation, mainly in plant leaves and buds (Fig. 12). Figure 13 shows ice nucleation and propagation in a bean leaf induced by a 2 μl droplet of a *Pseudomonas syringae* solution (left) and a drop of deionised water (right) (A). Since the water drop loses heat by evaporation it is cooler than the leaf and outside the chosen temperature scale (2 K) at the lower end (black). The droplet with the 'Ice^{+} bacteria' freezes first (B), gains a higher temperature outside the temperature range at the upper end (white) due to the exothermic heat of freezing and serves as centre for ice nucleation throughout the leaf (B to E). At the end, when the freezing of the leaf and the exothermic process came to an end, the leaf turned blue (in false colours), only then the water drop started to freeze and to obtain heat and a higher temperature (change from black to white).

A specially interesting field for IR thermography is that of the already mentioned thermogenic plants, a group of plants mainly consisting of species of the arum lily family and a few water lilies. Hanna Skubatz and colleagues presented an impressive collection of IR pictures of eight members of the *Arum* lily family, the temperature distribution along these plants (mainly in their appendices) and respiration data from the most thermogenic tissues [94]. The same authors thermographically followed the metabolic burst of the voodoo lily (*Sauromatum*

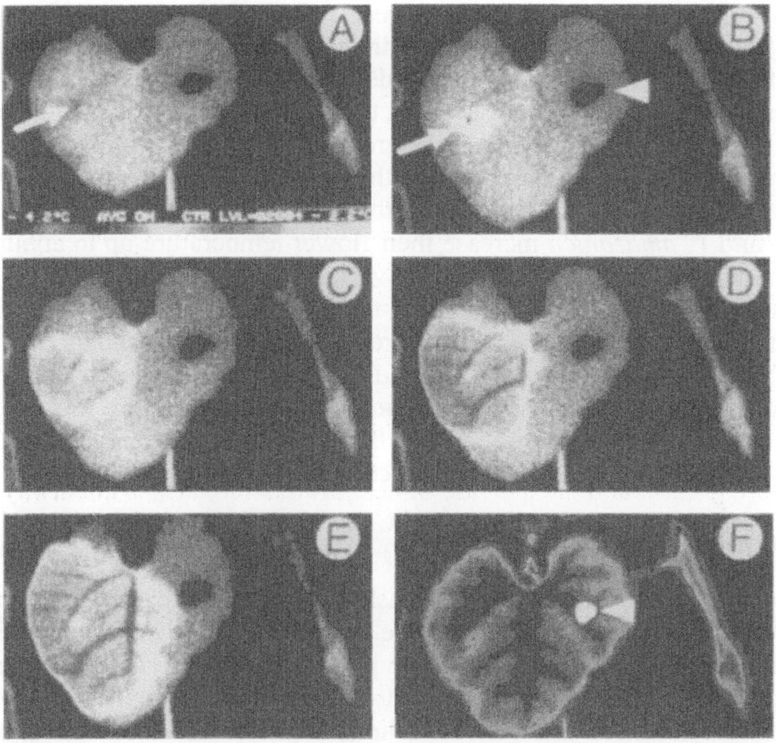

Fig. 13 Ice nucleation and propagation in a bean leaf shown by false-colour thermography. The temperature range was chosen 2 K. Black and white parts are out of range at the lower and the upper end, resp. For further explanations see text. With permission from [91] *(See colour section, p. 348).*

Fig. 14 Thermogenic active evening flower of the giant water lily *V. cruziana* in false colour. At air and water temperatures of 24.0 and 31.0°C, resp., the centre of the blossom shows a temperature from 30.9 to 33.5, significantly above the air temperature. The white area in the left upper corner represents the arm of the investigator. With permission from [23] *(See colour section, p. 348).*

guttatum) [95], as was done in the same year by one of the present authors (I. L.) [21, 22]. Similar *Arum* investigations on the temporal and spatial distribution of heat production were dedicated to the Lords and Ladies (*Arum maculatum*) showing two main centres of metabolic activity, the appendix and the male florets [96].

In the course of thermoanalytical greenhouse investigations on the giant tropical water lily *Victoria cruziana*, IR thermography was applied intensively, for flowers as well as for the structure of the huge floating leaves [23, 24, 97]. *V. cruziana* and *V. amazonica* increase their flower temperature up to about 10 K above ambient and release an attractive sweet odour of pineapple or fruit salad. The most thermogenic tissue is found in the inner stamina that are not directly visible from the outside. Thus, the infrared picture (Fig. 14) shows the hottest part (33.5°C) around the 'tunnel' that leads through the paracarpels into the floral chamber and that was closed in the moment when the picture was taken. As the flower was 10 cm above the water, the temperature difference against air (24.0°C) and not against water (31.0°C) counts.

Conclusion

In recent years, plant calorimetry has finished its sleeping beauty dormancy that lasted more or less continuously since the beginning of last century and realizes quite a number of scientifically attractive princes around it. They appeared with new armours and weapons and are eager to dedicate their life to her beauty. New sophisticated calorimeter allow investigations that were excluded before, combination with other highly specific instruments renders more information than gathered earlier by calorimetry alone, rigid and flexible light guides illuminate the otherwise dark vessels and open the field to photosynthesis experiments, infrared thermography provides a means to determine temperature distributions without injuring or even only touching the object. Imagination and technical skill are asked for further developments in plant calorimetry that will for sure come.

A special issue of *Thermochimica Acta* in 2004 will be dedicated to the outcomes of the XIIIth Conference of the International Society for Biological Calorimetry *Energetics of Adaptation and Development – From Molecular Mechanisms to Clinical Practice* that took place in Würzburg / Germany 2003. Many contributions deal with photocalorimetry and its application to plant systems as well as with conventional heat flow and combustion calorimetry for plant cells. Among other, topics like *A new calorespirometric instrument*, *Calorespirometric ratios and metabolic efficiency*, *Energy processes in model plant cells*, *Heat production and respiration of wheat roots*, *Calorimetric studies of vegetable tissue wounding*, *Life zones for key plant species can be predicted with calorespirometry and temperature measurements*; *Photo-Bio-Calorimetry of Chlorella vulgaris growth*, *Energetic evaluation of forest formations by bomb calorimetry* will be approached. This issue is recommended for further information.

References

1 Lamarck, J. B. A.: Flore Françoise ou Description Succinte de Toutes les Plantes. 2nd edition. Tome 3., H. Agasse, Paris 1778, p. 538.
2 Leick, E.: Untersuchungen über die Blütenwärme der Araceen. Bruncken & Co, Greifswald 1910.
3 Rodewald, H. Quantitative Untersuchungen über die Wärme- und Kohlensäure-Abgabe athmender Pflanzentheile. *Jahrbücher für Wissenschaftliche Botanik,* 18 (1887) 264–345.
4 Rodewald, H. Weitere Untersuchungen über den Stoff- und Kraftumsatz im Athmungsprocess der Pflanze. *Jahrbücher für Wissenschaftliche Botanik,* 20 (1889) 261–291.
5 Thornton, W. M. The relation of oxygen to the heat of combustion of organic compounds. *Philosophical Magazine, Sixth Series,* 33 (1917) 196–203.
6 Pierce G. J. A new respiration calorimeter. *Botanical Gazette,* 46 (1908) 193–202.
7 Wiedemann, H. G. Lamprecht, I. Wood. in: Handbook of Thermal Analysis and Calorimetry. Vol. 4: From Macromolecules to Man. R. B. Kemp (ed.). Elsevier Science B. V., Amsterdam 1999, 765–809.
8 Calvet, E. Prat, H.:Microcalorimétrie - Applications physico-chimiques et biologiques. Masson et Cie, Paris 1956.
9 Prat, H. Calorimetry of higher organisms. in: Biochemical Calorimetry, H. D. Brown (ed.) Chapter 9, Academic Press, New York 1969, 181–198.
10 James, (ed.) A. M.: Thermal and Energetic Studies of Cellular Biological Systems. Wright, Bristol 1987.
11 Criddle, R. S. Hansen, L. D. Calorimetric methods for analysis of plant metabolism. in: Handbook of Thermal Analysis and Calorimetry. Vol. 4: From Macromolecules to Man (Kemp, R. B. ed.), Elsevier Science B. V. Amsterdam 1999, 711–763.
12 Lamprecht, I. Application of Differential Scanning Calorimetry to questions of biological interest. Review No. 11, *TARard A,* 19(1) (1990) 1–5.
13 Lamprecht, I. Hemminger, W. Höhne. G. W. H. Calorimetry in the Biological Sciences. *Thermochimica Acta,* 193 (1991) (Special issue), p. 452.
14 Hansen, L. D. Hopkin, M. S. Criddle, R. S. Plant calorimetry: A window to plant physiology and ecology. *Thermochimica Acta,* 300 (1997) 183-197.
15 Wadsö, I. Microcalorimetric techniques for the investigation of living plant materials. *Thermochimica Acta,* 250 (1995) 285–304.
16 Criddle, R. S. Breidenbach, R. W. Hansen L. D. Plant calorimetry. Part 1. *Thermochimica Acta,* 193 (1991) 215–232.
17 Hansen, L. D. Hopkin, M. S. Taylor, D. K. Anekonda, T. S. Rank, D. R. Breidenbach, R. W. Criddle R. S. Plant calorimetry. Part 2. Modelling the differences between apples and oranges. *Thermochimica Acta,* 250 (1995) 215–232.
18 Leick, E. Die Erwärmungstypen der Araceen und ihre biologische Bedeutung. *Berichte der Deutschen Botanischen Gesellschaft,* 33 (1915) 518–536.
19 Seymour, R. Plants that warm themselves. *Scientific American,* (1997), March, 91–95.
20 Seymour, R. S. Schultze-Motel, P. Heat-producing flowers. *Endeavour,* 21 (1997) 125–129.
21 Lamprecht, I. Schaarschmidt, B. Thermographische Untersuchungen an einem Aronstabgewächs. *ThermoMed,* 7 (1991) 75–79.

22 Lamprecht, I. Drong, K. Schaarschmidt, B. Welge, G. Some like it hot - Calorimetric investigations of voodoo lilies. *Thermochimica Acta,* 187 (1991) 33–40.

23 Lamprecht, I. Schmolz, E. Blanco, L. Romero, C. M. Flower ovens: thermal investigations on heat producing plants. *Thermochimica Acta,* 391 (2002) 107–118.

24 Lamprecht, I. Schmolz, E. Hilsberg, S. Schlegel, S. A tropical water lily with strong thermogenic behaviour - Thermometric and thermographic investigations on *Victoria cruziana. Thermochimica Acta,* 382 (2002) 199–210.

25 Criddle, R. S. R. Breidenbach, W. Rank, D. R. Hopkin, M. S. Hansen, L. D. Simultaneous calorimetric and respirometric measurements on plant tissues. *Thermochimica Acta,* 172 (1990) 213–221.

26 Fontana, A. J. Hilt, K. L. Paige, D. Hansen, L. D. Criddle, R. S. Calorespirometric analysis of plant tissue metabolism using calorimetry and pressure measurement. *Thermochimica Acta,* 258 (1995) 1–14.

27 Bäckman, P. Breidenbach, R. W. Johansson, P. Wadsö, I. A gas perfusion microcalorimeter for studies of plant tissue. *Thermochimica Acta,* 251 (1995) 323–333.

28 Johansson, P. Wadsö, I. A photo microcalorimetric system for studies of plant tissue. *J. Biochem. Biophys. Methods,* 35 (1997) 103–114.

29 Petrov, V. Ye. Alyabyev, A. Ju. Loseva, N. L. Klementyeva, G. S. Tribunskich, V. I. A differential photomicrocalorimetric method for investigating the rate of energy storage in plants. *Thermochimica Acta,* 251 (1995) 351–356.

30 Loseva, N. Alyabyev, A. Ju. Rackimova, G. G. Estrina, R. I. Aspects of the energetic balance of plant cells under different salt conditions. *Thermochimica Acta,* 251 (1995) 367–362.

31 Lamprecht, I. Seymour, R. S. Schultze-Motel, P. Direct and indirect calorimetry of thermogenic flowers of the sacred lotus, *Nelumbo nucifera. Thermochimica Acta,* 309 (1998) 5–16.

32 Bonnier, G. Recherches sur la chaleur végétale. *Annales des Sciences Naturelles* / Botanique, 7. ser., tome 18, (1893) 1–32.

33 Sigstad, E. E. Schabes, F. I. Isothermal microcalorimetry allows detection of 'aquaporines' in quinoa seeds. *Thermochimica Acta,* 349 (2000) 95–101.

34 Britikov, E. A. Zholkevich, V. N. Borisova, T. A. Musatova, N. A. Kovaleva, L. V. Rye pollination studied by microcalorimetry and gasometry. *Fisiologija Rastenii,* 21 (1974) 320-328 (Russian).

35 Zholkevich, V. N. Borisova, T. A. Peisakhson, B. I. Energy redistribution in the cells directly before their division or elongation. *Fisiologija Rasteni,i* 19 (1972) 1245–1251 (Russian).

36 Zholkevich, V. N. Energy balance of respiring plant tissues under various conditions of water supply. *Fisiologija Rastenii* 8 (1961) 407–416. (Russian)

37 Zholkevich, V. N. Holler, V. A. Rogacheva, A. A. Relation between respiration and heat production in slightly withering plants. *Dokl. Akad. Nauk SSSR,* 169(3) (1966) 713–716. (Russian)

38 Zholkevich, V. N. Holler, V. A. Kushnirenko, S. V. Aftereffect of cooling on the efficiency of respiration of cucumber leaves. *Fisiologija Rastenii,* 9 (1962) 353–358. (Russian)

39 Bogie, H. E. Kreshek, G. C. Harmet, K. H. Calorimetric studies of the elongation of *Avena* coleoptile segments. *Plant Physiology,* 57 (1976) 842–845.

40 Hansen, L. D. Church, J. N. Matheson, S. McCarlie, V. W. Thygerson, T. Criddle, R. S. Smith, B. N. Kinetics of plant growth and metabolism. *Thermochimica Acta,* 388 (2002) 415–425.
41 Breidenbach, R. W. Saxton, M. J. Hansen, L. D. Criddle, R. S. Heat generation and dissipation in plants: Can the alternative oxidative phosphorylation pathway serve a thermoregulatory role in plant tissues other than specialized organs? *Plant Physiology,* 114 (1997) 1137–1140.
42 Ellingson, D. Olson, A. Matheson, S. Criddle, R. S. Smith, B. N. Hansen, L. D. Determination of the enthalpy change for anabolism by four methods. *Thermochimica Acta,* 400 (2003) 79–85.
43 Hansen, L. D. Woodward, R. A. Breidenbach, R. W. Criddle, R. S. Dark metabolic heat rates and integrated growth rates of coast redwood clones are correlated. *Thermochimica Acta,* 211 (1992) 21–32.
44 Anekonda, T. S. Adams, W. T. Genetics of dark respiration and its relationship with drought hardiness in coastal Douglas-fir. *Thermochimica Acta,* 349 (2000) 69–77.
45 Stoutemyer, M. R. Burger, D. W. Calorespirometric studies of *in vitro*-grown carnation (*Dianthus caryophyllus* L. var. 'Improved White Sim') shoot tips. *Plant Cell, Tissue and Organ Culture,* 53 (1998) 189–196.
46 Zotin A. I.: Thermodynamic Aspects of Developmental Biology. Monographs in Developmental Biology, vol. 5. S. Karger, Basel 1972, p. 159.
47 Vladimirova I. G.: The energetics of regeneration processes. in: Thermodynamics of Biological Processes (I. Lamprecht, A. I. Zotin, eds.). de Gruyter Berlin 1978, 243–255.
48 Smith, B. N. Hansen, L. D. Breidenbach, R. W. Criddle, R. S. Rank, D. R. Fontana, A. J. Paige, D. Metabolic heat rate and respiratory substrate changes in aging potato slices. *Thermochimica Acta,* 349 (2000) 121–124.
49 Smith, B. N. Criddle, R. S. Hansen L. D. Plant growth, respiration and environmental stress. *Journal of Plant Biology,* 27 (2000) 89–97.
50 Criddle, R. S. Hansen, L. D. Breidenbach, R. W. Ward, M. R. Huffaker, R. C. Effects of NaCl on metabolic heat evolution rates by barley roots. *Plant Physiology,* 90 (1989) 53–58.
51 Minibayeva, F. V. Gordon, L. Kh. Alyabyev, A. Ju. Rakhmatullina, D. F. Loseva, N. L. Heat production of root cells upon the dissipation of ion gradients on plasma membrane. *Thermochimica Acta,* 309 (1998) 139–143.
52 Lygin, A. Gordon, L. Alyabyev, A. Rakhmatullina, D. Loseva, N. Nikolayev, B. Heat production and oxygen consumption of root cells with blocked fatty-acid synthesis. *Thermochimica Acta,* 316 (1998) 155–158.
53 Smith, B. N. Monaco, T. A. Jones, C. Holmes, R. A. Hansen, L. D. McArthur, E. D. Freeman D. C. Stress-induced metabolic differences between populations and subspecies of *Artemisia tridentata* (sagebrush) from a single hillside. *Thermochimica Acta,* 394 (2002) 205–210.
54 Ordentlich, A. Linzer, R. Raskin, I. Alternative respiration and heat evolution in plants. *Plant Physiology,* 97 (1991) 1545–1550.
55 Moynihan, M. R. Ordentlich, A. Raskin, I. Chilling-induced heat evolution in plants. *Plant Physiology,* 108 (1995) 995–999.

56 Queiroz, C. G. S. Mares-Guia, M. L. Magalhaes, A. C. Microcalorimetric evaluation of metabolic heat rates in coffee (*Coffea arabica* L.) roots of seedlings subjected to chilling stress. *Thermochimica Acta,* 351 (2000) 33–37.
57 Hemming, D. J. B. Monaco, T. A. Hansen, L. D. Smith, B. N. Respiration as measured by scanning calorimetry reflects the temperature dependence of different soybean cultivars. *Thermochimica Acta,* 349 (2000) 131–134.
58 Rascón-Chu, A. Carvajal-Millán, E. García-Estrada, R. Siller, J. H. Martínez, J. J. Guerrero, V. M. Gardea A. A. Chilling injury in husk tomato leaves as defined by scanning calorimetry. *Thermochimica Acta,* 349 (2000) 125–129.
59 Gardea, A. A. Carvajal-Millan, E. Orozco, J. A. Guerrero, V. M. Llamas, J. Effect of chilling on calorimetric responses of dormant vegetative apple buds. *Thermochimica Acta,* 349 (2000) 89–94.
60 Thygerson, T. Harris, J. M. Smith, B. N. Hansen, L. D. Pendleton, R. L. Booth, D. T. Metabolic response to temperature for six populations of winterfat (*Eurotia lanata*). *Thermochimica Acta,* 394 (2002) 211–217.
61 Knutson, R. M. Plants in heat. *Natural History,* 88(3) (1979) 42–47.
62 Seymour, R. Schultze-Motel, P. Thermoregulating lotus flowers. *Nature,* 383 (1996) 305.
63 Siedow, J. N. Berthold, D. A. The alternative oxidase: A cyanide-resistant respiratory pathway in higher plants. *Physiologia Plantarum,* 66 (1986) 569–573.
64 Lytle, C. M. Smith, B. N. Hopkin, M. S. Hansen, L. D. Criddle, R. S. Oxygen-dependence of metabolic heat production in the appendix tissue of the voodoo lily (*Sauromatum guttatum* Schott). *Thermochimica Acta,* 349 (2000) 135–140.
65 Skubatz, H. Meeuse, B. J. D. Energy loss in tissue slices of the inflorescence of *Sauromatum guttatum* (Schott) analysed by microcalorimetry. *Journal of Experimental Botany,* 44 (1993) 493–499.
66 Skubatz, H. Tang, W. Meeuse, B. J. D. Oscillatory heat-production in the male cones of cycads. *Journal Experimental Botany,* 44 (1993) 489–492.
67 Seymour, R. S. Bartholomew, G. A. Barnhart, M. C. Respiration and heat production by the inflorescence of *Philodendron selloum* Koch. *Planta,* 157 (1983) 336–343.
68 Seymour, R. S. Analysis of heat production in a thermogenic arum lily, *Philodendron selloum,* by three calorimetric methods. *Thermochimica Acta* 193 (1991) 91–97.
69 Seymour, R. S. Schultze-Motel, P. Lamprecht, I. Heat production by sacred lotus flowers depends on ambient temperature, not light cycle. *Journal of Experimental Botany,* 49 (1998) 1213–1217.
70 Lamprecht I.: Combustion calorimetry. in: Handbook of Thermal Analysis and Calorimetry. Vol. 4: From Macromolecules to Man. R. B. Kemp (ed.). Elsevier Science B. V., Amsterdam 1999, 175–218.
71 Lamprecht I.: Combustion calorimeters. in: Handbook of Thermal Analysis and Calorimetry. Vol. 1: Principles and Practice. M. E. Brown (ed.). Chapter 14,10. Elsevier Science B. V., Amsterdam 1999, 657–675.
72 Núñez Regueira, L. Rodríguez-Añon, J. A. Proupín-Castiñeiras, J. Vilanova-Diz, A. Montero-Santoveña, N. Determination of calorific values of forest waste biomass by static bomb calorimetry. *Thermochimica Acta,* 371 (2001) 23–31.

73 Núñez Regueira, L. Proupín-Castiñeiras, J. Rodríguez-Añón, J. A. Energy evaluation of forest residues originated from *Eucalyptus globulus* Labill in Galicia. *Bioresource Technology,* 82 (2002), 5–13.
74 Núñez Regueira, L. Rodríguez Añón, J. A. Proupín Castiñeiras, J. Vilanova Diz, A. Romero Garzia, A. Using bomb calorimetry for determination of risk indices of wildfires originating from pine residues. *Thermochimica Acta,* 394 (2002) 291–304.
75 Rodriguez Añon, J. A. Calorific values and flammability for forest wastes during the seasons of the year. *Fuel and Energy Abstracts,* 36 (1995) 346.
76 Rodríguez Añón, J. A. Fraga López, F. Proupín Castiñeiras, J. Palacios Ledo, J. Núñez Regueira, L. Calorific values and flammability for forest wastes during the seasons of the year. *Bioresource Technology,* 52 (1995) 269–274.
77 Núñez Regueira, L. Rodríguez-Añón, J. A. Proupín-Castiñeiras, J. Design of risk index maps as a tool to prevent forest fires in the humid Atlantic zone of Galicia (NW Spain). *Thermochimica Acta,* 349 (2000) 103–119.
78 Loike, J. D. Silverstein, S. C. Sturtevant, J. M. Application of differential scanning microcalorimetry to the study of cellular processes: Heat production and glucose oxidation of murine macrophages. *Proceedings of the National Academy of Sciences of the USA* 78 (1981) 5958–5962.
79 Hansen, L. D. Criddle R. S. Determination of phase changes and metabolic rates in plant tissues as a function of temperature by heat conduction DSC. *Thermochimica Acta,* 160 (1990) 173–192.
80 Criddle, R. S. Hansen, L. D. Metabolic rate of barley root as a continuous function of temperature. *Journal of Plant Physiology,* 138 (1991) 376–382.
81 Wesolowski, M. Konieczynski, P. Thermoanalytical, chemical and principal component analysis of plant drugs. *International Journal of Pharmaceutics,* 262 (2003) 29–37.
82 Wesolowski, M. Konieczynski, P. Thermal decomposition and elemental composition of medicinal plant materials–leaves and flowers: Principal component analysis of the results. *Thermochimica Acta,* 397 (2003) 171–180.
83 Szabó, P. Várhegyi, G. Till, F. Faix, O. Thermogravimetric/mass spectrometric characterization of two energy crops, *Arundo donax* and *Miscanthus sinensis*. *Journal of Analytical and Applied Pyrolysis,* 36 (1996) 179–190.
84 Maldague, X. P. V. Nondestructive Evaluation of Materials by Infrared Thermography. Springer, London Berlin 1993, p. 207
85 Chaerle, L. Van Der Straeten, D. Seeing is believing: imaging techniques to monitor plant health. *Biochimica et Biophysica Acta,* (BBA) - Gene Structure and Expression, 1519 (2001) 153–166.
86 Chaerle, L. Van Caeneghem, W. Messens, E. Lambert, H. Van Montagu, M. Van Der Straeten D. Presymptomatic visualization of plant-virus interactions by thermography. *Nature Biotechnology,* 17 (1999) 813-816.
87 Chaerle, L. De Boever, F. Van Der Straeten D. Infrared detection of early biotic and wound stress in plants. *Thermology International,* 12 (2002) 100–106.
88 Van Der Straeten, D. Chaerle, L. Sharkov, G. Lambers, H. Van Montagu, M. Salicylic acid enhances the activity of the alternative pathway of respiration in tobacco leaves and induces thermogenicity. *Planta,* 196 (1995) 412–419.

89 Malamy, J. Hennig, J. Klessig D. F. Temperature dependent induction of salicylic acid and its conjugates during the resistance response to tobacco mosaic virus infection. *Plant Cell* 4 (1992) 359–366.
90 Pearce, R. S. Plant freezing and damage. *Annals of Botany,* 87 (2001) 417–424.
91 Wisniewski, M. Lindow, S. E. Ashworth E. N. Observations of ice nucleation and propagation in plants using infrared video thermography. *Plant Physiology,* 113 (1997) 327–334.
92 Workmaster, B. A. A. Palta, J. P. Wisniewski, M. Ice nucleation and propagation in cranberry uprights and fruit using infrared video thermography. *Journal of the American Society for Horticultural Science,* 124 (1999) 619–625.
93 Hamed, F. Fuller, M. P. Telli, G. The pattern of freezing of grapevine shoots during early bud growth. *Cryo-Letters,* 21 (2000) 255–260.
94 Skubatz, H. Nelson, T. A. Dong, A. M. Meeuse, B. J. D. Bendich, A. J. Infrared thermography of *Arum* lily inflorescences. *Planta,* 182 (1990) 432–436.
95 Skubatz, H. Nelson, T. A. Meeuse, B. J. D. Bendich, A. J. Heat production in the voodoo lily (*Sauromatum guttatum*) as monitored by infrared thermography. *Plant Physiology,* 95 (1991) 1084–1088.
96 Bermadinger-Stabentheiner, E. Stabentheiner, A. Dynamics of thermogenesis and structure of epidermal tissues in inflorescences of *Arum maculatum. New Phytology,* 131 (1995) 41–50.
97 Lamprecht, I. Schmolz, E. Blanco, L. and Romero, C. M. Energy metabolism of the thermogenic tropical water lily, *Victoria cruziana. Thermochimica Acta,* 394 (2002) 191–204.
98 Criddle, R. S. Breidenbach, R. W. Lewis, E. A. Eatough, D. J. Hansen, L. D. Effects of temperature and oxygen depletion on metabolic rates of tomato and carrot cell cultures and cuttings measured calorimetrically. *Plant, Cell and Environment,* 11 (1988) 695–701.
99 Meeuse, B. J. D. Raskin, I. Sexual reproduction in the arum lily family, with emphasis on thermogenicity. *Sexual Plant Reproduction,* 1 (1988) 3–15.
100 Knutson R. M. Heat production and temperature regulation in Eastern Skunk Cabbage. *Science,* 186 (1974) 746–747.
101 Nagy, K. A. Odell, D. K. Seymour, R. S. Temperature regulation by the inflorescence of Philodendron. *Science,* 178 (1972) 1195–1197.
102 Meeuse, B. J. D. Buggeln, R. G. Time, space, light and darkness in the metabolic flare-up of the Sauromatum appendix. *Acta Botanica Neerlandica,* 18 (1969) 159–172.

Chapter 9

Thermobiochemical studies of animal cell systems in vitro

Evidence of their nature from bioreactor experiments

R. B. Kemp*

University of Wales, Aberystwyth, Institute of Biological Sciences,Edward Llwyd Building, Aberystwyth, SY23 3DA, Wales, UK

Introduction

Over the last 20 years, there has been a steadily increasing use of animal cells, mostly from insects and mammals, to produce medically important protein-based macromolecules such as recombinant proteins from genetically engineered cells, monoclonal antibodies from hybridoma cells and vaccines from established cell lines. The cells are grown under precisely controlled conditions that are maintained by monitoring such parameters as temperature, pH and dissolved oxygen (DO) by on-line probes [1–3]. While control of the physical conditions is relatively easily achieved in this way, maintaining a suitable environment for the on-line measurement of metabolic variables is a problem [4]. This means that the heterologous products cannot be produced with the highest efficiency because cellular requirements for metabolites cannot be matched precisely with their availability. In off-line systems, for many years cells have been disrupted and the intensity of the catabolic flux measured from the specific activities of key allosteric enzymes, such as phosphofructophosphate and citrate synthase in glucose metabolism [5].

For some years the most common variables used to assess the metabolism of cells in a bioreactor were the rate of consumption of a substrate (e.g. glucose) and/or the oxygen uptake rate (OUR) and/or the production rate of a catabolite that leaves the cell (e.g. carbon dioxide or lactate) using biosensors (see, for instance, [6, 7]). Conventional polarographic OUR assessments [8, 9] are not sensitive enough for the dilute animal cell concentrations (~10^6 cells per cm^3) typical of a bioreactor [10]. With the exceptions of OUR for high-density cell suspensions and the carbon dioxide (CO_2) assay devised by Bonarius *et. al.* [11], in terms of their application most of the relevant measurement devices have at least

* rbk@aber.ac.uk

D. Lörinczy (ed.), The Nature of Biological Systems as Revealed by Thermal Methods, 215–249.

three problems [12]. First, because they use specific enzymes, they cannot be sterilised by chemical or thermal means. Safe and continuous sampling devices, including flow injection (FIA) techniques [13], are needed as well as the actual biosensor(s) [14]. These significantly increase the cost and the complexity of the total measuring system. Secondly, many of the different types of biosensor only measure the concentration of the metabolite, making it necessary to obtain at least two values with a time difference to calculate the metabolic rate. This interval must be relatively long to achieve reasonably accurate results. Then, the assay system cannot measure the metabolic rate in real time. Thirdly, enzyme-based biosensors are consumable and therefore have a high recurrent cost [14].

In recent years, there has been a marked increase in the use of on-line, non-invasive techniques with: (i) optical sensors such as midrange IR, UV, Raman, and fluorescence spectroscopy [7, 15, 16] to estimate the concentrations of substrates and products; (ii) frequency dielectric spectroscopy to measure the amount of biomass from on-line changes in capacitance [17–19]; and (iii) direct calorimetry to indicate increases in heat as animal cells grow with time [19, 20]. For the latter, two different calorimetric principles have been adopted for the heat measurements. For von Stockar and his colleagues (see [20]), the bioreactor itself is an in situ heat accumulation calorimeter in which the temperature of oil in the jacket surrounding the bioreactor reflects the heat produced by the cells [21]. At least in its early form, the modified Mettler-Toledo instrument could only detect relatively high cell concentrations ($>10^7$ cells cm^{-3}) whereas the starting inoculum in a tank, batch bioreactor is typically 2×10^5 cm^{-3} [22]. This meant that it was not until the third generation of the design that the Mettler-Toledo calorimeter was sufficiently sensitive to monitor the heat production of typically dilute cell suspensions [16]. In a different approach, Kemp and his colleagues [12, 19, 23] used an on-line, ex-situ heat conduction calorimeter that measures directly the heat flow *rate* ($\Phi = \mathrm{d}Q/\mathrm{d}t$, Watts, W) [21]. This means that, as well as thermodynamic information, the measurement gives the kinetics of the process. This flow is regarded as the rate of thermal (th) advancement, $\mathrm{d}_{th}\xi/\mathrm{d}t$ in the cellular energy transformations [24]. The advancement, or the extent of reaction as it is sometimes known, is an important concept in energy transformation because explicitly it is in terms of the stoichiometric coefficients, ν_i, of the *i*-th species in the reaction. For a reaction, the advancement, $\mathrm{d}\xi_B$, is,

$$\mathrm{d}\xi_B = \mathrm{d}n_i \, \nu_i^{-1} \tag{1}$$

where $\mathrm{d}n_i$ is the change in the amount of a reactant or a product, *i* and the subscript B denotes $|\nu_B| = 1$. Thus, $\mathrm{d}\xi_B$ is side-independent.

Biologists talk of 'metabolic activity' but, to the physical chemist, this can be substituted by a formal expression for the rate of advancement of the aggregated biochemical reactions in living cells, $\mathrm{d}\xi_B/\mathrm{d}t$, or their flux when the rate is expressed as being specific to mass, $(1/x)(d\xi_B/dt)$, where x represents the amount of biomass. The thermal advancement of energy transformation, $d_{th}\xi$, is related to $d\xi_B$ by the expression,

$$d_{th}\xi = v_i \Delta H_{Bi} d\xi_B \tag{2}$$

where ΔH_{Bi} is the molar enthalpy change of the reaction in terms of species I [25]. The change in thermal advancement, $d_{th}\xi$, is exactly equivalent to the change in heat, dQ. The relationship seen in Eq. (2) has been exploited in various ways, not least in the enthalpy balance method [26]. For it, the following information is required: (i) the experimental reaction flux, J_B (mol s^{-1} m^{-3}),

$$J_B = d\xi_B / x dt \tag{3}$$

This must be found for each species in a chemical reaction; and (ii) the calculated molar reaction enthalpy change, ΔH_B (J mol^{-1}), which is determned using the molar enthalpy of formation for each species I,

$$\Delta H_B = \sum v_i H_i \tag{4}$$

The combination of Eq. (3) and Eq. (4), gives the reaction enthalpy flux, = J_H, (W m^{-3}),

$$J_H = J_B \Delta H_B \tag{5}$$

where J_B is the reaction molar flux (mol s^{-1} per amount of biomass). J_H is directly comparable to the experimentally measured (scalar) heat flux, J_Q. This theoretical treatment demonstrates the fact that studies to obtain heat flux with heat conduction calorimeters gives thermodynamic information as well as kinetic data. Despite the well-established background detailed above that is summarised in Eq. (2), there has been little recognition among the fraternity of biochemical engineers that heat flow rate is potentially the perfect probe for cell cultures because it is a function of the metabolic rate. There are two approaches that can be made to introduce such a device to bioreactors. One is to make the vessel into a calorimeter and the other is to circulate the cells in suspension to a microcalorimeter that has a flow-through vessel. Both solutions will be illustrated in this review but more attention will be paid to the latter because it has been the subject of more research for bioengineered animal cells.

The bioreactor as a calorimeter

One of the problems in the design of industrial bioreactors and fermenters is to dissipate the heat produced by the cells. If, instead of simply removing/utilising it after an exergy analysis, the rate of heat flow were to be measured, then the rate of metabolism would be assessed continuously on-line. To this end, some years ago von Stockar's group [27] adapted a 2-L Mettler RC-1 heat accumulation calorimeter (see [21] for the classification of calorimeter types) originally designed for the safety testing of the chemical reactions encountered in industrial plants to measure the heat produced by cultures of animal and microbial cells. In this modified form, it has been employed mainly in microbial energetics (see review by Duboc *et al.*, [28]) but its potential for use with animal cell culture was explored briefly several years ago using a hybridoma cell line [29]. Two techniques for delivering air to the cell culture were compared, namely 'under surface aeration' and 'sparging by air'. When allied to data for the consumption of glucose and the production of lactate, the results indicated that the culture with under surface aeration had a yield of heat to glucose, $Y_{Q/S}$, of 41.49 kJ c-mol^{-1}, compared to $Y_{Q/S}$ for the sparged culture which was much higher at 66.6 kJ c-mol^{-1}. This meant that the latter culture was more aerobic, a finding underlined by the fact that it contained less lactate and gave a lower $Y_{lac/glc}$ yield of 0.96 c-mol/c-mol) than did the one gassed by under surface aeration ($Y_{lac/glc} = 1.24$ c-mol/c-mol). At that time when these results were obtained (1989), the detection limit for the reaction calorimeter was only ~2×10^6 cells per cm^3, generally an order of magnitude higher than the starting inoculum for animal cultures that in a standard batch culture generally rise to a maximum concentration of approximately 1×10^6 cm^{-3}. In the last 5 years, an ongoing series of improvements to the Mettler-Toledo RC1 reaction calorimeter [20, 28] resulted in the detection limit sufficiently low to follow the course of batch cultures of many different animal cell types. Recently, Garcia-Payo et al. [30] summarised the changes to the instrument in terms of limiting non-biological heat flows and altering the geometry of the glass reactors that has resulted in the improved resolution of it from ±50 mW l^{-1} five years ago [20] to 4 to 12 mW l^{-1} now. This is similar to that of the heat conduction flow microcalorimeter attached to a 3-L bioreactor [12]. While the improved Mettler-Toledo calorimeter allows model experiments on the bench scale with animal cells at the typically high dilutions found initially in the batch cultures of industrial applications, only the flow microcalorimeter can act as an on-line, ex-situ metabolic probe for industrial scale bioreactors.

Combination of a bioreactor with a flow microcalorimeter

By 1997, Kemp *et al* [19] had developed a solution for measuring the heat flow rate of animal cells growing under the controlled conditions of the bioreactor that is independent of the size of the culture vessel. It allows the measurement of the heat produced in industrial-scale plant by circulating the cell suspension from the bioreactor to the flow calorimeter and then returning it. As a model for a large-scale facility, a 3-L Applikon bioreactor containing a stirred suspension of Chinese Hamster Ovary cells (CHO320) genetically engineered to produce interferon-γ (IFN-γ) was connected by thermostated PEEK transmission tubing (see Fig. 1 [12]) to an ex situ, on-line Thermometric Thermal Activity Monitor (TAM) heat conduction, twin differential, flow microcalorimeter [31] (Thermo-

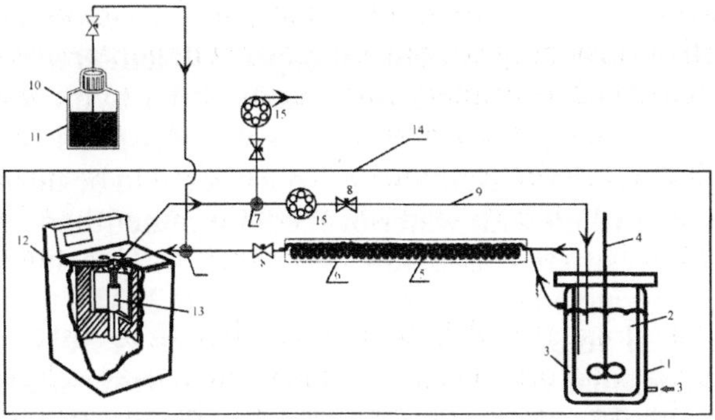

Fig. 1 A schematic diagram for the measurement of the heat flow rate of bioreactor-cultured animal cells using a flow microcalorimeter. 1 – bioreactor; 2 – cultured cells; 3 – jacket water for temperature control in the bioreactor; 4 – agitator; 5 – the outlet tubing of the jacket water is used for warming the cell suspension in the tubing leading to the microcalorimeter; 6 – a non-conductive sponge-plastic pipe is used to reduce heat dissipation along the outlet jacket water attached to the PEEK tubing transmitting the cell suspension to the calorimeter; 7 – PEEK T-piece; 8 – PEEK two-way valve; 9 – PEEK tubing (1 mm I.D.); 10 – glass bottle holding sterilized medium for washing the flow vessel of the microcalorimeter through short-time interruption of the heat flow measurement; 11 – sterile medium for cleaning the flow vessel free of possible accumulated animal cells, that would result in an overestimation of the heat flow rate for the cells in the bioreactor; 12 – a representation of the 4-channel Thermometric calorimeter (TAM); 13 – the gold flow vessel assembly; 14 – the dotted boundary, enclosing essentially the bioreactor and the flow vessel of the microcalorimeter, is an open thermodynamic system for enthalpy balance studies; 15 – the peristaltic pump (Reproduced from [12] with permission)

metric AB, Järfälla, Sweden) with a gold flow-through vessel that had been modified for downward flow to try and minimise blockage by cellular debris. The early experiments were performed with the commercially available vessel that has a nominal capacity of 0.6 cm^3 but a customised one of larger volume became available for later work. It is not always understood that the thermal volume of a calorimetric vessel may differ from the physical one for various reasons. Wadsö has frequently emphasised this point (for instance, in [32]) and advocated that, in addition to the daily electrical test, chemical calibration should be undertaken for a newly delivered vessel under the exact experimental conditions and repeated whenever there was an alteration in these conditions. He found that the slow imidazole-catalysed hydrolysis of triacetin is a suitable reaction [32]. Using this reaction, Guan *et al.* [33] found that the thermal volume of the standard gold flow-through vessel at pumping rate of 35 $cm^3\ h^{-1}$ was 0.77 cm^3, nearly 30% greater than the nominal volume. More recently, O'Neill *et al.* [34] showed that the kinetics of the triacetin hydrolysis reaction is too slow for the calibration of flow vessels with their relatively short residence time and suggested the use of the base catalysed hydrolysis of methyl paraben with faster kinetics as an alternative. In a subsequent paper, O'Neill *et al.* [35] showed that the substance that filled the reference vessel of the twin differential micro-calorimeter affected the thermal volume of the vessel, often in combination with the pumping rate. Indeed, these authors maintain it could be the only factor if the reference vessel was filled with water or medium rather than air, and the range of flow rates was restricted to below 50 $cm^3 h^{-1}$. This implies that, at the 35 $cm^3 h^{-1}$ pumping rate employed by Guan et al. [12, 33], the thermal volume equalled the physical volume. Occasionally, the flow vessel in their hands suffered from blockages that were probably due to the narrow internal diameter (ID = 1 mm) compounded by the comparatively slow pumping rate.

In more recent studies in Aberystwyth, a specially fabricated flow module kindly provided by Thermometric AB to Kemp's design (see Fig. 2) has been used as the standard [36]. Except for the peristaltic pump itself that has Viton tubing, the flow line was 1.5 mm (id) stainless steel throughout and was optimised for a pumping rate of 100 $cm^3 h^{-1}$, a speed that was achieved without undue noise in the signal. At this rate, chemical calibration with the tiracetin reaction gave an effective (thermal) volume of 1.05 cm^3, very close to the nominal volume of 1.00 cm^3 [36] Most recently, the vessel has been re-calibrated with the methyl paraben hydrolysis reaction and found to be within the specified vol-

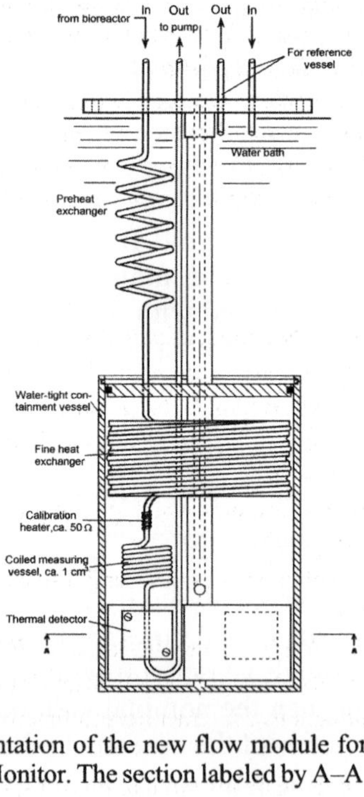

Fig. 2(A) A pictorial representation of the new flow module for the Thermometric 2277 Thermal Activity Monitor. The section labeled by A–A is illustrated in Fig. 2(B)

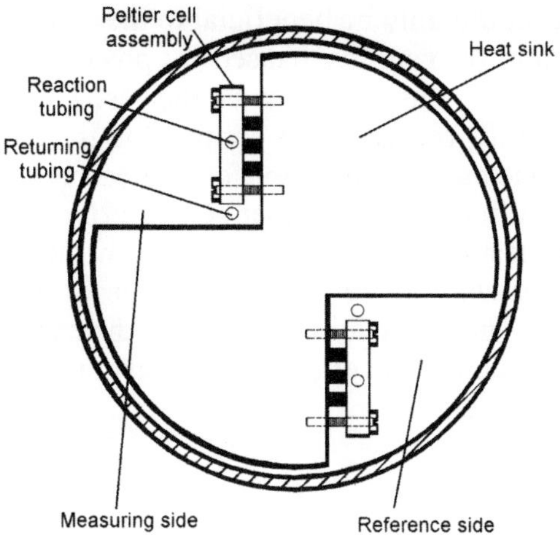

Fig. 2(B) A sectional view of the layout of the thermal detector in the new flow module for Thermometric 2277 TAM. The vertical position of this section is labeled in Fig. 2(A) as A–A. (Reproduced from [36] with permission)

ume [34]. Although Chisti [37] drew attention to the dangers of hydrodynamic damage to cells in such situations particularly with respect apoptosis, but also purely mechanical damage, exhaustive tests of prolonged exposure of the cells to pumping over several hours through the tubing failed to reveal any such deleterious changes [33, 36].

When the heat flow rate data are obtained simultaneously with the measurements of the concentrations of materials (substrates and products) in the bioreactor, the validity of the results can be tested by calculations described as the enthalpy balance method (see Section 5; also [26, 38]). Bearing in mind that the system is open in the thermodynamic sense [39], it is important for the validity of these calculations to (i) define the boundary of the system and (ii) ensure that the conditions are the same throughout that system. For the first requirement (see Fig. 1), it was decided that the infinitely thin boundary was in the walls of the glass bioreactor and the tubing. The nature of the latter depends on the flow system. For the initial studies, it was (in sequence) the PEEK transmission tubing, the Viton (or Aliprene) of the pump tubing and the gold in the wall of the flow vessel. For the customised vessel, the boundary was in the stainless steel and the Viton tubing. To help maintain similar conditions throughout the system, the transmission tubing was thermostated at the same temperature as the bioreactor and the calorimeter, while the materials for the transmission lines and the pump tubing were selected mainly for their low gaseous diffusivity. In the initial studies, Guan et al. [33] found experimentally that pumping the cell suspension at a rate of 35 cm^3 h^{-1} was sufficient to ensure that the cells did not consume so much of the dissolved oxygen in the medium to drive the concentration of it below normoxic levels before the cells were returned to the bioreactor. Even so, a higher rate could only be beneficial to the cell metabolism, providing that there was no effect on viability caused by physical trauma [37]. In trials with the customised vessel, this proved to be the case [36].

Kemp and Guan [40] recorded the continuous trace of the heat flow rate and compared it with the numbers of viable cells measured at discrete time intervals during the batch culture of recombinant CHO 320 cells. The important conclusion from these assessments was that the heat flow rate as the indicator of the metabolic rate declined while there were still increases in the cell number concentration (see Fig. 3). The implications were that the heat flow rate could be an even more sensitive reflection of cellular metabolism if it were made specific to biomass, i.e. scalar heat flux.

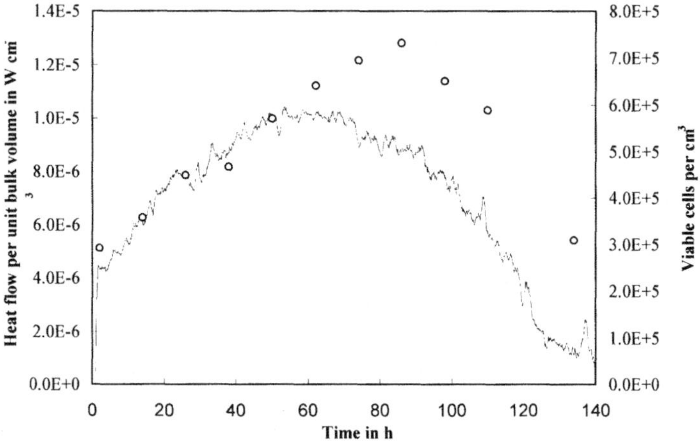

Fig. 3 The heat flow rate of growing cells measured on-line by the microcalorimeter and scaled to the unit bulk volume of RPMI 1640-based culture medium buffered with 20 mM HEPES and 4 mM bicarbonate (—). Estimates were made for the number of viable cells per cm^3 bulk volume (o) at discrete time intervals (Reproduced from [40] with permission)

Counting cell numbers by haemocytometry is a labour-intensive, time-consuming and a relatively inaccurate (± 10%) off-line method to assess biomass. Over the years, the Institute of Biological Sciences in Aberystwyth has been interested in the biological applications of dielectrics [41] and, after a series of trials, it was discovered that the animal cell biomass could be measured satisfactorily with the in situ, on-line probe of a radio frequency (0.5 MHz) dielectric spectrometer. Aber Instruments Ltd. (Aberystwyth, Wales, UK) then marketed it as the Viable Cell Monitor. The capacitance (*C*, in Farads, F) measurements made with this instrument have been shown both theoretically [41] and by the essential calibration of particle size with a flow cytometer [12] to record the volume fraction of the viable cells, providing there is no change in conductivity and also no alteration to the cell volume during the culture period. Then, monitoring the change in capacitance indicates the variation in biomass.

In the experiments, the signal from the spectrometer probe in the bioreactor was digitised and fed to the Applikon BioXpert software used as a log for the culture monitoring data [12]. The capacitance signal was blanked for the value obtained with the growth medium, both this signal (ΔC) and the one from the calorimeter were smoothed by the moving average technique in the software and heat flow rate was divided by the digitised capacitance signal (biomass) to give the flux. The resulting trace, $J_{\Phi/C}$, was comparable with the control in which the heat flow rate at discrete times during the culture was divided by the cell number concentration (N) obtained off-line with a Coulter counter, $J_{\Phi/N}$, (Fig. 4) and it had the advantage of being continuous, on-line [12, 23]. The on-line heat flux signal was then compared after the experiment with the discrete values for the changes in the concentrations of the major substrates, glucose and glutamine, measured by off-line methods (Fig. 5). It was highly significant that the heat flux monitor detected a decline in the metabolism of the CHO 320 cells before there was an apparent alteration in the consumption of the two substrates. This result confirmed the fact that, despite evidence of reduced metabolic flux, the cells continued to grow (see Fig. 3) and produce IFN-γ.

It should be recognised that others have also adopted dielectric spectrometry to measure animal cell biomass on-line but so far have not combined it with heat flux measurements. For instance, Ducommun et al. [42] indirectly determined the concentration of free suspended and immobilised CHO SSF cells by this on-line method and also applied it successfully to two industrial high density culture pro-

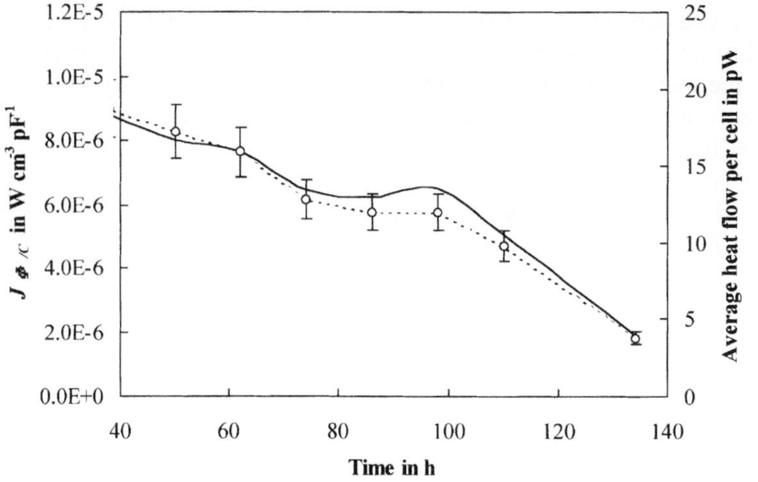

Fig. 4 On-line heat flux measurements adjusted to per cm^3 bulk volume (—) and heat flow per viable CHO 320 cell (o) over 140 h of a batch culture. In the expression of $J_{\Phi/C}$, the values are given in terms of 1 cm^3 bulk medium volume (Reproduced from [12] with permission)

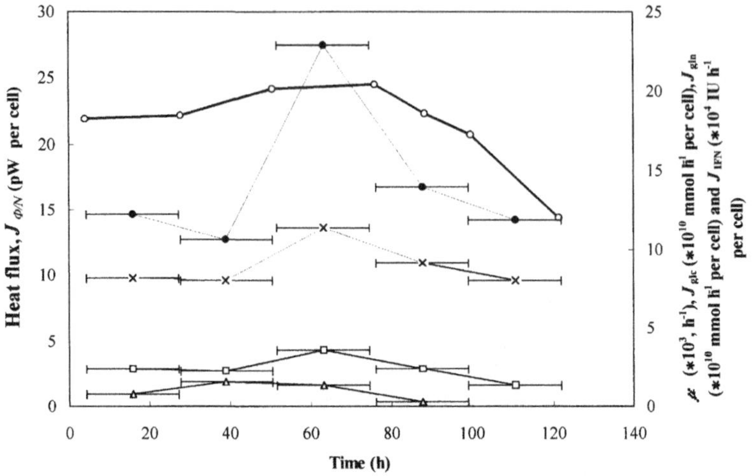

Fig. 5 Comparison of the heat flux ($J_{Ö/N}$) with the fluxes of glucose (J_{Glc}), glutamine (J_{Gln}) and IFN-γ (J_{IFN}), as well as the specific growth rate (μ) during the batch cultivation of CHO 320 cells in suspension. Heat flux (○), glucose flux (□), glutamine flux (Δ), IFN-γ flux (×) and specific growth rate (•). The bars indicate the period over which the discrete off-line measurements were made to give the individual average values for fluxes (Reproduced from [12] with permission)

cesses, using it to determine on-line the concentration of CHO cells immobilised on macroporous microcarriers in a stirred bioreactor and in a packed-bed of immobilised hybridoma cells on disk carriers [43]. For the latter, dielectric spectrometry was used as a tool to characterise the packed-bed process, showing for instance the maximum cell concentration that could be reached was 2.0×10^{11} cells per kg of disk carrier. From the future perspective of combining dielectric spectrometry with calorimetry, the more exciting finding from von Stockar's laboratory [44] was that the instrument could be modified to scan CHO perfusion cultures every 20 min over the excitation frequency range of 0.2 MHz to 10.0 MHZ to give both maximum and zero cell viability and signify the end of lactate consumption. The combination of scanning dielectric spectrometry (SDS) and calorimetry could give profound insight into the relation between the metabolic rate of the cells and their anabolic processes, particularly in the production of heterologous proteins.

Even with the putative power of SDS measurements combined with heat flux, it is often important to know the changes of crucial metabolites, such as glucose, glutamine and lactate as vital components in the overall process. Guan *et al.* [12] made discrete measurements of these materials but, of course, the profile of the concentration changes is of limited value compared with the fluxes. In order to reveal changes in the rate of change in material, normally the values for the concentrations of a given substance must be differentiated at two time points, t_1 and t_2 for each period of the cell culture. Although this procedure will give the rate, it is only the average, relative to that time period, as opposed to the instantaneous rate available from the calorimeter. Data for the material concentrations can be converted to the instantaneous rate by using the Tikhonov-Phillips integrative method of smoothing [45, 46].

Guan *et al.* [12] took the estimations for the changes in the catabolic substrates and products during the growth of CHO 320 cells, obtained the average relative rates and made them specific to cell number concentration. The calculations (see Fig. 5 and [23]) confirmed the impression that the specific growth rate (μ) and the fluxes of the major substrates and products, including IFN-γ, began to decrease at the relatively early time in the culture indicated by the heat flux biosensor. The exact mathematical relationship of the heat flux to the material fluxes was established by plotting the former fluxes against the collection of the latter. It was then clearly seen from Fig. 6 that heat flux is a function of the material fluxes in a monotonic relationship. From this, Kemp and Guan [25] concluded that heat flux can be utilised in cell culture as an on-line, real time probe which: (i) has a rigorous thermodynamic foundation to it; (ii) gives the instantaneous metabolic flux; (iii) is non-invasive as well as being non-destructive; and (iv) is robust with no consumable components. Therefore, it can be used to monitor cultures and improve the design of media as well as being made an effective control variable for fed-batch cultures.

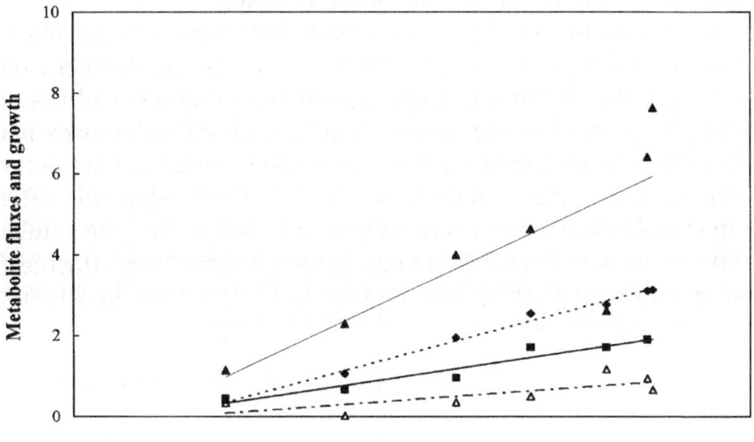

Fig. 6 Heat flux is plotted as a function of the specific growth rate (■) showing the monotonically decreasing relationship. This dependence extends to the fluxes for $10^4 \times$ IFN-γ production flux, IU s^{-1} per cell, (▲) and the major substrates, $10^7 \times$ glucose consumption flux, mol s^{-1} per cell, (◆) and $10^7 \times$ glutamine consumption flux, mol s^{-1} per cell (△) (Reproduced from [23] with permission)

Heat Flux and the growth reaction

Although there is, of course, a myriad of reactions in the eukaryotic cell, it transpires that the net formation of biomass (cell growth) is from a relatively few substrates and can be described by a comparatively simple chemical reaction accompanied by an enthalpy change [23]. Battley [47] outlined the general procedures for writing microbial growth-process equations and the these simple rules in principle can be extended to the information necessary for writing the growth reaction for the cells of higher organisms:

i) The elemental composition and the quantity of the substrates used as a source of carbon and energy for growth.
ii) The elemental composition and quantity of any organic products of the growth process. This includes the cells and, especially in the case of those producing heterologous proteins, any secondary product. The elemental composition of the cells (biomass) should consist of the ash-free proportions of C, H, O and N and is usually expressed as in terms of C-mol. The formula has the form $CH_{nH}O_{no}ON_{nN,}$ where nH, nO and nN represent the relative numbers of hydrogen, oxygen and nitrogen atoms, respectively, when the number of carbon atoms is taken as one.
iii) The elemental composition of the nitrogen sources. In animal cells, these are the amino acids.

The validity of the constructed growth reaction can be tested by the enthalpy balance method (see Eq. (3) to Eq. (5); and [19, 25, 48]), in which the calculated enthalpy change for the reaction is compared with the enthalpy change measured calorimetrically as the heat flux. The components of the growth reaction for animal cells can be simplified sufficiently to make the exercise practicable, providing that the cells are grown in a defined medium without serum [40], for two reasons. First, it has been proved experimentally [49] that the micronutrients can be ignored for the enthalpy balance. Secondly, no account need to be taken of the amino acids, except glutamine, because of evidence from microbial studies at least that the enthalpy of anabolism is negligible [47]. After taking these factors into account, Guan and Kemp [49] decided that the following stoichiometric equation was a justifiable basis for constructing the growth reaction of cells producing heterologous proteins,

$$\begin{aligned} &\nu_{\mathrm{Glc}}\mathrm{C_6H_{12}O(Glc)} + \nu_{\mathrm{G\,ln}}\mathrm{C_5H_{10}N_2O_3(G\,ln)}\nu_0\mathrm{O_2} \rightarrow \mathrm{CH_\alpha O_\beta N_\gamma(cell)} \\ &+\nu_{\mathrm{L}}\mathrm{C_3H_6O_3(Lac)} + \nu_{\mathrm{C}}\mathrm{CO_2} + \nu_{\mathrm{N}}\mathrm{NH_3} + \nu_{\mathrm{H}}\mathrm{H_2O} + \left(\int_1^2 J_{\mathrm{H,r}}\,dt\right) \end{aligned} \quad (6)$$

Eq. (6) is only valid with the following conditions: (i) molecular formulae are used to aid the subsequent material and energy balances; (ii) any target protein (e.g. IFN-g) must be combined with the biomass which is expressed by a C-molar formula as customarily utilised for microbial biomass (see [28, 39]); and (c) the stoichiometric coefficient of the cell mass is set at unity. Then, the enthalpy change of the growth reaction is based on the unit number of the C-molar biomass [49]. It should be remembered at this point that there is experimental evidence (see Fig. 6) of a one-to-one monotonic relationship between the metabolic flux (see Eq. (3)) and the stoichiometry of the growth equation (Eq. (6)). This can be expressed by:

$$(1/x)(\mathrm{d}\xi/\mathrm{d}t) \leftrightarrow v \tag{7}$$

According to Kemp and Guan [25], Eq. (7) means that: (i) the metabolic activity of the cell determines the stoichiometry of the growth equation; and (ii) a particular set of stoichiometric coefficients for this equation corresponds to a value of the metabolic activity for the same amount of viable cell mass. Thus, for Eq. (7), the growth reaction is characterised by its set of stoichiometric coefficients and it denotes that the cellular metabolic requirement can be determined by measuring the metabolic flux, specifically represented by the heat flux variable [25]. It was shown in Fig. 4 that the metabolic flux varies with the time in culture. Therefore, the stoichiometry expressed by Eq. (7) must change to reflect it. This means that the cellular nutritional demand must be reflected in the growth equation when the heat flux variable is measured for the culture on-line and in real time. Beforehand, however, it is necessary to incorporate the experimental data into the growth equation and then to validate it by constructing an enthalpy balance.

The growth reaction and the enthalpy balance method

The Mayer enthalpy balance method is the means to determine that the equation describing a given chemical reaction is correct. The complexity of the equation does not limit the method but the calculations must obey Hess's Law that the net heat evolved or absorbed in any chemical change depends only on the initial and final states, and is independent of the stages to reach the final state. Thus, in the present case study of CHO320 cells in batch culture, it is invaluable for validating the growth reaction, including the assumptions underlying it in terms of, for example, the effect of excluding micronutrients (see the Section 4). As a reminder of the current application, in terms of Eq. (5) the enthalpy change of the growth reaction can be calculated as the sum of the individual reaction enthalpy fluxes, J_H, for the constituents in Eq. (6) from their measured reaction fluxes and the molar reaction enthalpies using the material concentrations at any one time in the culture.

Many of the molar reaction enthalpies are well known but they must still be corrected for the side reactions current in the culture environment, for instance the enthalpies of neutralisation of the products. For calculating reaction enthalpies not available in the literature, the most comprehensive reference work is Wilhoit [50]. It may be necessary to calculate reaction enthalpies from balanced reaction stoichiometries and known enthalpies of formation, $\Delta_f H^o$, in dilute aqueous solution. There are, however, other useful sources of information in, for instance, Battley [47], Gnaiger and Kemp [51] and von Stockar *et al.* [39].

Enthalpy recovery is the term used for testing the validity of the enthalpy balance [24]. It is the ratio of the heat flux to the enthalpy flux, J_Q/J_H, is termed the enthalpy recovery (see review in Kemp *et al.* [19]). According to Gnaiger [24], if the ratio is unity, then the description of the (growth) reaction is correct. If the value is greater than one, then there is at least one unidentified reaction in the chemical system. If it is less than unity, there must be at least one unnoticed endothermic reaction.

The direct application of the enthalpy balance method to the growth reaction strictly requires knowledge of the enthalpy change for the combustion of the cell mass. Unfortunately, it has proved difficult to obtain a clean sample of animal cell biomass free of the exogenous protein and the other environmental macromolecules in the medium. This is because the plasma membrane is relatively delicate and easily lysed during washing procedures. In addition, the large amounts required for combustion, three dried samples of 0.5 g each, is a formidable proposition in terms of the numbers of animal cells, unlike for microbes (see procedures outlined by Gurakan *et al.* [52]). Because it is now possible to grow cells in completely protein-free medium, this is an experiment waiting to be done with a bomb calorimeter [52].

If the required instrumentation is not available for direct experimentation, the enthalpy of combustion, $\Delta_c H_i$, for compound *i*, in the present case biomass, can be calculated from the known Thornton regularity for the heat evolved per equivalent of oxygen, Q_O, [54] using the relationship [55],

$$\Delta_c H_i = Q_O \gamma_i \tag{8}$$

where γ_i is the degree of reductance for any compound *i* of the generalised C-molar formula $CH_{ei1}O_{ei2}N_{ei3}$ defined by,

$$\gamma_i = 4 + e_{i1} - 2e_{i2} - 3e_{i3} \tag{9}$$

This means that γ_i is four times the number of moles of oxygen required to oxidise one C-mole of compound *i* to CO_2, H_2O and N_2. The value for the heat released per mole O_2 during combustion is in a relatively small range [54], with the best estimate at -115 kJ deg^{-1} of reductance [54, 55]. This approach requires knowledge of the degree of reductance calculated from the elemental analysis of the biomass. In the case of animal cells, sample preparation is a problem for the reasons stated above but the number of cells required is smaller because the required size of the lyophilised sample is only ~3 mg. Empirical formulae for the elemental composition of biomass have started to appear in the literature. Bushell *et al.* [56] has given the formula for the murine hybridoma, PQXB1/2, as $CH_{1.7}O_{0.25}N_{0.25}$ which, from Eq. (9), means the degree of reductance of the biomass (γ_b) is 4.45. More recent determinations of the biomass formula are for the SP2/0-Ag 14 myeloma cell line at $CH_{1.78}O_{0.43}N_{0.25}$ [57] to give $\gamma_b = 4.17$ and for the *Zac3* hybridoma at $CH_{1.64}O_{0.36}N_{0.24}$ [58], $\gamma_b = 4.2$. Erickson [54] stated that an acceptable value for the generalised degree of reductance when there is no known elemental formula is 4.291. When the generalised value for Thornton's regularity and the calculated degrees of reductance from the above experimental elemental analyses were applied to Eq. (8), the resulting enthalpy of combustion for animal cell biomass is between –480 and –512 kJ mol^{-1} O_2, with the likelihood of it being towards the lower side. Because the main substrates (*s*), glucose ($\gamma_s = 4$) and glucosamine ($\gamma_s = 3.6$), are less reduced than the biomass, CO_2 is a net product of anabolism in the process as opposed to being a net reactant, e.g. in anaplerotic reactions.

Glucose in the culture medium is usually regarded as the source of energy and glutamine primarily for biosynthetic purposes. However, most culture media are not optimised for biomass production because the quantities of the amino acids are not present in the correct stoichiometric relationships [12, 59, 60]. The cells therefore must acquire some of the essential biosynthetic precursors from glucose via the glycolytic pathway. As an additional complication, a proportion of the glutamine is oxidised to provide energy by a pathway known as glutaminolysis [11, 59–62], with the rest of it directly (purine and pyrimidine synthesis) and indirectly (transamination to give other amino acids and amino sugars) incorporated into biomass.

In terms of irreversible thermodynamics, it is important to realise that the high quality energy – Gibbs energy – is in the electrons of the substrates. Many of these valence electrons are either (i) lost as H_2O in the oxidation to CO_2 or (ii) excreted from the cell as lactate. However, the degree of reductance found for animal cell biomass indicates that many of the electrons are conserved in the biomass. This implies that much of the substrate Gibbs energy is dissipated as chemical entropy in the biomass rather than as heat. This leads to the question of the efficiency of the biomass production that has been rather well addressed for microbial systems [27, 28, 39, 54, 63, 64]. However, this topic has not really been addressed at all for animal cells produced in industrial plants despite the essence of it having been solved for microorganisms. This is largely because the medical products from these cells are manufactured under patent and thus are not price-sensitive at this time.

The application of the growth reaction to medium design

It is typical in animal cell culture that one substrate is depleted from the growth medium before the others. As an example, glutamine was exhausted before glucose in cultures of CHO 320 cells (see Fig. 5). This is because, in the early days of cell culture, an empirical approach was adopted in formulating culture media, rather than a rational one based on the demand by the cell [23]. In order to take a more scientific approach, Xie and Wang [61] completed the total stoichiometric analysis of all the medium requirements for a particular type of hybridoma cell. This is a very lengthy and expensive exercise, however, and not possible in all laboratories. As an alternative method, Bonarius *et al.* [11] used a mass balance method to improve the medium for a second type of hybridoma cell. Another way is to use frequency statistics for medium optimisation by applying Plackett-Burmann equations [65]. Kemp and Guan [25] chose an alternative to these three methods by constructing the growth reaction to show the demands for substrates by the cells. Ideally, this would need either knowledge of the enthalpy of combustion for the particular cell type or, at least, sufficient confidence in the regularities detailed above to use a reasonable approximation to it. Even if this approach is not possible, Guan and Kemp [49] showed that the indirect application of the enthalpy balance method has advantages in this respect.

The growth of genetically engineered CHO 320 cells in a bioreactor [12] was the example chosen to explain the methodology [49]. For this purpose, the growth reaction given in Eq. (6) was divided into two half-reactions, namely the catabolic half-reaction, Eq. (10), and the anabolic half-reaction, Eq. (11),

$$\begin{aligned}&\left(\nu_{S_1}\right)_{cat}\text{Glucose} + \left(\nu_{S_2}\right)_{cat}\text{Glutamine} + \nu_O O_2 \rightarrow \nu_L\,\text{Lactate}\\&+\left(\nu_C\right)_{cat}CO_2 + \left(\nu_N\right)_{cat}NH_3 + \left(\nu_H\right)_{cat}H_2O + \nu_X\left(\int_1^2 J_{H,x}\right)_{cat}\end{aligned} \tag{10}$$

$$\begin{aligned}&\left(\nu_{S_1}\right)_{ana}\text{Glucose} + \left(\nu_{S_2}\right)_{ana}\text{Glutamine} \rightarrow \text{Biomass}\\&+\left(\nu_C\right)_{ana}CO_2 + \left(\nu_N\right)_{ana}NH_3 + \left(\nu_H\right)_{ana}H_2O\end{aligned} \tag{11}$$

The criteria for making this separation are well-rehearsed [49] but for this particular case the anabolic half-reaction included both (i) substrate degradation to form biosynthetic precursors and (ii) the subsequent syntheses from them to form the diverse macromolecules that constitute the biomass and the heterologous protein, IFN-γ. For the reason stated in Section 5, carbon dioxide was incorporated into the right hand side of this half-reaction, together with the consequent H_2O. It should be remembered from Battley's work [47] that Eq. (11) is dependent implicitly on the following relation for the enthalpy change of anabolism, $\Delta_r H_{ana}$ [49],

$$\Delta_r H_{ana} = 0 \tag{12}$$

Although most of the evidence supports this relationship, calculations for cultured *Saccharomyces cerevisiae* cells revealed a small but significant enthalpy of anabolism [66] but, even so, the assumption in Eq. (12) would only fail by a few percent.

By accepting Eq. (12), it follows that the heat flux of the growth reaction must have been due entirely to the catabolic half-reaction given in Eq. (10). The details of the method are given in Guan and Kemp [49] but, in essence, the enthalpy change of the catabolic half-reaction in Eq. (11) was calculated as stated in Section 5 from the experimental reaction flux data and, if the enthalpy change balances the observed heat flux, then the description of the growth reaction is correct. It should be remembered, however, that batch cultures are not in steady state [49]. Consequently, the metabolic flux and the heat flux that is a function of it must change to reflect alterations in the environmental conditions, including substrate availability and the accumulation of toxic products. Experimental evidence for this claim was seen for the CHO320 cells by the differing fluxes during the batch culture (see Fig. 5). Guan and Kemp [49] assayed all the major substrates and products at approximately 24-h intervals and then constructed the necessary equations for the metabolic, catabolic and anabolic reactions. The data reflected the alterations in metabolism with the dynamically changing environment. In particular, the reaction equations showed the demand for substrates with time. As an illustration [49], when cells were in early 'exponential' growth (4–28 h), the demand for glucose and glutamine was in the ratio of 3:1, and not 5:1 as provided in the medium. Thus, this is one of the bases for a strategy to supply nutrients in fed-batch culture (see later). More immediately, it was the impetus to design a new medium for improved growth of CHO320 cells and their production of a heterologous protein [67].

The role of the heat flux probe in fed-batch culture

In batch culture, the cell number increases until at least one of the substrates, even a single amino acid not involved in catabolism directly, is depleted, upon which the metabolic flux declines, the cells stop growing and eventually they die by necrosis and/or apoptosis [68]. This is an obvious disadvantage in terms of the overall productivity of cells used in the pharmaceutical industry to produce medically important substances because the cultures have to be started at frequent intervals. There is a limit to the concentrations of metabolites in the medium because unacceptably high osmotic pressure values would be created and so the generally adopted strategy is to retain the relatively low levels of metabolites in the batch culture but then to feed more nutrients to the cells when there are signs of a decline in the growth rate. The problem has been to detect this slowdown because of the paucity of on-line probes that can measure a variable indicative of the decreasing rate [69]. Heat flux is the perfect candidate for the role of detector because of the monotonic relationship between the metabolic flux and the stoichiometry of the growth reaction (see Eq. (7)). Since heat flux, J_{th}, is the correct expression for metabolic flux, Eq. (7) can be converted as,

$$v_i = f\left(J_{th}\right) \tag{13}$$

Guan and Kemp [49, 67] illustrated the strength of this relationship from data for cells grown in batch culture with an improved medium (Fig. 7). When the utilisation of each of the substrates, glucose, glutamine and oxygen, was calculated relative to the heat flux for the stoichiometric ratios, v_{Glc}/v_O and v_{Gln}/v_O, and the results

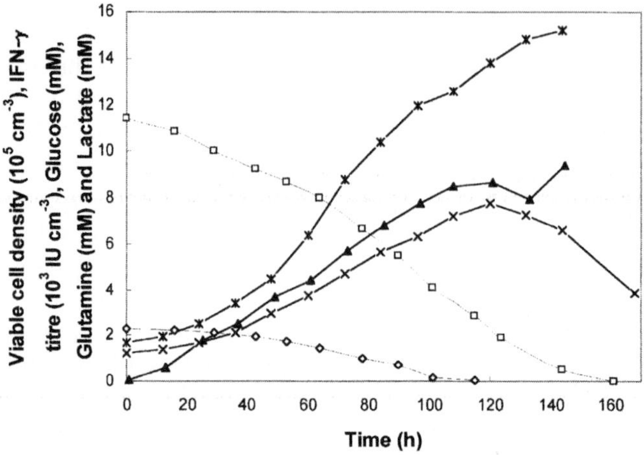

Fig. 7 Changes in the concentrations of the major substrates and products during the batch culture of CHO 320 cells grown in an improved medium (see text). Cell growth (×) is shown together with the concentrations of glucose (□), glutamine (◊) and lactate (▲) at each time. Also shown is the constitutive secretion of IFN-γ (∗) (Reproduced from [67] with permission)

expressed graphically, it can be seen in Fig. 8 that heat flux was indeed a monotonically increasing function of these ratios over the 120-h period of the culture. This was an indirect demonstration from experimental data that heat flux is a valid probe of the metabolic flux. Direct experimental evidence of the relationship, however, can only be achieved by using continuous cultures in which theoretically there is a metabolic steady state at each dilution rate [70].

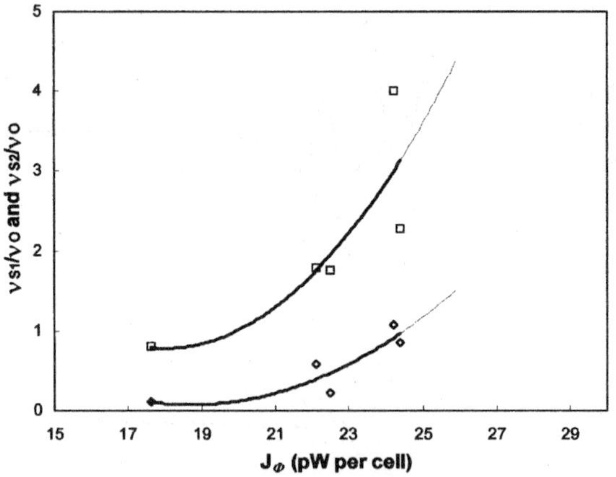

Fig. 8 The heat flux over a specified set of values is compared against the stoichiometric ratios for the consumption of glucose (s_1) and glutamine (s_2) to oxygen. The data for the two specific cases (see Eq. (15) and Eq. (16)) show that the relationship of heat flux to glucose (□) and glutamine (◊) fluxes (Reproduced from [67] with permission)

In contrast to the good steady states achieved in continuous cultures of microbial cells with their relatively simple metabolism, a given dilution rate for animal cell culture can still allow variation in the concentrations of substrates and products because of the metabolic complexity of the cells [25]. A measuring system to monitor on-line the heat flux directly related to the metabolic flux [12] has a distinct advantage in this situation. A constant value for the former indicates the metabolic flux is at steady state. It is also a good reflection of the quantity of viable cell mass.

Guan and Kemp [71] examined the heat flux plateaux for different dilution rates of CHO320 cells in continuous culture. They showed the metabolic flux was at steady state and that its value increased with increasing dilution rate, see Fig. (9). For a given steady state, in most cases glucose was at a low concentration and glutamine was depleted entirely. The concentration of lactate was constant apart from $D = 0.016\ h^{-1}$ where there appeared to be variation. As pointed out by Kemp and Guan [25], however, the fact that the value for the heat flux was constant for each of the selected periods (usually each period lasted from 4 to 7 days) denotes that the steady states were properly achieved without long periods of transition. This was the first time that the direct measurement of steady state had been made for cultured animal cells. It was also pointed out [25] that this instructive result opened up a new direction to explore the changes that occur in the metabolic fluxes in response to alterations in the cellular environment. It is as important to realise that Fig. 9 gives the required proof of the strict relationship between the material stoichiometric coefficients and their equivalent in terms of heat because it validates the use of heat flux as a control variable.

From the above results, it is suggested that, once a culture system has been defined in terms of cell metabolism, the heat flux monitor is the ideal tool to signal the early intervention to control feeding by an automated on-line technique. Thus, fed-batch experiments were undertaken, using the averaged decrease in on-line heat flux over successive 1-h periods as the signal in the Applikon BioXpert software to

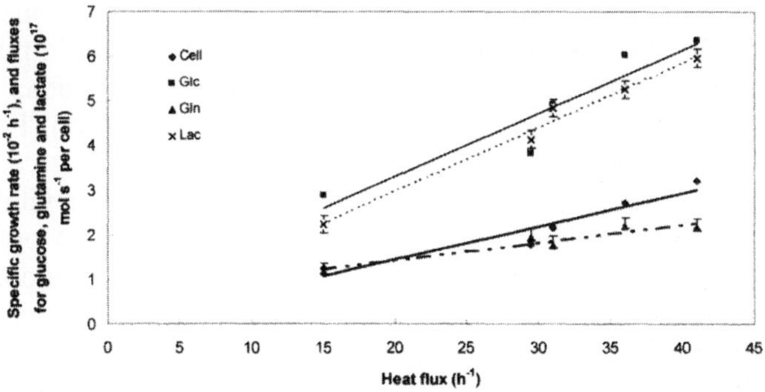

Fig. 9 The cell specific growth rate, and fluxes for glucose, glutamine and lactate were correlated to heat flux at different dilution rates in a continuous culture (Reproduced from [71] with permission)

trigger the feeding of a nutrient cocktail (glucose, 50 mM; glutamine, 16 mM) to the cells. The results depicted in Fig. 10 showed that calorimeter-controlled nutrient feeding effectively restored the metabolic activity at cell concentrations below ca. 10^6 cm^{-3}. Above this level, feeding slowed the deterioration in metabolism to a degree dependent on the time in culture. These results gave incontrovertible evidence that heat flux is a robust and reliable probe for use to control nutrient feeding in batch culture.

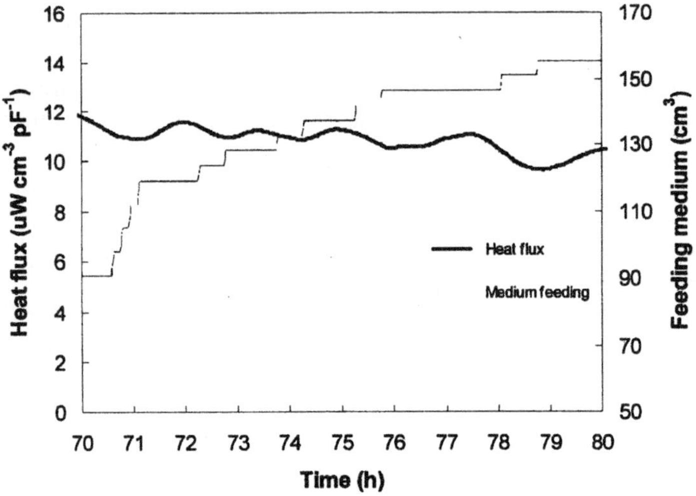

Fig. 10 This shows a small section of the heat profile for a fed-batch culture (from 70 to 80 h) to illustrate that the medium feeding was triggered by the declining heat flux values over 1-h assessment periods. The heat flux was restored, to a varied extent, by this feeding strategy (Reproduced from [71] with permission)

Measurement of Oxygen

Most animal cell bioreactors contain an electrode to measure the amount of DO in the culture medium with the data being fed to a biocontroller set to maintain the required level of oxygen saturation. Thus, oxygen is one of the parameters that are kept constant during the period of the culture. A second reason, however, to measure oxygen is because respiration, estimated from the OUR, is central to the metabolism of animal cells. Estimating OUR is often regarded as a vital factor for the future success in the biotechnology industry of strategies to optimise cell cultures for the most efficient production of high quality heterologous proteins. Yet it has remained difficult to measure OUR for the relatively dilute number of animal cells in the typical bioreactor culture, often starting at 2×10^5 cells per cm^3. The problem is due to the relatively small oxygen flux by animal cells [10]. It is exacerbated by (i) the low solubility of the hydrophobic gas in the culture medium and (ii) the salting out effect, which lowers the solubility by as much as 10% compared with the value in pure water [9]. These two factors have meant that the polarographic measurement of OUR, as opposed to DO, has severe limitations. The alternative of using very expensive mass spectrometers (MS) in order to obtain the required sensitivity might seem excessive [10].

For the MS measurements, the necessary oxygen mass balance for the whole bioreactor, termed the global balance has been written by, for instance, Heinzle's group [72, 73]. A more pragmatic solution in terms of cost would be to base the balance on the liquid phase of the culture, particularly with a constant DO concentration. This course was pursued by Ramírez and Mutharasan [74] and it enabled them to calculate the molar fraction of oxygen very accurately from the set point of the oxygen, nitrogen and carbon dioxide gas flow rate adjusted automatically to control DO by the so-called stationary liquid phase balance (SLPB). There was still the problem shared by all methods involving only the liquid phase [75] of obtaining the volumetric mass transfer coefficient, k_La. The value for it can be very variable, especially in sparged systems which obviously are very disturbed, but Ruffieux *et al.* [10] found a good solution to the problem in the use of hydrophobic polytetrafluoroethylene (PTFE) tubing (W.L. Gore and Associates GmbH, Putzbrunn, Germany) that acts as a porous membrane for aeration.

Ducommun *et al.* [76] recently described the instrumentation that has enabled the measurement of OUR in dilute cell suspensions, mainly because aeration was achieved through the use of the porous PTFE tubing rather than simply relying on the surface of the medium [74]. In brief, the bioreactor was aerated through a metered mixture of oxygen, nitrogen and carbon dioxide (the latter initially at 5% v/v) delivered through a set length of the tubing. Because this has a very high oxygen diffusion coefficient, the oxygen concentration on the surface in contact with the medium was constant and solely dependent on the oxygen molar fraction in the tubing. By avoiding gas bubbles, the k_La was much more stable but to maintain it constant the DO had to be measured with high precision. This was made possible by a low drift, sterilisable Ingold oxygen electrode, A/D interfaced to the appropriate software, which held the DO at the pre-determined percentage saturation throughout the experiments. The total gas flow rate of nitrogen (with CO_2) and oxygen was also constant. This was achieved because the mass flow controllers (Brooks, Veenendaal, The Netherlands) for the three gases were connected via their electronic control module to the acquisition and control software in order to measure and control the flow rates of the nitrogen, carbon dioxide and oxygen. The DO control was made automatically through the software using an on/off control program that maintained it to within 1% of the set point. The k_La at the gas-liquid interface of the PTFE tubing was small, so the value of it depended solely on the length of the tubing. This is much easier and more reliable than having to estimate the k_La at the surface of the medium [74]. In this case, it has to be determined experimentally at a given time from the inverse of the time constant of the system using the dynamic response curve technique [75]. It then has to be assumed that the value remains constant over the complete time of the culture.

The principle of the measurement is that, because the dissolved oxygen concentration, C_L, is maintained constant, the oxygen transfer rate (OTR) must be equal to the OUR. Therefore an oxygen balance in the liquid phase yields,

$$\mathrm{OTR}=k_{L}a[C_{L}^{*}-C_{L}] \tag{14}$$

where C_L* is the DO concentration in equilibrium with the gaseous phase in the PTFE tubing. The gas flow in the tubing is maintained at a very high rate so only <0.5% of the oxygen is consumed and thus it can be assumed the oxygen is equilibrated between the gas and liquid phases [10,74,75]. So, C_L* can be calculated as

$$C_{L}^{*}=B(Pn_{M}F_{o})/(HF_{T}) \tag{15}$$

where F_O and F_T are the oxygen and total gas flow rates respectively, H is the apparent Henry's constant for oxygen in the medium, P is the total pressure, n_M is the molar concentration of the medium (assuming water), and B is a conversion factor equal to 1000 when C_L* is expressed in millimolars. Henry's constant can be obtained assuming that the oxygen concentration in the medium saturated with air is 0.194 mM [77].

It is the typically low number concentration of animal cells in a traditional batch culture that makes it necessary to use of the global balance or the SLPB necessary for OUR measurements. Under other circumstances with whole animals/tissues/organs in aqueous media or in cultures with higher cell density, the polarographic measurement of OUR is perfectly satisfactory [9]. OUR of cells at low number concentration is also possible but only for small volumes (~2–5 cm^3) using the high-resolution Oroboros Oxygraph (Oroboros Bioenergetics and Biomedical Instruments GmbH, Schoepfstr. 18, A-6020 Innsbruck, Austria) with large-diameter, highly sensitive electrodes [78]. If the (net) catabolic process is totally aerobic without an anaerobic component such as glycolytic lactate formation, then the OUR measurements constitute a form of indirect calorimetry. When it is necessary for thermodynamic and energetic reasons to express the results in terms of the heat flux and/or molar reaction enthalpy, the appropriate oxycaloric equivalent, $\Delta_k H_O$, is applied to the data for oxygen (O) flux (J_O),

$$J_{Q}=\Delta_{k}H_{O}J_{O} \tag{16}$$

Oxycaloric equivalent s are the theoretical values for the enthalpy changes of the catabolic part of the metabolic cycle from the substrates to the products in terms of the biomass and, in biotechnology, the heterologous proteins. They correspond to the term, Thornton's regularity, the values for which are obtained by combustion calorimetry [53, 54]. The catabolic half-cycle, e.g. from glucose to HCO_3^- and H^+, does not include any coupled processes such as ATP production [51]. This means that no work is done, so the net efficiency is zero [79, 80]. These equivalents mostly are calculated from the standard enthalpies of formation (see Section 5) and, for given substrates, are the same as the values obtained by bomb calorimetry [53]. Combustion of material in a calorimeter is dissipative and thus totally inefficient, not being coupled to an energy conserving mechanism. It was found by experiment that the oxycaloric equivalents for all substrates are similar [51, 79] (see Section 5) because of the regularity for the heat evolved per equivalent of oxygen consumed in terms of the available electrons [54, 55], sometimes known as Thornton's regularity or rule. Gnaiger and Kemp [51] calculated the theoretical oxycaloric equivalents for a variety of substrates and conditions and found a narrow range from –430 to –480 kJ mol^{-1} O_2; on average $\Delta_k H_O = -450$ kJ $mol^{-1} \pm 15\%$.

The Calorimetric-respirometric (CR) ratio

For measurements by direct and indirect calorimetry, the theoretical oxycaloric equivalent is the same as the expected ratio of the calorimetric heat flux and the respirometric oxygen flux, the CR ratio,

$$\text{CR ratio} = J_Q \,/\, J_O \tag{17}$$

For whole animals in normoxic conditions, their homeostatic mechanisms ensure that the CR ratio is the same as the above generalised oxycaloric equivalent [81] with the Cori cycle in the liver converting to glucose any lactate in the circulating blood after much of it has been oxidised in the muscle by the appropriate isoform of lactate dehydrogenase. It was noted some years ago [82, 83] that it is rare indeed for cultured, growing cells derived from vertebrate tissue to have a CR ratio close to the generalised oxycaloric equivalent. The only authenticated report of it is for hamster mature brown adipocytes that have a CR ratio of –490 kJ mol^{-1} O_2 when treated with noradrenaline [84]. This hormone is known to stimulate thermogenesis in the brown fat by acting on the 33-kDa uncoupling protein (UCP; also known as thermogenin) that resides in the inner mitochondrial membrane to open channels between the cytosol and the matrix [85]. In consequence, it caused the dissipation of the protonmotive force in the adipocytes. Similarly to action of respiratory couplers, the electron transport from NADH to O_2 then proceeded at a maximal rate not coupled to the demand for ATP [80]. There was a proportionate increase in both heat production and OUR as a result of the relatively uncontrolled catabolic flux. This is analogous to the situation in the bomb calorimeter (see Lamprecht [53]) because uncoupling resulted in a totally inefficient process that permits no energy conservation, i.e. there was no oxidative phosphorylation to produce ATP [80]. Without a source of ATP other than the mitochondrion, there was the rapid death of the uncoupled brown adipocytes.

Gnaiger and Kemp [51] established that the highly exothermic CR ratios of vertebrate cells are due to the integration of anaerobic pathways with aerobic metabolism. They showed that the net production of the most common anaerobic end product, lactate (Lac), is accompanied by a dissipative catabolic enthalpy change, $\Delta_k H_{Lac}$, of –80 kJ mol^{-1} when the acid is buffered in the cytosol. However, the plasma membrane has an outward pump for lactate and then the enthalpy change depends on the nature of the buffer in the medium. When lactate is excreted into a bicarbonate buffer, the enthalpy change is –80 kJ mol^{-1}. It is –59 kJ mol^{-1} in a phosphate buffer and –77 kJ mol^{-1} in 20 mM HEPES buffer. The molar amount of it produced per unit amount of oxygen consumed (Lac/O_2) is a good indication of the relative extent of the aerobic glycolysis [51]. The catabolic (k) heat change per mol O_2, $\Delta_k H_{(ox+anox)}$ (CR ratio), is then calculated as,

$$\Delta_k H_{(ox+anox)} = \Delta_k H_O + \mathrm{Lac} / O_2 \cdot \Delta_k H_{Lac} \tag{18}$$

In many cases of highly exothermic CR ratios, the enthalpy balance approach in Eq. (2) proved that aerobic glycolysis to produce lactate was the cause [25, 38, 40, 51, 82, 83, 86]. The situation is not straightforward for many types of cultured vertebrate cell, however, because the growth media almost always contain glutamine. Although this is used for purine and pyrimidine synthesis in the anabolism of nucleic acids, in some cases a proportion of it is completely oxidised to provide energy. As stated in Section 5, it can also be partially oxidised by glutaminolysis to lactate [45, 46], with an ATP stoichiometric coefficient of 6ATP/O_2 [23, 25]. The operation of this pathway causes an overestimate of aerobic glycolysis from estimates of lactate secretion because, of course, the oxycaloric equivalent for glutaminolysis is within the normal range [51] and thus cannot be distinguished from the oxidation of other carbon sources. For the HEPES-buffered medium used in the culture of 2C11-12 mouse macrophage hybridoma cells [87], the metabolic reaction and corresponding enthalpy change is [88],

$$\begin{aligned} &{}^{+}\mathrm{H_3N(CHCH_2CH_2CONH_2)COO^{-1}(aq)} + 1\tfrac{1}{2}\mathrm{O_2(g)} + 3\mathrm{H_2O(l)} \\ &+\mathrm{HEPES(aq)} \rightarrow \mathrm{CH_3CH(OH)COO^{-}(aq)} + 2\mathrm{HCO_3^{-}(aq)} \\ &+2\mathrm{NH_4^{+}(aq)} + \mathrm{HEPES-H^{+}(aq)} \qquad -695\,\mathrm{kJ} \end{aligned} \tag{19}$$

to give the oxycaloric equivalent of glutamine oxidation to lactate as –463 kJ mol^{-1}. The extent of the overestimation can only be determined by radioisotopic experiments as performed by Kemp *et al.* [87]. They found that 38% of the glutamine utilised by the immunologically activated cells was oxidised to lactate (9 pmol lac s^{-1} per 10^6 cells).

Kemp and Guan [89] advised that, if the enthalpy recovery was not achieved for lactate, then it is preferable to look for possible methodological deficiencies before resorting to further metabolic analyses. On the principle that it is preferable to avoid systematic errors, it is vital to set the conditions for both the heat and oxygen measurements to be as similar as possible because the use of ratios considerably increases experimental error. In general, the polarographic measurement of OUR is performed on stirred cell suspensions whereas there is only a limited availability of stirred calorimetric vessels [21, 32]. If suspensions are not stirred, then the sedimented layers of cells in the unstirred layer rapidly exhaust the DO and exhibit the so-called 'crowding' effect (25, 26, 38, 40, 48, 51, 82, 83, 86, 87, 89] with increased glycolysis due to the fact that the Pasteur effect results in the production of more lactate. It should also be remembered that polarographic measurements are suspect if there is no movement of the aqueous phase to prevent the formation of an unstirred layer close to the membrane. As another methodological difference, in many cases, the calorimetric vessel has a gaseous headspace that allows recruitment of oxygen, whereas OUR is usually measured without one.

Guan and Kemp [67] stressed the fact that normal mammalian cells in culture produce lactate in many cases due to poor design of the medium. These were formulated in each laboratory on the empirical basis of a buffered physiological saline with glucose at the level in blood, the essential amino acids and serum. It is common knowledge that, in the glycolytic pathway, amino sugars arise from fructose 6-phosphate and the amino donor, glutamine; the amino acids serine (for fatty acids) and alanine are formed from 3-phosphoglycerate; the heterocyclic acids, phenylalanine, tyrosine and tryptophan from phosphoenolpyruvate; and oxaloacetate from pyruvate and carbon dioxide (the anaplerotic reaction). These reactions mostly require the production of three-carbon units at a rapid rate. As a result, a large quantity of pyruvate is available that is surplus to energy requirements and is reduced to lactate to pay back NAD^+ [67] The partial oxidation of glutamine to lactate catalysed by enzymes of the Krebs' cycle and followed by transamination, enables other amino acids to be formed from α-ketoglutarate and oxaloacetate.

In order to optimise the cell growth in batch culture and to minimise the production of the toxic product lactate (and ammonia), it was mentioned in Section 6 that Xie and Wang [61] made an exhaustive stoichiometric analysis of all the metabolic requirements for the growth of a hybridoma cell type. The reformatted medium was successful in reducing the cellular lactate production by 90%. The more empirical approach was to monitor by heat flux measurements, the steps taken to improve the culture medium of recombinant CHO 320 cells [67]. The redesigned medium improved the specific growth rate and the flux of IFN-γ (Fig. 11) while decreasing the catabolic flux, especially of glucose and lactate (Table 1).

The interesting relationship between the highly negative CR ratios and the cell growth can be seen in Fig. 12 from studying the growth of recombinant CHO 320 cells [12]. The oxygen flux remained constant over the whole culture

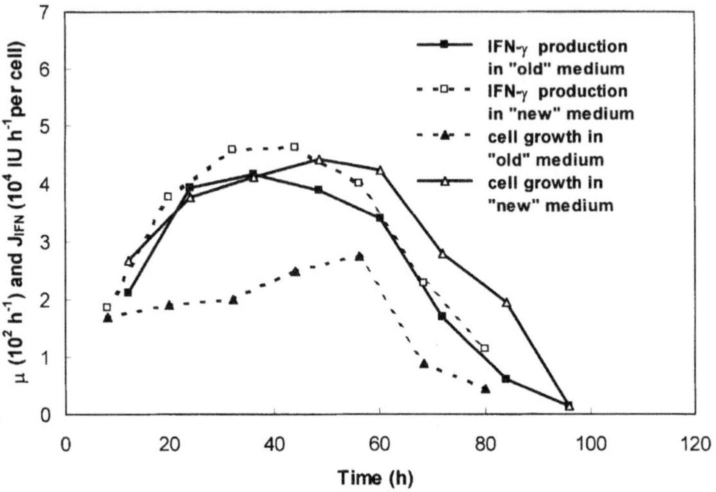

Fig. 11 The effects of the semi-empirically improved growth medium that is based on RPMI 1640 on both the specific cell growth rate and the production flux of interferon-γ during a batch culture of CHO320 cells in a controlled bioreactor (Reproduced from [67] with permission)

Table 1. Representative fluxes of the major metabolites for CHO 320 cells growing in the original medium and the improved version of it at 48 h during batch culture in the bioreactor

Medium	Fluxes			
	m (h^{-1})	Glucose (mol s^{-1} per cell)	Glutamine (mol s^{-1} per cell)	Lactate (mol s^{-1} per cell)
Improved	0.049	4.11×10^{-17}	1.67×10^{-17}	5.28×10^{-17}
Original	0.028	5.81×10^{-17}	1.78×10^{-17}	1.08×10^{-16}

period even when there was no net cell growth. In contrast, the CR ratio was highly exothermic (~ –700 kJ mol^{-1} O_2) only during the growth phase, reducing to a level (–443 kJ mol^{-1} O_2) indicative of solely oxidative metabolism when there was a decline in cell numbers. This was good evidence that measuring the on-line lactate concentration in the culture medium could give an accurate approximation of cell growth. An inexpensive alternative might be formally to monitor on-line the amount of NaOH required to neutralise lactate when, typically, pH is a controlled parameter.

Guan *et al.* [12] thought there was some evidence that the amount of excreted lactate decreased after the two substrates, glucose and glutamine, were fully exhausted in the medium (Fig. 13). They rationalised that, since there was still an oxygen flux (see Fig. 12), it is possible that the lactate was oxidised as the source of energy. The mechanism for the oxidation of the lactate to pyruvate requires the

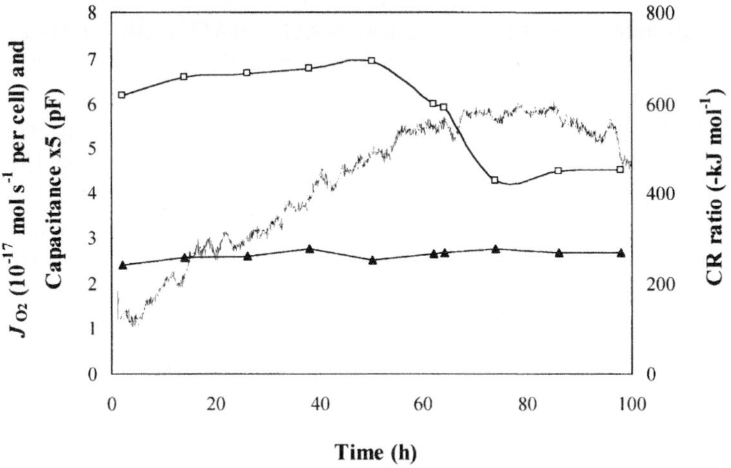

Fig. 12 Comparison of the viable cell concentration measured by the on-line capacitance signal with oxygen flux and the calorimetric-respirometric (CR) ratio in a batch culture of growing CHO320 cells. The medium was buffered by 20 mM HEPES and 4 mM sodium bicarbonate. The curves are □—□ for the averaged oxygen flux in terms of an individual viable cell on average, and ▲—▲ for CR ratio. The trace —— stands for the smoothed capacitance signal in which the value for the cell-free medium is deducted by the Applikon BioXpert software (Reproduced from [86] with permission)

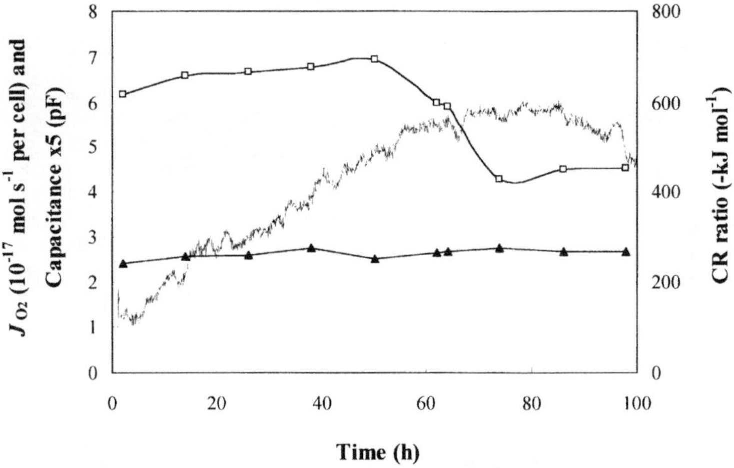

Fig. 13 Heat flux ($J_{\varphi/x(c)}$) compared with the changes in concentrations of glucose, lactate, ammonia, glutamine for a typical batch culture of CHO 320 cells. Symbols are —— for specific heat flow rate, □—□ for glucose, ■—■ for lactate, ▲—▲ for ammonia, and Δ——Δ for glutamine (Reproduced from [12] with permission)

presence of the isozyme H_4 of the lactate dehydrogenase complex in association with the mitochondria, together with a low $NADH/NAD^+$ ratio. These conditions occur in liver cells for the operation of the Cori cycle to convert lactate to glucose but it is not clear that the typical tissue cell can catabolise lactate in this way. It seems likely that pyruvate carboxylase (PYC) is more important to the operation of this oxidative pathway than the dehydrogenase because there is evidence that transfecting BHK cells with a *PYC* construct dramatically decreased lactate production and improved cell growth [90]. In any case, there is an advantage for the control of cells in bioreactors to combine the on-line estimation of OUR with continuous heat flux measurements [12]. For this purpose, Feng *et al.* [91] has performed experiments combining the flow microcalorimeter to give the heat flow rate with SLPB to determine OUR (see Section 8).

Genetically engineered NS1-derived, TB/C3 hybridoma cells that produces antibodies against human Immunoglobulin G (IgG) and overexpresses the anti-apoptotic protein bcl-2 (pEF bcl2-MC1neopA plasmid) [92, 93] were grown on RPMI 1640 medium supplemented with 5% (v/v) fetal calf serum and 2 mM L-alanyl-L-glutamine (Glutamax©, Gibco) in an Applikon 3-L tank bioreactor. The on-line curves for heat flow rate and OUR of the cultured cells are shown in Fig. 14. Initially these metabolic variables increased at a rate comparable to the growth rate. As with other cell types, the CR ratio was highly exothermic, indicating the intensity of aerobic glycolysis, which was identified as being due to lactate production [91]. From approximately 44 h onwards, the heat flow rate and OUR decreased to different degrees, although the cells were still increasing in number to a certain extent. The change in the metabolic rate indi-

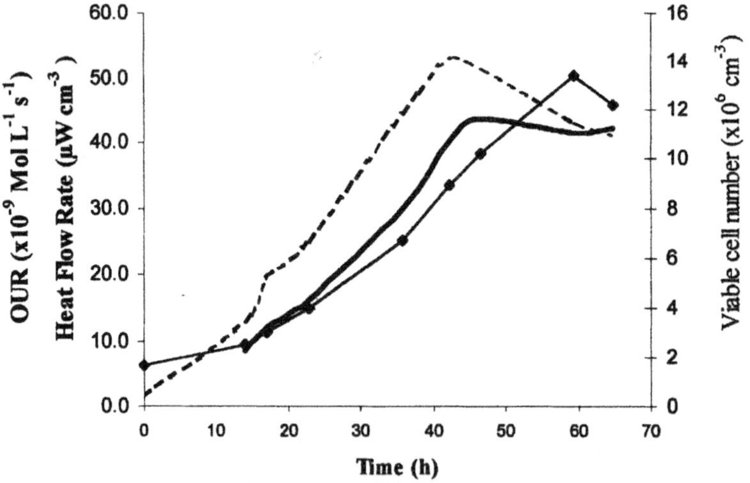

Fig. 14 Typical on-line traces for heat flow rate (— —) and oxygen uptake rate (—) compared with off-line counts of the viable cell density (o), during the batch culture of TB/C3 bcl-2 cells in the Applikon tank bioreactor (Reproduced from [91] with permission)..

cated by the on-line continuous measurements at an earlier stage than the decrease and eventual cessation of cell growth was similar to that first reported for batch cultures of CHO 320 cells (see Fig. 4) in which the heat flow rate reduced in value before the peak of cell growth. This finding was interpreted as meaning that the cells in G1 phase of the cell cycle were still able to divide once more despite the fact that the metabolic rate was in decline, probably because of substrate exhaustion and/or the accumulation of toxic metabolic by-products [12]. However, it will be seen in Fig. 14 that the heat flow rate declined more steeply than the OUR, so that the CR ratio more closely approximated to the oxycaloric equivalent for respiration without anaerobic catabolism. Feng *et al.* [91] concluded that the highly exothermic CR ratio during much of the batch culture was evidence for the inadequacy of the medium to provide all the appropriate biosynthetic precursors. Late in the batch culture, the cells were not growing and the CR ratio reflected the reduced glycolysis. In spinner cultures there was some evidence of apoptosis with 10% apoptotic bcl-2 cells at 48 h [91]. According to Kroemer [94], among others, one of the early molecular signs of the apoptotic cascade is the mitochondrial permeability transition (MPT) which should seriously affect respiration. OUR and HFR measurements in the tank bioreactor unfortunately gave no direct indication of the putative MPT (see Fig. 14). Although it is reasonable to assume from the spinner cultures that apoptosis was prevalent in the bioreactor at 48 h, it is possible that the averaging effect of many cells in the 3-L volume would mask the detection of it by MPT. Nevertheless, both OUR and HFR, together with the CR ratio, have potential in discovering how changes in such medium ingredients as amino acids, including glutamine, dissolved oxygen, Mg^{2+}, Ca^{2+} and K^{+} ions and vitamins induce the apoptotic cascade [95].

Conclusions

Although it is the case that only a few cell biology laboratories possess a calorimeter, it is to be hoped that this paper has demonstrated that thermobiochemistry already has had a significant impact in several areas associated with cell physiology and particularly in the applied fields of biotechnology and pharmacology. The present author is convinced that its importance will increase dramatically in the coming post-genomic age, when everyone will be wishing to know the functions of all those myriads of expression systems and how they interact to give life. Calorimetry has the distinct advantage of a rigorous theoretical foundation in thermodynamics that will ensure its precise contribution to cell biology and physiology.

Acknowledgements

The author is grateful to the Biotechnology and Biological Sciences Research Council for financial support in grants numbered 2/3680, 2/TO3789 and 2/E10985. Mr Dayo Olomolaiye in collaboration with Mr Anthony Pugh drew the figures – many thanks.

References

1 Cartwright, T. (1988). *Animal Cells as Bioreactors*, Cambridge University Press, Cambridge.

2 Hesse, F. Ebel, M. Konisch, N. Sterlinski, R. Kessler, W. and Wagner, R. (2003). Comparison of a production process in a membrane-aerated stirred tank and up to 1000-L airlift bioreactors using BHK-21 cells and chemically defined protein-free medium, *Biotechnology Progress* 19, 833–843.

3 Langheinrich, C. and Nienow, A. W. (1999). Control of ph in large-scale, free suspension animal cell bioreactors: Alkali addition and pH excursions, *Biotechnology and Bioengineering*, 66, 171–179.

4 Sauerwald, T. M. Betenbaugh, M. J. and Oyler, G. A. (2002). Inhibiting apoptosis in mammalian cell culture using caspase inhibitor XIAP and deletion mutants, *Biotechnology and Bioengineering*, 77, 704–716.

5 Koshland, D. E. (1987) Evolution of catalytic function, *Cold Spring Harbor Symposia on Quantitative Bio*logy, 52, 1–7.

6 Omesa, T. Higaskiyama, K. i. Schoya, S. and Suga, K. i. (1993). Effects of lactate concentration on hybridoma culture in lactate-controlled fed-batch operation, *Biotechnology and Bioengineering*, 39, 556–564.

7 Ulber, R. Frerichs, J. G, and Beutel, S. (2003). Optical sensor systems for bioprocess monitoring, *Analytical and Bioanalytical Chemistry*, 376, 342–348.

8 Gnaiger, E. (2001). Bioenergetics at low oxygen: dependence of respiration and phosphorylation on oxygen and adenosine diphosphate supply, *Respiration Physiology*, 128, 277–297.

9 Gnaiger, E. and Forstner, H. (1983). *Polarographic Oxygen Sensors: Aquatic and Physiological Applications*, Springer-Verlag, Berlin.

10 Ruffieux, P.-A. von Stockar, U. and Marison, I. W. (1998). Measurement of volumetric (OUR) and determination of specific (qO_2) oxygen uptake rates in animal cell cultures, *Journal of Biotechnology*, 63, 85–95.

11 Bonarius, H. P. J. de Gooijer, C. D. Tramper, J. and Schmid, G. (1995). Determination of the respiration and the respiration quotient in mammalian-cell culture in bicarbonate buffered cells, *Biotechnology and Bioengineering*, 45, 524–535.

12 Guan,Y. Evans. P. M. and Kemp, R. B.: Specific Heat Flow Rate (1998). An on-line monitor and potential control variable of the specific metabolic rate in animal cell culture that combines microcalorimetry with dielectric spectroscopy. *Biotechnology and Bioengineering*. 58, 464–477.

13 Male, K. B. Gartu, P. O. Kamen, A. A. and Luong, J. H. T. (1997) On-line monitoring of glucose in mammalian cell culture using a flow injection analysis (FIA) mediated biosensor, *Biotechnology and Bioengineering*, 55, 497–504.

14 Kumar, M. A. Thakur, M. S. Senthuran, A. Senthuran, V. Karanth, N. G. Hatti-Kaul, R. and Mattiasson, B. (2001). An automated flow injection analysis system for on-line monitoring

of glucose and L-lactate during lactic acid fermentation in a recycle bioreactor, *World Journal of Microbiology and Biotechnology*, 17, 23–29.
15 Rhiel, M. Ducommun, P. Bolzonella, I. Marison, I. and von Stockar, U. (2002). Real-time in situ monitoring of freely suspended and immobilized cell cultures based on mid-infrared spectroscopic measurements, *Biotechnology and Bioengineering*, 77, 174–185.
16 von Stockar, U. Valentinotti, S. Marison, I. Cannizzaro, C. and Herwig, C. (2003). Know-how and know-why in biochemical engineering. *Biotechnology Advances*, 21, 417–430.
17 Ducommun, P. Bolzonella, I. Rhiel, M. Pugeaud, P. von Stockar, U. and Marison, I. W. (2001). On-line determination of animal cell concentration, *Biotechnology and Bioengineering*, 72, 515–522.
18 Olomolaiye, D. Guan, Y. H. Carvell, J. P. and Kemp, R. B. (2001). Measurement of the viable cell density: Validation and integration of on-line and off-line capacitance biomass monitors for cell culture processes, in E. Lindner-Olsson, N. Chatzissavidou, and E. Lüllau (eds), *Animal Cell Technology: Proceedings of the 17th ESACT Meeting*, Kluwer Academic Publishers, Dordrecht, pp. 455–458.
19 Kemp, R. B. Evans, P. M. and Guan, Y. (1997). An Enthalpy Balance Approach to Studies of Metabolic Activity in Mammalian Cells. *Journal of Thermal Analysis,* 49, 755–770.
20 Marison, I. Liu, J.-S. Ampuero, S. von Stockar, U. and Schenker, B. (1998). Biological reaction calorimetry: Development of high sensitivity bio-calorimeters, *Thermochimica Acta*, 309, 157–173.
21 Kemp, R. B. (1998). Nonscanning calorimetry, in P. Gallagher (ed.), *Handbook of Thermal Analysis and Calorimetry*, Vol.1, M. Brown (ed.), *Principles and Practice,* Ch. 14, Elsevier, Amsterdam, pp. 577–675.
22 Freshney, R. I. (2000). Culture of animal cells: *A manual of basic technique*, Wiley, New York, 4th edition.
23 Kemp, R. B. and Guan, Y. (1998). Probing the metabolism of genetically-engineered mammalian cells by heat flux. *Thermochimica Acta*, 309, 63–78.
24 Gnaiger, E. (1993). Nonequilibrium thermodynamics of energy transformations, *Pure and Applied Chemistry*, 65, 1983–2002.
25 Kemp, R. B. and Guan, Y. H. (1999). Microcalorimetric studies of isolated animal cells, in R. B. Kemp (ed.) *Handbook of Thermal Analysis and Calorimetry*, Vol.4, *From Macromolecules to Man,* Ch. 11, Elsevier, Amsterdam, pp. 557–656.
26 Kemp, R. B. (1993). Developments in cellular microcalorimetry with particular emphasis on the valuable role of the energy (enthalpy) balance method. Th*ermochimica Acta*, 219, 17–41.
27 von Stockar, U. and Marison, I. W. (1989). The use of calorimetry in biotechnology, *Advances in Biochemical Engineering/Biotechnology*, 40, 93–136.
28 Duboc, P. Marison, I. and von Stockar, U. (1999). Quantitative calorimetry and biochemical engineering, in R. B. Kemp (ed.) *Handbook of Thermal Analysis and Calorimetry*, Vol.4, *From Macromolecules to Man,* Ch. 6, Elsevier, Amsterdam (1999), pp. 267–365.
29 Randolph, T. W. Marison., I. W. Berney, C. and von Stockar, U. (1989). Bench-scale calorimetry of hybridomas in suspension culture, *Biotechnology Techniques*, 3, 369–374.
30 Garcia-Payo, M. C. Ampuero, S. Liu, J. S. Marison, I. W. and von Stockar, U. (2002). The development and characterization of a high resolution bio-reaction calorimeter for weakly exothermic cultures, *Thermochimica Acta*, 391, 25–39.

31 Suurkuusk, J. and Wadsö, I. (1982). A multichannel micro-calorimetry system, *Chemica Scripta*, 20, 155–163.
32 Wadsö, I. (1994). Microcalorimetry of aqueous and biological systems, in K.N. Marsh, P.A.G. O'Hare (eds.), *Experimental Thermodynamics*, Vol. IV, Blackwell Scientific Publishers, Oxford, pp. 267–3–1.
33 Guan, Y. Evans, P. M. and Kemp, R. B. (1997). A modified flow microcalorimeter for measuring the heat dissipation by mammalian cells in batch culture. *Journal of. Thermal Analysis*, 49, 785–794.
34 O'Neill, M. A. A. Beezer, A. E. Labetoulle. C. Nicolaides, L. Mitchell, J. C. Orchard, J. A. Connor, J. A. Kemp, R. B. and Olomolaiye, D. (2003). The base catalysed hydrolysis of methyl paraben: A test reaction for flow microcalorimeters used for determination of both kinetic and thermodynamic parameters. *Thermochimica Acta*, 399, 63–71.
35 O'Neill, M. A. A. Beezer, A. E. Kemp, R. B. Olomolaiye, D. Volpe, P. L. O. and Oliveira, D. (2004). Practical and theoretical consideration of flow-through microcalorimetry: Determination of 'thermal volume' and its flow rate dependence. *Thermochimica Acta*, in press.
36 Guan, Y. H. Lloyd, P. C. and Kemp, R. B. (1999). A calorimetric flow vessel for measuring the metabolic activity of animal cells. *Thermochimica Acta,* 332, 211–220.
37 Chisti, Y. (2001). Hydrodynamic damage to animal cells, *Critical Reviews in Biotechnology*, 21, 67–110.
38 Kemp, R. B. (1991). Calorimetric studies of heat flux in animal cells, *Thermochimica Acta*, 193, 253–267.
39 von Stockar, U. Gustafsson, L. Larsson, C. Marison, I. Tissot, P. and Gnaiger, E. (1993). Thermodynamic considerations in constructing energy balances for cellular growth, *Biochimica et Biophysica Acta*, 1183, 221–240.
40 Kemp, R. B. and Guan, Y. (1997). Heat flux and the calorimetric-respirometric ratio as measures of catabolic flux in mammalian cells. *Thermochimica Acta*, 300, 199–211.
41 Harris, C. M. Todd, R. W. Bungard, S. J. Lovitt, R. W. Morris, J. G. Kell, D. B. (1987). The dielectric permittivity of microbial suspensions at radio-frequencies: A novel method for the real-time estimation of microbial biomass. Enzyme and Microbial Technology, 6, 181–186.
42 Ducommun, P. Bolzonella, I. Rhiel, M. Pugeaud, P. von Stockar, U. and Marison, I. W. (2001). On-line determination of animal cell concentration, *Biotechnology and Bioengineering*, 72, 515–522.
43 Ducommun, P. Kadouri, A. von Stockar, U. and Marison, I. W. (2002). On-line determination of animal cell concentration in two industrial high-density culture processes by dielectric spectroscopy, *Biotechnology and Bioengineering*, 77, 316–323.
44 Cannizzaro, C. Gugerli, R. Marison, I. and von Stockar, U. (2003). On-line biomass monitoring of CHO perfusion culture with scanning dielectric spectroscopy, *Biotechnology and Bioengineering*, 84, 597–610.
45 Phillips, D. L. (1962). A technique for the numerical solution of certain integral equations of the first kind, *Journal of the Association for Computing Machinery*, 9, 84–97.
46 Tikhonov, A. N. and Arsenin, V. Y. (1977). *Solutions of Ill-posed Problems*, V. H. Winston and Sons, Washington, DC.
47 Battley, E. H.: *Energetics of Microbial Growth*, Wiley-Interscience, New York, 1987.
48 Kemp, R. B. (2000). 'Gae me ae spark o' Nature's fire' (R. Burns): An insight to cell physiology from calorimetry. *J. Thermal Analysis and Calorimetry*, 60, 831–843.

49 Guan, Y. H. and Kemp, R. B. (1999). Dynamic medium optimization by on-line heat flux measurement in mammalian cell culture. *Journal of Biotechnology*, 69, 95–114.
50 Wilhoit, R. C. (1969). Selected values of thermodynamic properties, in H.D. Brown (ed.), *Biochemical Microcalorimetry*, Academic Press, New York, Appendix, pp. 305–317.
51 Gnaiger, E. and Kemp, R. B. (1990). Anaerobic metabolism in aerobic mammalian cells: information from the ratio of calorimetric heat flux and respirometric oxygen flux, *Biochimica et Biophysica Acta*, 1016, 328–332.
52 Gurakan, T. Marison, I. W. von Stockar, U. Gustafsson, L. and Gnaiger, E. (1990). Proposals for a standardised sample handling procedure for the determination of elemental composition and enthalpy of combustion of biological material, *Thermochimica Acta*, 172, 251–266.
53 Lamprecht, I. (1999). Combustion Calorimetry, in R. B. Kemp (ed.), *Handbook of Thermal Analysis and Calorimetry*, Vol. 4, *From Macromolecules to Man*, Ch. 4, Elsevier, Amsterdam, pp. 175–218
54 Erickson, L. E. (1987). Energy requirements in biological systems, in A. M. James (ed.), *Thermal and Energetic Studies of Cellular Biological Systems*, Wright, Bristol, pp. 12–33.
55 von Stockar, U. and Marison, I. W. (1991). Large-scale calorimetry and biotechnology. *Thermochimica Acta*, 193, 215–242.
56 Bushell, M. E. Bell, S. E. Scott, M. F. Spier, R. E. Wardell, J. N. and Sanders, P. G. (1994). Enhancement of monoclonal antibody yield by hybridoma fed-batch culture, resulting in extended maintenance of viable cell population, *Biotechnology and Bioengineering*, 45, 1099–1106.
57 Vriezen, N. and van Dijken, J. P. (1998). Fluxes and enzyme activities in central metabolism of myeloma cells grown in chemostat culture, *Biotechnology and Bioengineering*, 59, 28–39.
58 Ruffieux, P.-A. (1998). *Determination of Metabolic Fluxes for Animal Cells in Continuous Culture*, Ph.D. Thesis, EPFL, Lausanne.
59 Ardawi, M. S. M. and Newsholme, E. A. (1985). Metabolism in lymphocytes and its importance in the immune response, *Essays in Biochemistry*, 21, 1–44.
60 McKeehan, W. L. (1986). Glutaminolysis in animal cells, in M.J. Morgan (ed.), *Carbohydrate Metabolism in Cultured Cells*, Plenum Press, New York, pp. 111–150.
61 Xie, L. and Wang, D. I. C. (1994). Stoichiometric analysis of animal cell growth and its application in medium design, *Biotechnology and Bioengineering*, 43, 1175–1189.
62 Xie, L. and Wang, D. I. C. (1996). High cell density and high monoclonal antibody production through medium design and rational control in a bioreactor, *Biotechnology and Bioengineering*, 51, 725–729.
63 Roels, J. A. (1983). *Energetics and Kinetics in Biotechnology*, Elsevier, Amsterdam.
64 Heijnen, J. J., and van Dijken, J. P. (1992). In search of a thermodynamic description of biomass yields for the chemotrophic growth of microorganisms, *Biotechnology and Bioengineering*, 39, 833–858.
65 Castro, P. M. L. Hayter, P. M. Ison, A. P. and Bull, A. T. (1992). Application of a statistical design to the optimisation of culture medium for recombinant interferon-γ production by Chinese hamster ovary cells, *Applied Microbiology and Biotechnology*, 38, 84–90.
66 Battley, E. H. (1995). A reevaluation of the thermodynmics of growth of *Saccharomyces cerevisiae* on glucose, ethanol, and acetic acid, *Canadian Journal of Microbiology*, 41, 388–398.

67 Guan, Y. H. and Kemp, R. B. (1999). On-line Heat Flux Measurements Improve the Culture Medium for the Growth and Productivity of Genetically Engineered CHO cells, *Cytotechnology*, 30, 107–120.
68 Al-Rubeai, M. and Singh, R. P. (1998). Apoptosis in cell culture, *Current Opinion in Biotechnology*, 9, 152–156.
69 Zhou, W. and Mulchandani, A. (1995). Recent advances in bioprocess monitoring and control, *ACS Symposium Series*, 613, 88–98.
70 Pirt, S. J. (1975). *Principles of Microbe and Cell Cultivation*, Blackwell Scientific Publishers, Oxford.
71 Guan, Y. H. and Kemp, R. B. (1999). A model for the on-line scrutiny of metabolism: It's application to the changing nutritional demands of cultured animal cells, in A. Bernard, B. Griffiths, W. Noé, F. Wurm (eds.), *Animal Cell Technology: Products from Cells; Cells as Products. Proceedings of the 16th ESACT Meeting*, Kluwer, Dordrecht, p. 131–134.
72 Oeggerli, A. Eyer, K. Heinzle, E. (1995). On-line gas analysis in animal cell cultivation: I. Control of dissolved oxygen and pH. *Biotechnology and Bioengineering*, 45, 42–53.
73 Eyer, K. Oeggerli, A. and Heinzle, E. (1995). On-line gas analysis in animal cell cultivation: II. Methods for oxygen uptake rate estimation and its application to controlled feeding of glutamine, *Biotechnology and Bioengineering*, 45, 54–62.
74 Ramírez, O. T. and Mutharasan, R. (1990). Cell cycle- and growth phase-dependent variations in size distribution, antibody productivity, and oxygen demand in hybridoma cultures, *Biotechnology and Bioengineering*, 36, 839–848.
75 Higareda, A. E. Possani, L. D. and Ramírez, O. T. (1997). The use of culture redox potential and oxygen uptake rate for assessing glucose and glutamine depletion in hybridoma cultures, *Biotechnology and Bioengineering*, 56, 555–563.
76 Ducommun, P. Ruffieux, P.-A. Furter, M.-P. Marison, I. and von Stockar, U. (2000). A new method for on-line measurement of the volumetric oxygen uptake rate in membrane aerated animal cell cultures. *Journal of Biotechnology*, 78, 139–147.
77 Miller, W. M. Blanch, H. W. and Wilke, C. R. (1988). A kinetic analysis of hybridoma growth and metabolism in batch and continuous suspension culture: Effect of nutrient concentration, dilution rate, and pH, *Biotechnology and Bioengineering*, 32, 947–965.
78 Gnaiger, E. Steinlechner-Maran, R. Méndez, G. Eberl, T. and Margreiter, R. (1995). Control of mitochondrial and cellular respiration by oxygen, *Journal of Bioenergetics and Biomembranes*, 27, 583–596.
79 Gnaiger, E. (1990). Concepts on efficiency in biological calorimetry and metabolic flux control, *Thermochimica Acta*, 172, 31–52.
80 Gnaiger, E. Méndez, G. and Hand, S. C. (2000). High phosphorylation efficiency and depression of uncoupled respiration in mitochondria under hypoxia, *Proceedings of the National Academy of Sciences USA*, 87, 11080–11085.
81 Dauncey, M. J. (1995). From whole body to molecule: an integrated approach to the regulation of metabolism and growth, *Thermochimica Acta*, 250, 305–318.
82 Kemp, R. B. (1987). Heat dissipation and metabolism in isolated animal cells and whole tissues/organs, in A. M. James (ed.), *Thermal and Energetic Studies of Cellular Biological Systems*, Wright, Bristol, pp. 147–166.

83 Kemp, R. B. and Gnaiger, E. (1989). Aerobic and anaerobic energy flux in cultured animal cells, in W. Wieser and E. Gnaiger (eds.), *Energy Transformations in Cells and Organisms*, Georg Thieme Verlag, Stuttgart, pp. 91–97.
84 Nedergaard, J. Cannon, B. and Lindberg, O. (1977). Microcalorimetry of isolated mammalian cells, *Nature*, 267, 518–520.
85 Palou, A. Pico, C. Bonet, M. P., and Oliver, P. (1998). The uncoupling protein, thermogenin, *International Journal of Biochemistry and Cell Biology*, 30, 7–11.
86 Kemp, R.B. (2000). 'Fire burn and Cauldron bubble' (W. Shakespeare): What the calorimetric-respirometric (CR) ratio does for our understanding of cells? *Thermochimica Acta*, 355, 115–124.
87 Kemp, R. B. Belicic-Kolsek, A. Hoare, S. Schmalfeldt, M. Townsend, C. and Evans, P. M. (1995). A thermochemical study of metabolic pathways in activated and triggered 2C11-12 mouse macrophage hybridoma cells, *Thermochimica Acta*, 250, 259–276.
88 Bäckman, P. Kimura, T. Schön, A. and Wadsö, I. (1992). Effects of pH-variations on the kinetics of growth and energy metabolism in cultured T-lymphoma cells, Journal of Cell Physiology, 150, 99–105.
89 Kemp, R. B. and Guan, Y. (2000). The application of heat flux measurements to improve the growth of mammalian cells in culture, *Thermochimica Acta*, 349, 23–30.
90 Schulz, C. Irani, N. Wirth, M. van den Heuvel, J. and Wagner, R. (1999). Decreased lactate production and increased cell growth in BHK cells transfected with a pyruvate carboxylase construct, in: A. Bernard, B. Griffiths, W. Noé, F. Wurm (eds.), *Animal Cell Technology: Products from Cells; Cells as Products. Proceedings of the 16th ESACT Meeting*, Kluwer, Dordrecht, pp. 73–80.
91 Feng, Y. Olomolaiye, D. and Kemp, R. B. (2004). Thermobiochemical evidence for the rapid metabolic rate in hybridoma cells genetically engineered to overexpress the anti-apoptotic protein bcl-2 in batch culture. *Thermochimica Acta*, in press.
92 Singh, R. P. Emery, A. N. and Al-Rubeai M. (1996). Enhancement of survivability of mammalian cells by overexpression of the apoptosis-suppressor gene *bcl-2*. *Biotechnology and Bioengineering*, 52, 166–175.
93 Simpson, N. H. Milner, A. E. and Al-Rubeai, M. (1997). Prevention of hybridoma cell death by *bcl-2* during suboptimal conditions, *Biotechnology and Bioengineering*, 54, 1–16.
94 Kroemer, G. (2003) Mitochondrial control of apoptosis: An introduction, *Biochemical and Biophysical Research Communications*, 304, 433–435.
95 Ishaque, A. and Al-Rubeai, M. (2002). Role of vitamins in determining apoptosis and extent of suppression by bcl-2 during hybridoma cell culture, *Apoptosis*, 7, 231–239.

Chapter 10

Thermal investigations on social insects

E. Schmolz[1*] and I. Lamprecht[2]

[1]*Free University of Berlin, Institute for Biology/Zoology, Königin-Luise-Straße 1-3, D-14195 Berlin, Germany*
[2]*Free University of Berlin, Institute for Biology, Ehrenbergstraße 26-28, D-14195 Berlin, Germany*

Abstract

Social insects (honey- and bumblebees, wasps, hornets, ants and termites) are interesting in many aspects, among them the energetic advantages of social life and conquering of unfavourable territories. Own investigations and data from literature deal with the energy metabolism of these insects (except termites because of experimental difficulties), with locomotor activities, energy balances of foraging, energy saving by insulation of wasp nests compared with the afford to construct the wooden envelope, bee cluster strategy for surviving at low temperatures, and rearing of brood.

The energy and heat flow data were obtained by indirect and isoperibol direct calorimetry, bomb calorimetry, experiments with a customer constructed carousel flight calorimeter, thermometry, and false colour thermography.

Introduction: energetics of insects

Of all animal species on this planet, more than one third are insects. Roughly 750 000 insect species have been described so far, but only about 13 500 or 2 % of them are eusocial with a division of labour between reproductives like drones and queens and sterile forms like workers. Nevertheless, these 2 % of insect species make up more than half of the total insect biomass [1]. The energetical impact of social insects on ecosystems can not be underestimated, as insects themselves represent more than half of the animal biomass on earth. Their ecological success is not only due to their remarkable diversity, but also to their abundance in all ecosystems.

Eusocial insects are exceptional in the animal kingdom, as only a few members of a colony will produce offspring (reproductives: queens and drones), whereas the

* eschmolz@zedat.fu-berlin.de

D. Lörinczy (ed.), The Nature of Biological Systems as Revealed by Thermal Methods, 251–283.

majority is more or less sterile (workers) and only helps to rear the progeny of the reproductives. Both reproductives and helpers form a complex society, often with division of tasks and labour and advanced modes of communication between them. Eusociality has been evolved independently in at least 7 taxa of insects, in ants, bees, wasps, and termites, but also in some beetles, thrips and aphids [2]. Here, we will only deal with the first three, as our energetical knowledge about the latter ones is very poor. Social insects are most abundant at low latitudes and low elevations. Their activities are most conspicuous in summer in temperate areas, and throughout the year in subtropical to tropical climates. In temperate regions, honeybees (*Apis mellifera*) and some ant species like the red wood ant *Formica polyctena* hibernate socially with workers and reproductives together in winter clusters or nest hills. Normally, these colonies persist for some years (perennial colonies), whereas in bumblebees, wasps and most ants only the young queens, which have mated in autumn, overwinter alone to establish a new colony in the preceding season solitarily (annual colonies).

The evolutionary success of social insects can be attributed to two major advantages over solitary insects. Firstly, tasks like feeding and caring for the brood or foraging can be performed simultaneously by workers, instead of sequentially like in solitary species. This increases the overall work efficiency drastically. Secondly, colonies of insects which are able to communicate (even sometimes in a very primitive way) and react to each others behaviour are able to perform tasks which are impossible for solitary insects, like building of huge nests in sometimes unfavourable places or conditions, and defend these against large predators.

There is a striking analogy between an eusocial insect colony and a multicellular organism, which led entomologists from the beginning of the 20th century to invent the term 'superorganism' [3]. From an energetical point of view, superorganisms are fascinating objects, as it is easy to investigate their energy budgets, metabolic rates and physiology of their subunits, i.e. individuals, as well as the energy metabolism of complete colonies. In this chapter, we will first concentrate on the energetics of solitary insects as well as colony members and the physiology of single workers, queens and drones. We will then have a look on the energetics of superorganisms and on the action of parasites or enemies and the defence against them.

Energy consumption for development

Principally two different modes of development can be found in insects: an incomplete (hemimetabolous) and complete (holometabolous) development. Hemimetabolic insects do not have larvae, which are defined as juveniles possessing at least one trait which can not be found in adults. Juveniles with hemimetabolic development resemble their parents, except that they have no wings. A life cycle with two distinct morphs, the larva and the adult (sometimes denoted as 'imago'), is characteristic for holometaboluous insects. In fact, the appearance of these life forms differs so much that in some cases the connection between both can only be concluded from direct observations. The adult stage is normally rather short in many winged insect species (with the most extreme case of *Dolania americana*; a mayfly in which adults live no longer than 5 minutes in average [4]). The reason for this is the reduction of moultings (ecdysis) in adults, which prevents further growth as well as regeneration of lost body parts, which is common in many arthropods like spiders and crustaceans. Adult Pterygota do not moult their cuticle mainly for one reason: it is not possible or would be at least extremely cumbersome to shed their exoskeleton which includes the wings (the only known exception from this rule are mayflies (*Ephemeroptera*)). The adaptive advantage of the development of larvae lies in an avoidance of competition between adult insects and their young for space and food resources. For this end, both life forms colonise different habitats. Simplified, larvae serve as body mass and energy accumulators, whereas adults serve for only one purpose: reproduction. As the latter is sometimes achieved in a much shorter time than the first, larvae are the predominant life form in many holometaboluous insects. Both larvae and adults are connected through a pupal stage where the insect undergoes a complete metamorphosis, which can be regarded in many aspects as a second embryogenesis. During this stage, most larval organs and structures are degraded and the adult body is formed completely new.

Energetically, we can divide the life of a holometaboluous insect in three stages: (i) the larval stage, which serves mainly for the acquisition and accumulation of biomass and energy reserves. (ii) The pupal stage where the stored biomass is converted into an adult, and (iii) the adult which reproduces itself. In the adult stage, we can expect very high metabolic rates during activity, and very low rates during rest. The pupa should spent as less energy as possible for metamorphosis as it cannot take up food. The larva should have a highly efficient metabolism for rapid growth. As all three life stages have different lifestyles and serve for different purposes, the relationship between heat production rates and age or body mass cannot expected to be linear. Holometaboluous insects therefore represent an unique taxon for investigations on different aspects of animal growth, activities and reproduction.

ENERGETICS OF DEVELOPMENT IN *GALLERIA MELLONELLA*

The greater wax moth *Galleria mellonella* is a parasite of the honeybee *Apis mellifera* and is typical in many aspects for the Holometabola (sensu Ax [5]). Adult moths invade honeybee colonies, mate and lay their eggs. The growing moth larvae are the actual parasites: they eat away the combs, and feed on wax (hence the name), pollen, honey and bee brood. The period of larval growth endures about 40 days, depending on ambient conditions. Subsequently, the larvae pupate and undergo a metamorphosis. The hatching adult possesses only reduced mouth parts and is unable to take up food for the rest of its life. Such a situation is not exceptional among insects [2]. *Galleria mellonella* is an interesting model species for investigations on developmental energetics of holometaboluous insects because of two reasons. Over a short period, they gain a multiple of their body mass [6]. This does not only enhance experimental investigations, but means also that wax moth larvae underlie a high selective pressure for rapid growth, which should be as efficient as possible. From an energetic point of view, the inability of adult moths to take up food is most interesting. As early as the prepupal stage, they must have accumulated all energy reserves for metamorphosis as well as for their whole lifespan as adults.

Wax moth larvae exhibit a peculiarity during their development: they exhibit unusually high body temperatures, which can easily be detected when one touches accumulations of larvae [7]. Such conglomerations occur in nature on combs of highly infested bee hives. Principally, two reasons (phenomena) could be responsible for such high body temperatures: an increased individual heat production rate or a quasi-adiabatic agglomeration-effect, when many larvae crowd together and decrease their surface-volume ratio in order to prevent heat loss. The latter is known for some lepidopteran larvae [8]. In a series of calorimetric investigations, Schmolz and Schulz [9] were able to demonstrate that 5th and 6th instar larvae exhibit very high heat production rates. L_5 - instars have specific heat production rates of up to 160 mW/g wet weight (w.w.) (Fig. 1). As the calorimetric measurements were made on isolated individuals, agglomeration as cause for high body temperatures could be ruled out. Nevertheless, due to

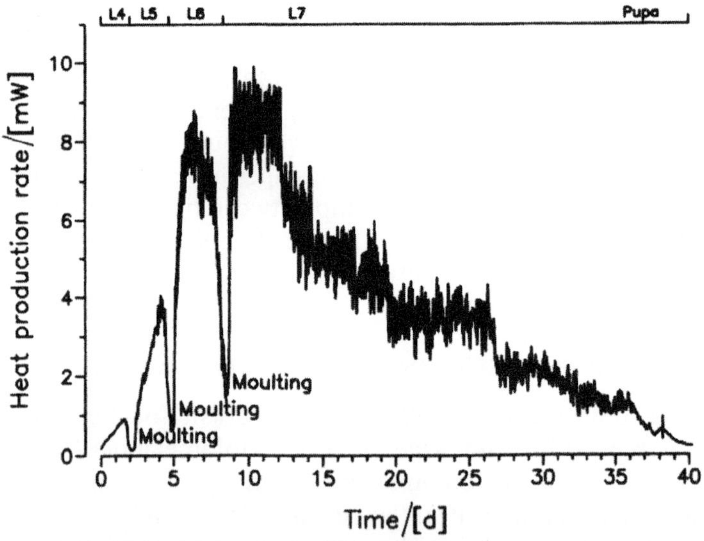

Fig. 1 Power-time curve of a wax moth larva during a long-time experiment. The larva was placed in the measuring vessel at the *L4* stage and continued to develop until the pupal stage. Note the decrease in heat production during ecdysis (moulting). Figure from [9]

their relatively large body mass, L_5 - instars may contribute an additional amount of heat in populations with high densities of wax moth larvae. Up to now, the adaptive significance of the increased heat production in L_5 and L_6 - larvae is not known. An interesting new concept in nutritional physiology of insects suggests a diet-induced thermogenesis, which has been already described for mammals [10, 11]. Insects are able to uncouple metabolic pathways when they are forced to take up unbalanced (e.g. low - protein) diets in order to dispense surfeits of energy and nutrients above metabolic needs [12]. Detailed studies on wax moth nutrition [13, 14] point to such an effect for *Galleria*, which would explain excessive heat production in old larvae.

ENERGETICS OF DEVELOPMENT IN HONEYBEES

The development time of honeybee larvae from egg to beginning of pupal metamorphosis lasts 9 days. Food uptake ends at larval day 7, external food supply from nurse bees ends with the sealing ('capping') of the brood cell at day 5. In this short time, the larva increases its body mass from 0.32 mg (first larval instar) to 173 mg (seventh larval instar). The rapid growth rate of honeybees should require an intense metabolism with relatively high heat production rates. Nevertheless, the heat production rates of honeybee brood are lower compared to *Galleria*. Nurse bees produce a nourishing fluid ('food sap') for feeding the larvae, which is rich in carbohydrates and proteins and more balanced than wax moth food. Diet-induced thermogenesis is highly unlikely to occur in honeybees. Further-

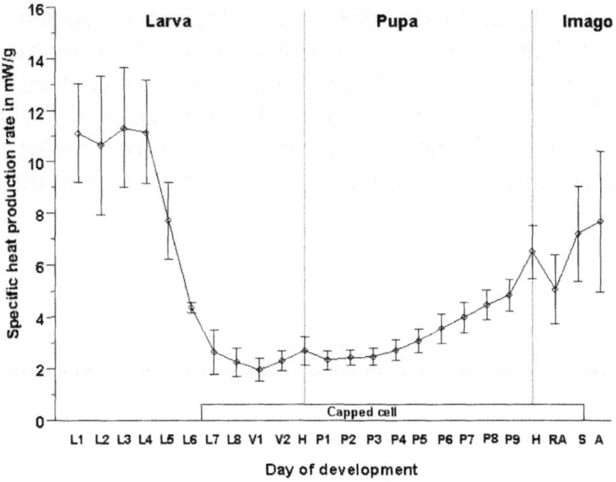

Fig. 2 Specific heat production rates of honeybee drones (*Apis mellifera*) during development. Bars indicate standard deviations. Abbreviations: *L*2, *L*3, ... *L*7: 2nd, 3rd,...7th day of larval development; *V*1, *V*2: day 1 and 2 of prepupal stage; *H*: moulting from larva to pupa; *H*: moult from pupa to adultus; *P*1-*P*8: Day 1-8 of pupal development; *RA*: resting adult between moult to adult and hatch from the capped cell; *S*: hatching adult; *A*: young adult (age < 24h). Unpublished data from the authors

more, growth is enhanced by maintaining a high and constant ambient temperature of 35–36°C on the brood combs [15]. Metabolic processes during development of the young bees are adapted and optimised for this temperature, leading to a rapid development, but as a drawback the larvae are extremely vulnerable to temperature changes. Bee pupae which were experimentally raised at lower temperature (32–34.5°C) have weaker learning abilities and their performance as foragers is significantly reduced [16]. Besides these peculiarities, the course of heat production during ontogenesis of bee larvae is normal (Fig. 2), with high heat production rates of about 10 mW/g during the first 4 days of development, and an decrease down to 2 mW/g in the last days before pupation. This decrease, which can also be observed in *Galleria,* is partly a result of a higher percentage of fat body mass, which is metabolically less active than other tissues and partly a means of energy saving as preparation for metamorphosis.

METAMORPHOSIS AND MOULTING

The typical time course of heat production rates during metamorphosis is a U-shaped power-time (*p-t*) curve. First, after ecdysis (moulting), larval organs are degraded. During this process, the heat production rates decrease. Subsequently a so-called 'second embryogenesis' takes place, when, starting from the imaginal cells, the body is reconstructed with all adult organs. The growing complexity of the pupal body is reflected by an increase of heat production. The U-shaped *p-t* pattern can be found in wax moths [6, 17] and the beetle *Tenebrio molitor* [18]. In honeybees, the course of the metabolic rate is not U- but rather J-shaped with no decrease of heat production at the beginning of metamorphosis

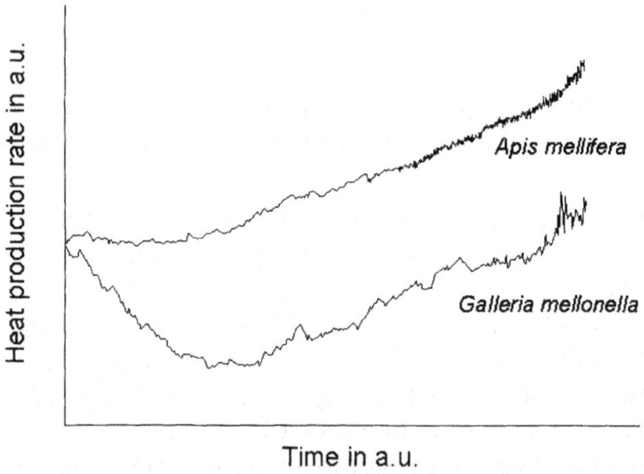

Fig. 3 Heat production rates of pupae of a wax moth (*Galleria mellonella*) and a honeybee worker (*Apis mellifera*). Both curves are not identical in scale; for better comparison of the curve structures, units are omitted. The honeybee curve is J-shaped, the wax moth curve shows a U-shape, which is typical for holometaboluous insects. Taken from [19]

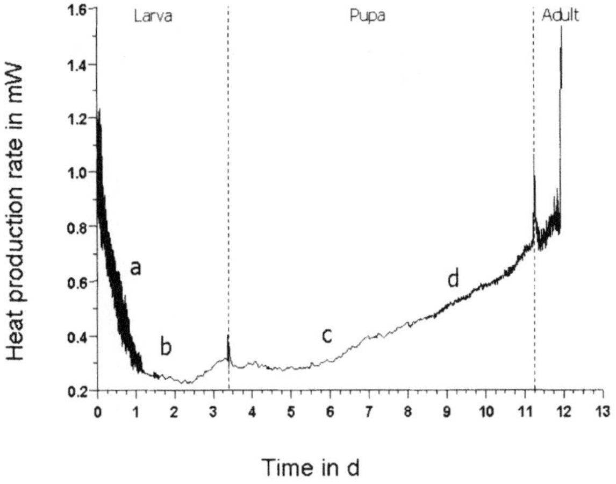

Fig. 4 Heat production rate of a honeybee worker pupa during metamorphosis. a–d refer to Fig. 5, where curve structures are shown in more detail. Unpublished data from the authors

[19] (Figs 3 and 4). As honeybees are less complex in their body plan compared to free-living larvae like those of wax moths or mealworms (*Tenebrio*), the degradation of organs is less complicated, which is reflected in the *p-t* curves. Consequently, the time needed for metamorphosis is shortened and takes only 8 days in worker bees. It is interesting to have a more detailed look on the curve

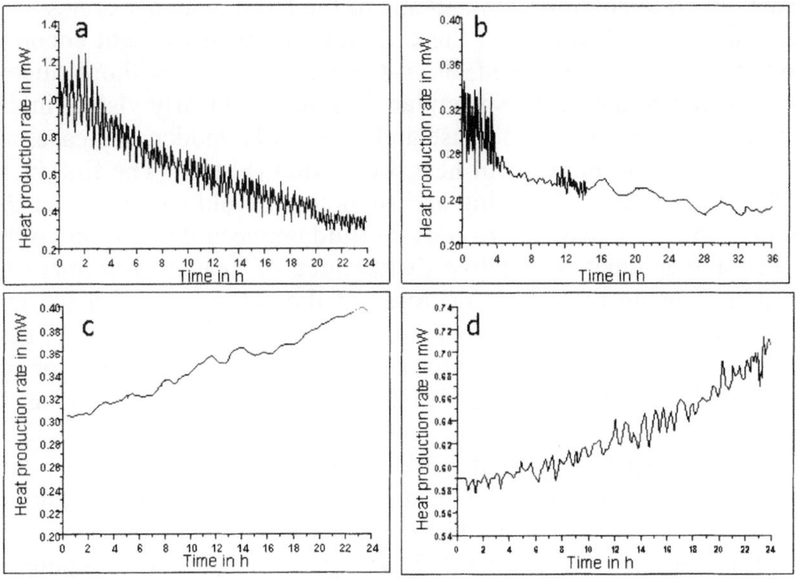

Fig. 5 Details of the heat production rate of a honeybee pupa, taken from Fig. 4. For details see text. Unpublished data from the authors

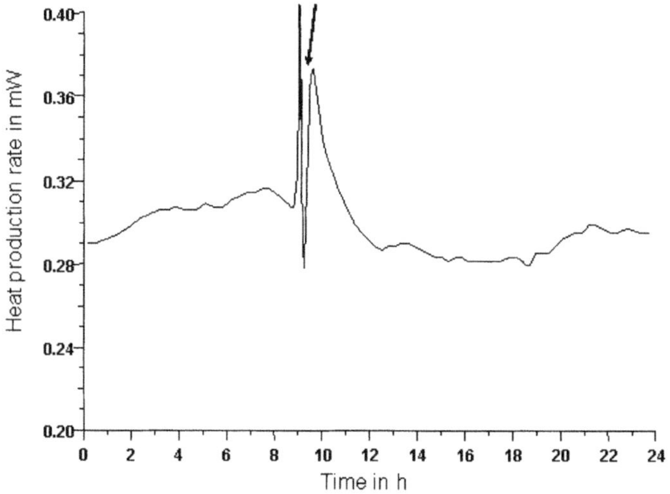

Fig. 6 Power-time curve of a honeybee prepupa during moult to pupa. The arrow indicates the release of ecdysial fluid and the decrease of heat flux signal as a consequence of evaporation. Unpublished data from the authors

structures during the pupal phase in honeybees (Fig. 5 a-d). Shortly before pupal ecdysis, the curve exhibits regular oscillations with an amplitude from about 0.5 mW in the beginning down to 0.08 mW before the larva becomes motionless and enters the prepupal stage. The fluctuations, first intensive and then reduced, can be attributed to spinning movements of the larva, which secretes a silk-like substance that is wallpapered to the cell wall and forms a light cocoon. After spinning movement has stopped, the *p-t* curve is smooth without fluctuations. Like in wax moths, the ecdysis from larva to pupa is clearly visible in the calorimeter curves, with a sharp increase and a similarly sudden decrease, which is again followed by an increase of heat production (Fig. 6). The first increase is due to the movement for breaking and shedding the old cuticle. After the pupa has moulted the old exoskeleton, ecdysal fluid between the new and old cuticle evaporates and induces a negative disturbance of heat measurement through evaporation, causing the abrupt decrease of the *p-t* curve. After the moulting fluid has evaporated, the artificial disturbance of heat measurement ceases and the curve increases. This pattern in the calorimeter curves is typical for all insect moultings and can also be found in *Galleria mellonella* [9, 20] as well as *Tenebrio molitor* [18].

During the first five days of worker pupa metamorphosis, the *p-t* curve is smooth with no abrupt changes. The curve then again exhibits oscillations with an amplitude of about 0.01 mW. The pupa is now coloured and has a dark cuticle. The fluctuations can be attributed to slight movements of the pupa, mainly of its legs. The imaginal ecdysis ends the honeybee metamorphosis with a strong peak of heat production, which is also typical for *Galleria*. As the amount of ecdysal fluid produced for the moult from pupa to adult (imago) is small, the disturbance is not clearly distinguishable from fluctuations caused by body movements.

ENERGY CONTENT DURING DEVELOPMENT

In principle, the main role of insect larvae is accumulation of body mass and energy for pupal metamorphosis. It is trivial to say that together with the growth of a larva its energy content (Q) increases. But the energy density (q), defined as energy content divided through dry body mass may change considerably during development and is different in honeybee drones and workers (Fig. 7). At the end of their larval development, drones have a higher energy density than workers. In relation to their energy content at the beginning of pupation, drones lose more energy than workers. The energy loss of holometaboluous insects during pupal metamorphosis amounts to roughly 50 % of the prepupal energy content (see Table 1) in all investigated species. The energy density of worker bee pupae remains more or less constant and the percentage body fat content does not change much over time, whereas in drones the energy density decreases dramatically. In *Galleria mellonella*, the energy density increases during the pupal phase, because more protein and carbohydrates are metabolised than fat, which remains as an energy depot of higher caloric value for the adult [6]. In contrast to this, worker bee pupae do not use up their energy depots selectively and have nearly no fat reserves for their adult life. They have a lower energy density compared to *Galleria* throughout their whole life. Adult wax moths have q values of 37.5 J/mg, whereas an adult honeybee worker has an average q value of 23.2

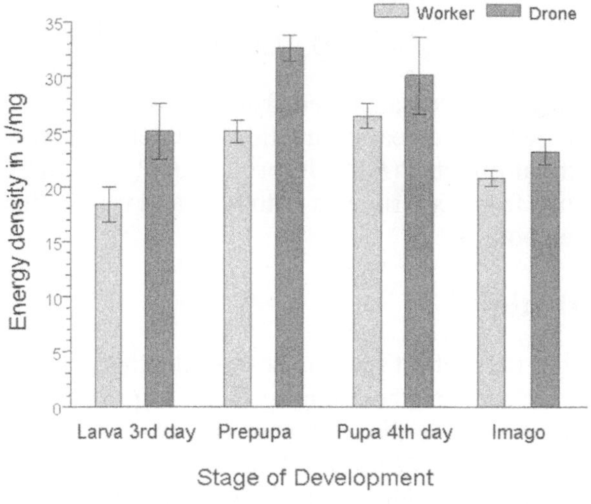

Fig. 7 Energy density of some representative developmental stages of honeybee workers and drones. Bars indicate standard deviations. Unpublished data from the authors

Table 1 Average energy content Q, energy density q and body mass M_B (dry weight) during metamorphosis of some insects

Species		Q/ J	Q/ J/mg	M_B/ mg	Δ_{tp}/ d	ΔQ/ J	ΔQ % of Q_P	Ref.
Galleria mellonella male *(Wax moth)*	Prepupa Adult	2940 1538	32.3 37.5	89 41	11.4	1402	47.7	[6]
Apis mellifera (Honeybee) worker	Prepupa Adult	958 499	25.0 23.2	38 22	7.9	459	48	[21]
Apis mellifera drone	Prepupa Adult	2934 1186	32.6 20.8	90 57	9.5	1748	59.6	[21]
Vespa crabro (Hornet) worker	Prepupa Adult	3800 2451	21.6 17.9	176 137	16.3	1349	55	[22]
Tripoxylon politum (solitary wasp)	Prepupa Adult	1987 1267	21.6 19.8	92 64	32 - 55	720	56.8	[96]

J/mg [6, 21]. Young hornet (*Vespa crabro*) workers have even lower values of 17.9 J/mg [22].

It is possible to calculate the amount of energy needed for pupal development (Q_M) by subtraction of the energy content at the end of metamorphosis (Q_A) from that of the pupa at the beginning (Q_L):

$$Q_M = Q_L - Q_A$$

Q_M is spent in a given time needed for metamorphosis (Δt_P), and one can calculate the mean heat production rate of the pupa (P_{CC}) as:

$$P_{CC} = Q_M / \Delta t_P$$

The heat production rate calculated from the decrease of energy content (P_{CC}) can be compared to the heat production rate determined on living pupae by means of direct calorimetry (P_{DC}). In principle, P_{CC} should equal P_{DC}, and empirical data gained from experiments with bees and wax moths render a good accordance between both.

Locomotor activities

Flying insects exhibit the highest specific heat production rates known in the animal kingdom [23]. Nearly all heat during flight is produced by the thoracal muscles, which have an extraordinary power output. The reason for this lies in the relatively small body mass of insects. In this case inertia plays nearly no role during locomotion. Gliding flight is thus known only for a few species (some butter- and dragonflies) and most flying insects have to continuously beat their wings with high frequencies (up to 500 Hz) to stay in the air.

The difference between the heat production rate of a resting animal and that of a maximally active one during exercise is called the metabolic scope. In terrestrial vertebrates, the metabolic scope is up to 10, i.e. the difference in the heat production rate during locomotion and rest is 10-fold. As resting insects have heat production rates which lie in the normal range expected for small invertebrates (between 0.1 and 10 mW/g) but exceptionally high rates during flight, the difference between heat production rates at rest and activity is up to 100 - fold. The amount of energy spent for flight activities is therefore the biggest part of their individual energy budget. Numerous accounts have been undertaken to measure energy consumption of insects during flight [24–28]. Nearly all of them followed indirect approaches like respirometry or thermometry. Nevertheless, it is possible to determine the heat production rates of some insects directly in an isoperibolic heat flow calorimeter [29]. For the investigation of heat productions during flight under controlled conditions, the insects have to be tethered to a gibbet in a wind channel or to a roundabout construction. In the first case, the insect remains at its position, and air drag simulates appropriate flight conditions, in the second approach the insect flies rounds in a carousel. Up to now, only one study has been published about the measurement of oxygen consumption of free-flying insects in a wind channel [30]. The results of these experiments remained controversial [31] and only a very few animal at all did fly. For obvious reasons, the direct measurement of heat production in wind channels is impossible, regardless if the insects are tethered or not. In a flight calorimeter, the insect is fixed to a small carousel arm. The calorimeter volume is about 10 L, big enough to allow roundabout flight in a sufficiently large radius, and small

Fig. 8 IR thermography of a flying hornet (*Vespa crabro*) worker in a wind channel experiment. The inserted photo shows a hornet in the same flying position for comparison. Unpublished data from the authors *(See colour section, p. 349).*

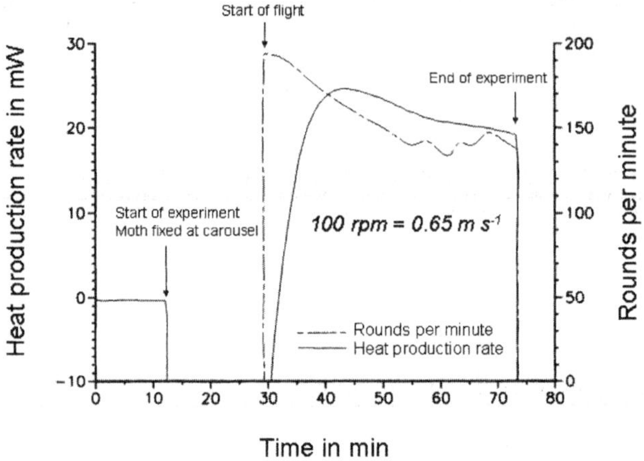

Fig. 9 Typical calorimetric curve of a flying wax moth. Straight line: heat production rate; dotted line: flight speed. After fixation of the wax moth to the carousel arm, the calorimeter signal is disturbed due to the influx of air from the surrounding, but increases rapidly after the moth begins to fly. Taken from [35]

enough to keep the time constant of the calorimeter low. A main drawback of direct flight calorimetry is the fact that illumination, which is necessary for most insects to induce flight behaviour, has to be limited due to the heat production of any stronger light sources. These obstacles can be overcome by the use of light guides connected to a cold light source, introducing only a small amount of extra heat. The second possibility is the use of chemical cold light, with an even lower heat production than the electrical light [29]. Up to now, the heat production rates of flying hornets and wax moth have been measured in such a calorimeter. The specific heat production rates of hornet workers and drones were temperature-dependend, decreasing from about 160 mW/g at 20°C to 100 mW/g at 30°C in workers [32]. A more detailed study on the flight behaviour of hornets in a bigger roundabout outside the calorimeter revealed that the reduction of heat production is accompanied by a reduction of flight speed as a means of thermoregulation during flight [33]. Like most insects, hornets regulate their body temperature during flight, either through forced convection and physical increase of heat loss, e.g. through a droplet of fluid between the mouthparts or, as new findings imply, through reduced heat production, which is the normal means of thermoregulation in birds and mammals [34]. The temperature of flying hornets can be followed in a wind channel with IR-thermography (Fig. 8). Such investigations [25] reveal the temperature distribution in flying insects without physical disturbance through thermocouples. The thorax as main source of heat is significantly warmer than the abdomen and the head.

Adult wax moths are good study objects for insect flight calorimetry, as they are easy to rear in the laboratory and fly at low illuminances (Fig. 9). Although the flight velocity of wax moths is rather slow (about 1m/s), they have

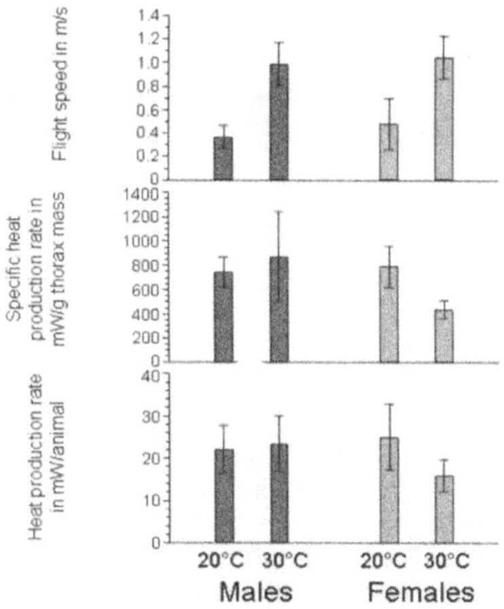

Fig. 10 Mean heat production rates and flight speeds of wax moth in relation to sex and ambient temperature. Bars indicate standard deviations. Note that the specific heat production rate is referred to the thorax mass, as the flight muscles of the thorax produce nearly all heat during flight and the abdomen mass of a female is unproportionally higher than that of a male. Taken from [35]

specific heat production rates of up to 870 mW/g thorax mass at 30°C [35]. Nevertheless, although a temperature-induced increase of flight speed with higher temperatures was observed (Fig. 10), the heat production rates did not change as observed in hornets and honeybees [32–34].

Sleep and hibernation

Most social insects are diurnal foragers. When they stay inside their nests at night, they perform nest duties like cleaning cells, care for the young or thermoregulation. Nevertheless, in honeybees the most active foragers pause in a state which closely resembles sleep. The phenomenon of sleep is not restricted to mammals, but can also be found in other vertebrates like birds or turtles and even in invertebrates [36, 37]. It may therefore not be too surprising to find sleep-like behaviour in honeybees [38], but the adaptive value of sleep for organisms is still puzzling and has not been clarified up to now. It has been proposed that sleep facilitates restorative processes [39]. Growth hormones are released during sleep, and it maintains the integrity of the immune system [40, 41]. Important neuronal functions like memory capabilities depend on sufficient amounts of sleep [42]. On the other hand, sleep can be a means of saving energy for organisms during inactive phases [43]. Honeybees are good test objects for the latter hypothesis, as they switch between different modes of thermoregulation and energy expenditure [44]. Workers maintain constant body temperatures when foraging for food. They can elevate their body temperature through simultaneous contraction of flight muscle antagonists ('shivering') and

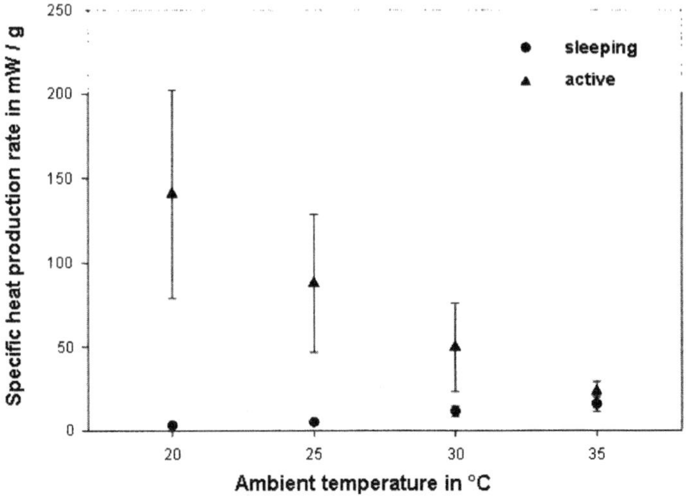

Fig. 11 Specific heat production rates of sleeping and active bees at different ambient temperatures. Active bees are endothermic, sleeping bees ectothermic. Data for sleeping bees were taken from [45], for active bees from [77]. Both data sets were obtained from measurements with the same calorimeter

are thus endothermic**, whereas resting honeybees are sometimes ectothermic or thermoregulatorically inactive and their body temperature equals the ambient temperature. Indeed, the heat production rates of sleeping bees increase with ambient temperature (ectothermic state). These rates are generally lower than in active bees at daytime, were the heat production decreases with increasing temperatures (Fig. 11) [45]. Nevertheless, sleeping or resting bees do not seek the coldest place on the combs in the hive, which would yield the biggest benefits in energy saving. Instead, they prefer mediate temperatures of about 28°C, as a slightly increased metabolic rate faciliates restaurative processes. In this case, energy saving is balanced with the needs for regeneration.

A situation were energy has clearly to be saved in insects is overwintering. Insects hibernate in different lifestages, as eggs, prepupae, pupae or adults. The first three mentioned have already stored energy reserves (for embryogenesis or metamorphosis), whereas adults have to provide themselves with food reserves before they undergo diapause (i.e. a controlled state of inactivity and rest during unfavourable conditions). Diapause metabolism is not necessarily equal to resting metabolism, as the physiology of diapausing insects changes considerably, e.g. as adaptation to sub-zero temperatures. During diapause, heat production rates are generally low and the insects are ectothermic. As calorimetric investi-

** The term 'endothermic' is used here in a physiological/biological sense and denotes a process where an organism actively elevates its body temperature and heat is being produced and not consumed, as the psysicochemical meaning of this term would imply

gations on beetles could show, the heat production rates and energy expenditure can be different between subspecies [46]. *Chrysomela lapponica* is a leaf beetle with at least to subspecies, one living in Northern Europe, and the other in Central Europe. The specific heat production rate during overwintering is higher in the northern population at a temperature of 5°C when compared to the Central European. An explanation for this is the smaller body mass of the northern beetles which renders higher specific heat production rates compared to the Central European group. Moreover, the average and maximum temperatures during overwintering in the leaf litter layer in the Central Europe are more variable and on average higher than under the snow cover in Northern Europe. The higher temperatures that the Central European beetles encounter during overwintering are detrimental because an increase of the environmental temperature means an increase of body temperature and thus also an increase of metabolic costs for the beetle. In order to reduce such metabolic costs at elevated overwintering temperatures, a low specific metabolic rate can be an adaptation of the Central European beetles to the climatic conditions in their habitat. In insects, as in all animals, the specific heat production rate is negatively correlated with body mass [47–49]. The large body mass of Central European beetles leads to a lower specific heat production rate and thus a further reduction of their energy expenditure. Nevertheless, one has to consider that bigger beetles may have a higher proportion of metabolically less active tissue (fat body) or even inactive extracellular material like the cuticle which contributes to their mass, thus leading to an underestimation of their real specific heat production rates.

In social insects living in colder climates, we can find two different strategies for hibernation. Honeybees and some ants overwinter as an intact colony. For this end, they have to accumulate energy reserves for the winter period (e.g. honey). Most other social insects like bumblebees, wasps and many ant species do not overwinter as a colony. Instead, only the young and inseminated queens, which were reared at the end of the summer season, leave their home colonies and seek for hibernation places. They solely rely on their internal energy reserves for overwintering and initiate their own, new colonies in the following spring. The old colony with its foundress dies out at the end of each season.

As we have seen in the case of leaf beetles, a main problem for insect overwintering are mild temperatures. Bumblebee and wasp or hornet queens have evolved strategies to survive sub-zero temperatures down to at least −18°C. Calorimetric investigations on hornets were also able to demonstrate that warm winters will not endanger overwintering queens as the energy reserves gained in the mother colony before their leave of the nest are sufficient for hibernation even at average temperatures up to 15°C [22]. In this case, social insects have a great advantage over solitary species, which have to seek for energy reserves themselves without help of numerous foragers. Nevertheless, desiccation remains a main problem for overwintering queens. One means to overcome the risk of water loss is discontinuous ventilation of the tracheae (tube-like breath-

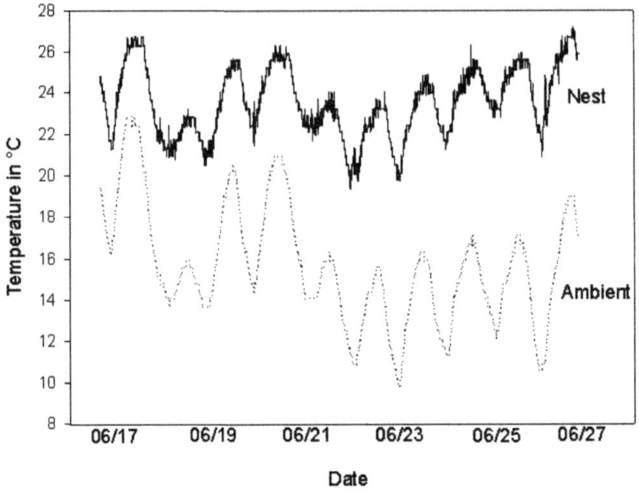

Fig. 12 Nest temperature of an embryo hornet nest with only the queen present. The workers had not hatched at the time of measurement. The nest had an approximate diameter of 10 cm and was placed in a bird nest box. The nest temperature was measured with thermocouples directly at the comb. Measurements ware made from 17th to 27th juny 1999 by one of the authors (E.S.). Unpublished data

ing organs of insects), as the main portion of water gets lost over the respiratory organs. Discontinuous ventilation appears as a distinct breathing cycle which conserves water in the respiratory organs. Such cycles of discontinuous ventilation can be observed in *p-t* curves of calorimetric measurements and have been described in detail elsewhere [19].

The most dangerous phase in the colony cycle of annual insect societies is spring, when young queens emerge from hibernation and found nests. At this time, the queen has no worker around her to help in nest construction or foraging. As the queens forage for larval food solitarily, the nest temperature can not be maintained at a constant level (Fig. 12) and the brood has to be insensitive to temperature changes and exposition to the cold. The queen spends a high amount of energy and time in warming the brood to accelerate larval growth [50, 51]. The estimated probability for a queen to survive the colony founding phase is about 10% [22]. If we consider this, it is not surprising that in honeybees and some ants the queen overwinters together with her workers. The greater probability to survive spring and the rapid growth of these colonies after winter suggest a huge advantage for such perennial colonies. The drawbacks are the need for stocking food during the summer season for successful overwintering of hundreds of workers. To illustrate this, we may look at the following example: an average honeybee colony consumes about 20 kg of collected honey during winter, which equals to an energy resource of ca. 260 MJ. For collection of the honey store, workers have to undertake 170 000 foraging flights (they have to collect about 50 kg nectar to produce 20 kg honey, and each forager can carry a load of 0.3 g). If each foraging flight lasts an average of 10 min, and the heat production rate of a flying honeybee amounts to 50 mW [28], the energy spent on such a flight is 30 J, which makes a total of only 5.1 MJ the honeybee colony has to invest for the collection of a sufficient amount of honey to survive winter. In winter, the bees are inactive, reduce their metabolism [52] and form clusters. The heat produced by the individual bees is conserved due to the decreased surface/volume ratio. Moreover, the bees are able to regulate the density of the cluster by packing worker bees more or less densely together. The heat conductivity of the cluster is altered in such a way that the core temperature amounts to 30 to 32°C. The temperature at the periphery of the cluster is always > 8°C, the critical temperature for worker bees at which they fall in chill coma. A theoretical model [53] explains the temperature distribution in a bee cluster and fits well to empirical data obtained from measurements in bee clusters [54].

Colonies of social insects

Social insect colonies were since long particularly fascinating for biologists because of the ability to maintain social homeostasis. Honeybees are able to regulate winter cluster or brood comb temperature precisely, ants control humidity in their nest hills, and even at cold days the temperature in most insect colonies is above the ambient. As early as 1837, Newport measured the temperature inside bumblebee nest with mercury thermometers [55]. But as we will see, the ability for maintenance of social homeostasis in respect to control of nest climate has been overestimated sometimes. A closer look reveals the dependency of regulatory processes on several parameters. This holds true especially for annual insect societies were colony development includes a solitary phase of nest foundation, production of an increasing number of workers during the social phase and a subsequent production of reproductive forms (drones and young queens). Colony demography in bumblebee, wasp and hornet nests changes substantially throughout the season, and it is not surprising that the ability for nest thermoregulation is weak at the beginning and the end of the colony cycle, when only a few workers are present [56, 57].

Thermoregulation in animals is closely connected to energy metabolism. Only few attempts have been published to measure the metabolic rates of wasp colonies [58–60]. The experimental setup of the first two mentioned studies was simple: after the end of the season, when all wasps had died, an electrical resistor was placed inside the nest. The energy which was necessary to heat up the nest interior by means of the resistor to those nest temperatures measured during summer was regarded as the metabolic rate of the colony. Colony metabolism was therefore determined only for very short periods.

A more sophisticated method followed calorimeter construction principles first published by Wesolowski and colleagues [61], who used modified camping cold boxes as robust heat flow calorimeters with large active volumes of up to 24

L. These camping cold boxes carry a Peltier element at the backside or bottom wall which serves in its original function as a heat pump when a DC current of 12 V is applied. Reversely, Peltier elements can be also used as heat flux sensors, because a heat gradient between cooling box inside and outside creates a thermoelectric voltage proportional to the temperature difference and thus the heat flow.

The first investigation on social insects with this calorimeter type was made on the bumblebee *Bombus lapidarius* [62]. Foragers of the bumblebee colony had free access to outdoor environment through a long exit tunnel, which limited cold draught from outside. The calorimeter itself was placed in a climatized laboratory room, and colony development could be followed calorimetrically throughout the season. *B. lapidarius* had a brood temperature between 27 and 32 °C, and the mean rate of heat loss of the colony varied between 0.3 and 1.4 W. Both temperature and the heat production rate decreased together with the declining number of foragers in the nest at the end of summer. A winter cluster of the honeybee *Apis mellifera* was also investigated for heat production during winter. The heat production rate of this cluster, which had a biomass of about 1 kg, was negatively correlated to ambient temperature with specific heat production rates from 20.5 mW/g at 0°C to 11.5 mW/g at 5°C [60]. To investigate the effects of colony size on the energetics and thermoregulation of hornet (*Vespa crabro*) nests, a detailed calorimetric study on this species was made with a similar experimental setup as that for bumblebees [22, 60]. Moreover, the biomass of the colonies could be determined, and thus specific heat production rates calculated. The level of heat production of the colonies was clearly dependent on colony size, and maximum heat production rates (i.e. the highest measured heat

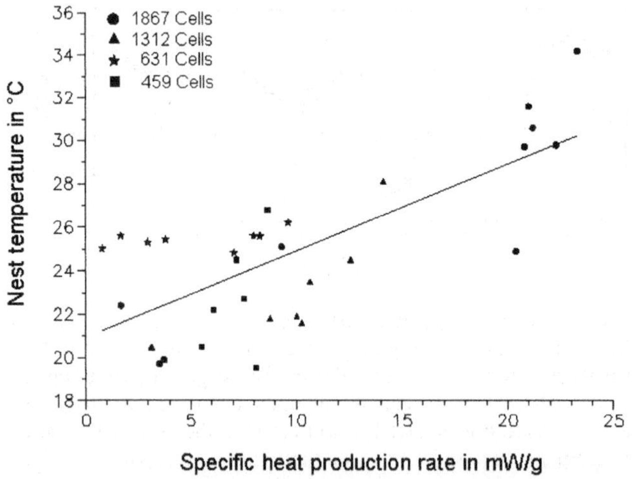

Fig. 13 Nest temperature and specific heat production rates of 4 hornet colonies. Different symbols denote each nest. The number of cells in the nests was determined at the end of the season in the empty nests as parameter for colony size. For details of measuring method see text

production rate of each nest during the season) varied between 12.5 W or 23.3 mW/g for the largest and 0.738 W or 9.6 mW/g for the smallest colony, respectively [22]. Specific heat production rate increased with colony mass. Figure 13 shows the relation between the specific heat production rates and nest temperature in 4 hornet colonies. These data are higher than values for heat production in *Dolichovespula*-nests as reported by Gibo and coworkers [58] showing values between 3.3 and 6.4 mW/g at an ambient temperature of 30°C. The difference may result from (i) different ambient temperatures in the experiments (20°C for the hornet nests compared to 30°C in the wasp nests), (ii) the indirect measuring method (heating up the nest by means of an electrical resistor after the demise of the colony) compared to the direct calorimetric measurements and (iii) the different nesting biology of *Dolichovespula* wasps, which build free hanging nests with probably better insulation capacities compared to *Vespa crabro*-nests, which are usually placed in cavities [63].

Although the ambient temperature in hornet nests remained stable throughout the experiment, the nest temperatures varied to a greater degree in all colonies and thermoregulation was imperfect. During the season, the curve of the mean daily nest temperatures showed considerable variations. The level of the nest temperature depended on colony size in the same way as the heat production rates. The largest colony had the highest nest temperature with 34.2°C, and the smallest colony reached only 25.3°C even at the point of maximum colony size.

In some aspects of their social physiology, hornet colonies may be regarded as superorganisms. It is interesting to compare their heat production rates, which represent the energy expenditure of active, and not of resting animals, to field metabolic rates of endotherm organisms like birds or mammals. The maximum heat production rate of a large hornet colony with a weight of 530 g amounts to 12.5 W. Multiplied with time, this renders a daily energy expenditure of 1080 kJ d^{-1}. This may be compared with a small nest which has a daily energy expenditure of 63.8 kJ d^{-1} and a mass of 63 g. Nagy [64] presents some allometric equations for field metabolic rates of various bird and mammal groups. According to his data, the field metabolic rate (FMR) of a placental mammal can be calculated as

$$\mathrm{FMR}[\mathrm{kJ}\cdot\mathrm{d}^{-1}]=3.35\cdot\mathrm{M}[\mathrm{g}]^{0.813}$$

and for passerine birds as

$$\mathrm{FMR}[\mathrm{kJ}\cdot\mathrm{d}^{-1}]=8.88\cdot\mathrm{M}[\mathrm{g}]^{0.749}$$

Using these equations, one can calculate that a mammal with a body mass of 530 g has a FMR of 549 kJ d^{-1}, and a bird of equal body mass a FMR of 974 kJ d^{-1}. Small mammals and birds with a weight of 63 g will have a FMR of 97 kJ d^{-1} and 198 kJ d^{-1}, respectively. Hence, the energy expenditure of a large hornet colony is nearly double that of the FMR of a mammal with equal weight, and nearly the same as the FMR of a bird. Small hornet colonies spend less energy than birds or mammals with equal weight. This is explained by the fact that birds and mammals regulate their body temperature precisely in a relatively narrow range, regardless of their body size, which makes them homeotherms. Hornets, in contrast, do not regulate their nest temperature to a specific temperature. The nest temperature of annual colonies is not only dependent on the stage of colony development, but also on colony size. Small or medium sized colonies never reach the level of nest temperatures which can be observed in big nests, and most annual insect societies in temperate regions are far from being homeotherm superorganisms. The time needed for development in holometaboluous insects is a function of temperature [65–67]. High nest temperatures will therefore shorten brood development, but workers must make a compromise between time and energy spent for thermoregulation (which includes not only the warming of the brood but also the building of the nest envelope) and for foraging. A colony of an annual social insect species can be best described as an incubator which is heated up whenever the energy budget of the colony allows it, and the brood is presumably adapted to a broad range of temperatures. Perennial insects like honeybees are able to maintain their brood comb temperatures at a much more narrow range. The development time for the larvae and pupae is shortened in bees compared to bumblebees, hornets and wasps, but adaptation to a specific temperature needed for development leads to a higher vulnerability of bee brood too cold temperatures. Both in bees and in hornets, workers were observed to warm pupae directly. A study using IR thermography was able to show the warming behaviour of the bees with a high energy expenditure [68]. In hornets and wasps, a similar behaviour was observed and a thermoregulatoric pheromone,

Fig. 14 IR thermography of a small hornet nest. The site of the combs inside the nest is clearly indicated by the yellowish area in the middle (about 22.4°C). The warm red spots are hornets walking on the nest envelope. Taken from [72] *(See colour section, p. 349).*

which was released by old pupae, was identified as z-9 pentacosene [69, 70]. Nevertheless, recent analysis of this phenomenon was not able to confirm these observations. z-9 pentacosene is a rather ubiquitous substance, which occurs in nearly all nest structures, and it did not induce warming behaviour [71].

An interesting feature of hornet and wasp colonies is their nest construction. The combs, which are oriented horizontally, are covered with a multi-layered nest envelope. The nest material consists of wood which is mixed with saliva to a paper-like pulp mass. Indeed, physical properties like the areal density of the nest envelope amounts to 76.5 g/m^2 and is equal to that of paper (80 to 110 g/m^2) [72]. Investigations on hornet nests, in which empty nests were heated up to steady-state internal temperatures with an electrical resistor inside the nest revealed a thermal conductivity of 0.08 to 0.20 W/(°C m), which is again about the same as the thermal conductivity of paper (0.14 W/(°C m)). During the season, the combs are used subsequently from the most upper one to the lowest and youngest comb for rearing brood. Old combs are abandoned during the season and workers start to seal off theses nest parts with nest material in order to reduce 'dead space' in the colony which has to be warmed. Figure 14 shows an IR thermography of an inhibited hornet colony. In the empty upper parts of the nest envelope the surface temperature is lower than in the inhabited parts, indicating a successful thermal insulation of this nest area.

The nest material is being collected by foragers, which carry small pieces of pulp to the nest. A typical foraging flight for wood lasts about 15 min, and the pulp load has an average weight of 7.6 mg, which corresponds to an envelope area of 1 cm^2. The envelope of a small nest has a mass of 23 g, built from about 3000 pulp loads collected in 750 foraging hours. A hornet worker flies at a power of about 72 mW (see above, locomotor activities) at an ambient temperature of 20°C. The energetic investment for pulp foraging is thus 72 mW · 750 h = 54 000 mWh or 194 kJ, an amount of energy equivalent to the energy required for heating the nest for 5 days at moderate summer temperatures. This calculation demonstrates that the building of a nest envelope makes up a surprisingly small part of the total energy budget of a colony [72].

A quite different but also impressive nesting structure can be found in red wood ants (*Formica polyctena*). Wood ants build large hills made of pine needles and other organic matter. The temperatures in these nest mounds are above the ambient and remain stable over longer periods. It has long been questioned how wood ants are able to heat up their nest hills, as ant workers are strictly ectothermic and do not heat themselves up at low ambient temperatures like bees or wasps do to heat their nests. Two main sources of internal thermogenesis have been proposed: microbes which produce heat through decay of organic nest material (similar to a compost pile) and/or the heat production of ants, which may produce metabolic heat through their activities, even if they are not endothermic. Calorimetric investigations on heat production of nest material revealed specific heat production rates of 0.18 mW/g at 20°C, whereas ants have specific heat production rates of 2.6 mW/g [73]. But when one calculates the total heat production of an ant hill, the proportion of biomass from ants and nest material has to be considered. As there is much more mass of nest material than of ants in an ant hill, the total heat production of nest material amounts to about 16 W compared to 2.6 W produced by the ants. Nevertheless, the heat production rates of nest material are highly dependent on humidity. At relative humidities below 80 %, the heat production of nest material is near to zero, and most heat is produced by the ants [74]. The heat source in an ant hill is probably changing with ambient and internal conditions [75]. Most likely, the ants themselves are able to control factors like humidity or even growth of microorganisms in the nest material.

Like in the case of overwintering leaf beetles (section 5) we can find different metabolic rates and lifestyles in subspecies of social insect species. A prominent case for this is the Western Honeybee *Apis mellifera*. Western honeybees are naturally distributed throughout the Middle East, Parts of Western Asia, Europe and Africa [76]. The behaviour and physiology of African and European honeybee subspecies differs drastically. Honeybee races from Africa are more aggressive, produce less honey, form no winter clusters and are more immune against many bee diseases. European bees are generally more gentle in their behaviour and build up large stocks of honey for overwintering, which makes them much more suitable for beekeeping. A multitude of calorimetric measurements on heat production rates of European honeybees (e.g. *Apis mellifera carnica*) has been published up to now [77, 78 (both with extensive compilations of data from other authors)], but only few studies about the metabolism of their African cousins. The Egyptian honeybee *Apis mellifera lamarckii* belongs to the African races and is significantly smaller than *Apis mellifera carnica,* has shorter legs and wings and a lower body mass [76]. Through the introduction of European bee races nowadays in Africa both bee races are kept by Egyptian beekeepers, but due to their aggressive temper *lamarckii* bees become less frequently used and play only a minor part in Egyptian honey production. The occurrence of new bee parasites in Egypt may change this picture in the future, as *lamarckii* bees are more resistent against a major bee pest, the brood mite *Varroa destructor* (see section 7). A calorimetric investigation of both subspecies revealed higher specific heat production rates in *lamarckii* bees [79]. Egyptian honeybees have highest heat production rates at 20°C (75 mW/g) whereas the highest value for *Apis mellifera carnica* at 25°C amounts to 47 mW/g. At temperatures between 30 and 40°C, the specific heat production rates of *lamarckii* bees do not change, which means that they obviously achieve thermoregulation by physical means (evaporation of fluid droplets, see section 3) and not by reduction of the metabolic rate, which is found in *carnica* bees. An interesting study on the metabolic rates of African-European bee hybrids determined with respirometry revealed low metabolic capacities. The metabolism of European as well as African bees is higher than that of hybrids probably due to disruption of coadapted enzymatic complexes in hybrids [23].

Social insects regulate the nest temperatures mainly to facilitate brood development. The amount of energy invested in brood rearing makes up the biggest part of the energy budged of a colony, and the main goal of an insect society is to produce and tend as much offspring as possible. A large hornet colony raises about 2000 workers during a season and produces about 3500 reproductives (drones and queens) [22]. The energy content of worker prepupae amounts to 3800 J (see Table 1) and is slightly lower than that of reproductives with 4300 J. In total, a colony invests about 7600 kJ in workers and about 15 000 kJ in the production of drones and young queens. Compared to this, the energy investment of honeybees in brood rearing is much higher. A well-developed honeybee colony with 80 000 workers in early summer (peak of colony strength) produces about 500 000 workers over the whole season. This number seems to be surprisingly high, but the average life span of an adult worker is only 2 to 3 weeks. The total energy content of all workers produced in one season amounts to about 480 000 kJ.

Colony defense

As we have seen in the previous section, social insect colonies contain large amounts of energy, either as brood or as stock (honey, pollen). The energy resources stored in a nest are alluring to predators and parasites which leads to a strong selection to evolve means of colony defence. The most annoying aspect of social insect life to humans is that the vast majority of all species have evolved poisonous stings. Even without stingers, which have been lost in many ants and some bees, the poison glands remain. In social insects, the defence of the colony is one of the most important tasks for all colony members, but the action of one sting against a large predator (e.g. many honey-hunting mammals, including humans) is not sufficient to protect a colony. A strong selection for rapid communication to recruit nestmates against predators or intruders led to the evolution of alarm pheromones. These pheromones, which can be found in honeybees, ants, wasps and termites, induce aggressive behaviour. An alarm pheromone consists typically of several components. Chemical analysis of alarm pheromones of social insects has proven to be difficult due to their high volatility and only a few components have been identified. In honeybees, isopentylacetate and 2-heptanone are the main components of the alarm pheromone. When confronted with these compounds, honeybees become extremely aggressive. Nevertheless, their alarm pheromone contains a total of 20 components. For most of them, the function and significance is unknown [80]. In the hornet *Vespa crabro*, 4 components have been identified so far. The main active substance seems to be 2-methyl-3-butene-2-ol [81].

The action of alarm pheromones can be investigated with ethological assays, which are cumbersome and time-consuming, because they can not be performed in the laboratory but only in the field, and then only in secure areas as provoked bees, wasps or hornets may represent a hazard to humans in the vicin-

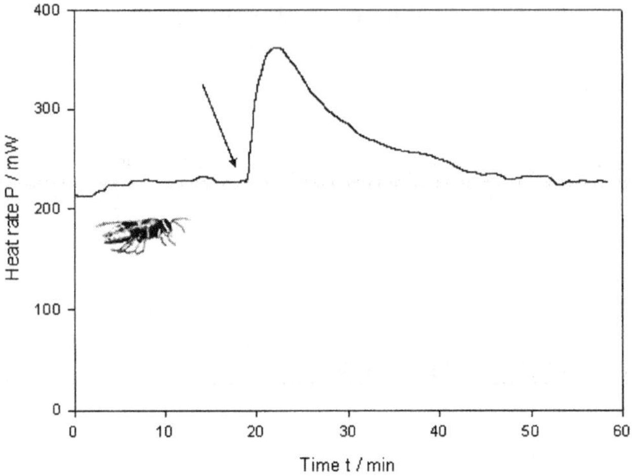

Fig. 15 Heat production rate of five hornet workers before and after application of 200 µl 3-methyl-2-butene-1-ol, an alarm pheromone component. Arrow indicates pheromone application. Unpublished data from E.S. and C. MacLean. Inserted sketch drawn by C. MacLean

ity of the test site. An alternative to behavioural field tests are studies on the physiological reaction of provoked insects. Earlier studies demonstrated the dramatic increase of honeybee heat production rates when exposed to alarm pheromones, which is mainly caused by strongly increased locomotive activities. In most cases, the chosen method was respirometry [82], but direct calorimetry has also been tested successfully for this purpose [19, 71, 83]. Physiological tests, which investigate the increase in heat production rates, cannot be used as biotests in a narrow sense, because not only alarm pheromones may induce an increase of heat production, but also other pheromones (e.g. sexual pheromones, brood pheromones), and the increase of heat production rates is not always necessarily an aggressive reaction. Nevertheless, such tests can be appropriate to quantify specific parameters of an alarm response when a substance has already been proven to be alarm-inducing. For calorimetric determination of the action of pheromones, an air stream containing the pheromones is led through the calorimeter vessel with bees or wasps (Fig. 15) [71, 83]. In honeybees, the metabolic rates increase by about 30 % up to values of 85 mW/g, depending on the pheromone substance. Isopentylactetate, 2-heptanone, 2-heptanol and 1-hexanol were identified as inducers of strong and rapid metabolic reactions, whereas other pheromone components like benzylacetate, 9-octadecen-1-ol and 2-nonanol were totally inactive [83]. A study on the action of alarm pheromones on hornets revealed that 2-methyl-3-buten-2-ol, 3-methyl-3-buten-1-ol, 3-methyl-2-buten-1-ol and 4-penten-1-ol were equally active in the metabolic tests. Previous ethological studies only identified 2-methyl-3-buten-2-ol as active component. Moreover,

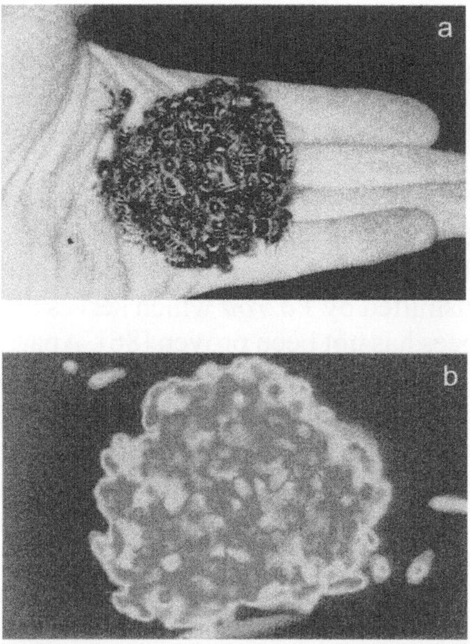

Fig. 16 a) IR thermograph showing a hot defensive ball of the Eastern Honeybee *Apis cerana*. b) Defensive ball with about 400 bees which engulf a predatory hornet. (Courtesy of Masato Ono) *(See colour section, p. 350)*.

even an alarm pheromone component of the honeybee, isopentylacetate, provoked strong metabolic responses of the hornets [71]. Studies on the action of alarm pheromones will be of increasing interest in the future, as not only the number of allergic reactions to insect stings is increasing but evidence is accumulating that some of the alarm pheromone components can be found in perfumes or cosmetic products [84].

Besides stinging an intruder to death or at least to drive him back, the eastern honeybee *Apis cerana* has evolved an unique method to fight their main predator, the giant hornet *Vespa mandarinia*. When a hornet appears at the bee hive entrance, honeybees aggregate around the intruder and more than 500 bees quickly engulf the hornet in a ball. IR-thermography shows that the ball temperature is very high (about 47°C), which proves lethal to the hornet but not to the bees [85] (Fig. 16).

Energetics of social insect parasites

Another type of enemies of social insects are parasites. As honeybees are economically and ecologically important due to their honey production and pollination activities, research on bee parasites focuses on the treatment of these bee pests. From an energetical point of view, only two bee parasites have been investigated in some detail, the bee mite *Varroa destructor*, which is a recently introduced brood parasite of great economical importance, and the wax moth *Galleria mellonella*.

Varroa destructor invades brood cell of honeybees shortly before pupation to feed on the body fluids (hemolymph) of the brood and adults of *Apis mellifera* L. and causes damage on the latter. Adult bees from *Varroa*-infected brood cells are severely malformed, and the extent of harm is directly proportional to the degree of infestation. However, the causality between viral, bacterial and fungal infections probably transmitted by *Varroa* which serves as a vector and the malformation of hatching bees has not been proven [86]. Apart from its role as a disease vector the damage caused by infestations even with a few numbers of mites on the ontogenesis and weight at hatching of the adult bees was demonstrated [87]. The weight loss of a bee infested at the brood stage, compared to non-infested bees, is directly proportional to the number of infesting mites. In order to also assess the energetic impact of *Varroa* mites on their host, the energetics of bee development and metabolic activities of *Varroa* were investigated [88, 89]. Adult mites sit on worker bees, feed on their hemolymph and are transported to the brood combs. Here, the mites enter brood cells shortly before pupation of the bees. They prefer drone brood cells, as drones have a longer development time than workers. Inside the cells, which are sealed off for metamorphosis of the bees, the mites feed on the pupal hemolymph and reproduce. Typically one mite (female) will produce one son and four daughters. Interestingly, the metabolic rates of mites, which parasitize adult workers, are significantly higher than those for bee brood (mites from adult workers 17.44 μW/mg, worker brood 14.32 μW/mg, drone brood 15.6 μW/mg; see Fig. 17). The weight loss of mites without constant food supply was determined during starvation experiments and

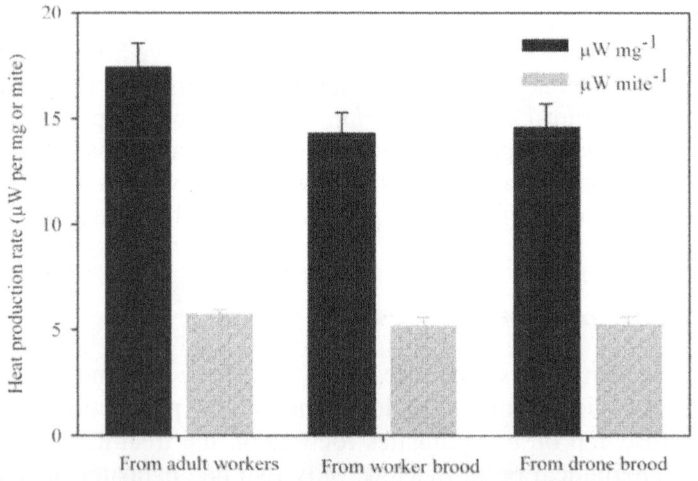

Fig. 17 Heat production rates of *Varroa* mites collected from adult bee workers, worker brood and drone brood. Bars indicate standard deviations. Taken from [88]

amounted to 0.7 mg/d. Total weight loss of pupae due to parazitation with mites was calculated as 25.7 mg for workers (body mass of prepupa: 145 mg) with an infestation with 2 mother mites, which produce 1 daughter mite each. The high mass loss of the bees and the low metabolic rates of the mites lead to the conclusion that mites can utilize only a small amount of the energy, which they take up as food [88, 89, 90]. Infestation with *Varroa* mites does not alter the energy density of bee pupae, which is a hint that mites feed unselectively on bee pupae, whereas it has a significant influence on the total energy content at the end of metamorphosis. Worker pupae infested with 4 to 6 mites have an energy content of about 15 % below the normal value at the end of pupation. Unparasitized young worker bees ready to hatch have an average body mass of 107 mg, which is practically the weight of an emerged adult [78]. This means that bees have no energy reserves left after metamorphosis, as they do not need them for their life in a social colony were food is readily provisioned by nestmates. As a consequence, they have nearly no safety margin in their energy reserves during pupation when they can not take up any external food and are more vulnerable to starvation than other solitary insects like wax moths. Therefore, an energy loss of about 15 % due to mite parasitation may have serious effects on the development of the bees. Mite infestation does not affect the heat production rates of bee pupae. The energy loss can therefore be directly linked to the hemolymph robbing of the mites [89].

Wax moths normally invade weak colonies, which are no longer able to defend themselves. The amount of destruction by wax moth larvae is tremendous, and after heavy infestations the bee combs are often barely recongnizable as combs at all. *Galleria mellonella* is an ideal organism to demonstrate that social insect colonies are a rich food source for any predator or parasite. Wax moths themselves are most interesting when one investigates how the energy they take up from their hosts is used during their own life cycle. As we have stated in section 2, the larvae consume energy from the bee hives, whereas neither pupae nor adults take up any external food. Table 2 gives an overview of wax moth heat production rates during their life cycle. The highest specific heat production rates can be found in the 5th larval stage and in flying adults. The latter means that wax moth spent a high amount of energy in a life phase when they are unable to restore energy. The energy content of a male wax moth prepupa amounts to about 3000 J. A dead male adult still has an energy content of 1300 J [6], i.e. a wax moth can utilize 1700 J for metamorphosis and adult life. The duration of the pupal phase is 11 days, the average heat production rate in this time is 1 mW. The moth consumes therefore about 950 J for its metamorphosis from larva to adultus. Consequently, only an energy of 750 J is left over for adult life. With a flight speed of 1 m/s and a heat production rate in flight of 25 mW, the male moth is theoretically able to fly a distance of about 30 km. Of course, it does not spent all energy in flying activities, but undoubtely locomotion and the active finding of a sexual partner consumes most of its energy, as the resting metabolic rate is low

Table 2 Calorimetrically determined specific heat production rates p and body mass M_B of the wax moth *Galleria mellonella* during its life cycle

Developmental stage	p (mW/g)	M_B (mg)	Reference
Larval stage 1 + 2	55	0.5	[9]
Larval stage 5	160	230	[9]
Larval stage 7	18	178	[9]
Pupa	4.1		[6]
Adults			
Resting male	5.6	84	[6]
Resting female	5.3	146	[6]
Flying male	277	84	[35]
Flying female	181	146	[35]

(0.47 mW). A simple experiment demonstrates this: if one keeps male and female wax moths either in mixed groups or as isolated individuals, the life duration alters dramatically. Isolated males live for 24 days, whereas males together with females have a much shorter life of only 3 days [6].

Honeybees have evolved some means to get rid of parasites, among them the use of propolis. Propolis is a sticky, wax-like substance collected by bee foragers from tree buds of various species, depending on the locality. The antimicrobial effects of propolis are known since long in natural medicine, but a series of investigations has revealed that it can be also effectively used against bee parasites like *Varroa destructor* or *Galleria mellonella* [91, 92]. Ethanolic extracts of propolis induce damaging effects against *Varroa* mites. Besides classical bioassays for toxicity of propolis, calorimetric investigation were able to demonstrate sublethal effects on mites, which could not be detected with traditional methods [93, 94]. Especially the action of propolis against wax moths, which was mainly in a sublethal range, could be detected by isoperibolic calorimetry. The development of *Galleria* pupae was negatively affected when they were treated with ethanolic propolis extracts [92].

Of medical interest are investigations on the action of propolis extracts on bacteria and fungi. Again, besides conventional petri dish assays, investigations on microbial cultures with flow calorimeters yielded important results about the effectiveness of these extracts [95].

Conclusion

Social insects have been investigated in many aspects. Evolutionary biologists are fascinated by their altruism, and the genetical basis for this is still under investigation. Ecologists have to deal with insect societies as they are key species in most ecosystems. Behavioural biologists investigate their means of communication and selforganized decision-making. For a physiologist dealing with animal energetics, social insects are particularly interesting as they can be viewed from two different perspectives, the individual one and that of a colonial superorganism. Social insects are ideal for investigation of the physiology of individuals, as they can be easily bred in large numbers. On the other hand, experiments on superorganismal energetics will improve the understanding of their ecological role as well as their evolution.

References

1 Hölldobler, B. Wilson, E. O.: A Journey to the Ants. Harvard University Press, Cambridge Massachusetts, 1994.

2 Gullan, P. J. Cranston, P. S.: The Insects. An Outline of Entomology. Chapman and Hall, London 1994.

3 Moritz, R. F. A. Southwick, E. E.: Bees as Superorganisms. Springer, Berlin 1992.

4 Sweeney, B. W. Vannote, R. L.: Population synchrony in mayflies: a predator satiation hypothesis. *Evolution,* 36 (1982) 810–821.

5 Ax, P.: Das System der Metazoa II. G. Fischer Verlag, Mainz 1999.

6 Schmolz, E. Drutschmann, S, Schricker, B. Lamprecht, I.: Calorimetric measurements of energy content and heat production rates during development of the wax moth *Galleria mellonella. Thermochim. Acta,* 337 (1999) 83–88.

7 Buchmann, S. L. Spangler, H. G.: Thermoregulation of the Greater Wax Moth *Galleria mellonella. Am. Bee J.,* 131 (1991) 772.

8 Mosebach-Pukovski, E.: Über die Larvengesellschaften von *Vanessa io* und *Vanessa urticae. Z. Morphol. Ökol. Tiere,* 33 (1937) 358.

9 Schmolz, E. Schulz, O.: Calorimetric investigations on thermoregulation and growth of wax moth larvae *Galleria mellonella. Thermochim. Acta,* 251 (1995) 241–245.

10 Kleiber, M.: The Fire of Life. Wiley, New York 1961.

11 Bachman, E. S. Dhillon, H. Zhang, C. Cinti, S. Bianco, A. C.,Kobilka, B. K. Lowell, B. B.: AR signaling required for diet-induced thermogenesis and obesity resistance. Science 297 (2003) 843–845.

12 Trier, T. M. Mattson, W. J.: Diet-induced thermogenesis in insects: a developing concept in nutritional ecology. *Enivron. Entomol.,* 32 (2003) 1–8.

13 Jindra, M. Sehnal, F.: Larval growth, food consumption, and utilization of dietary protein and energy in *Galleria mellonella. J. Insect Physiol.,* 35 (1989) 719–724.

14 Jindra, M. Sehnal, F.: Linkage between diet humidity, metabolic water production and heat dissipation in the larvae of *Galleria mellonella. Insect Biochem.,* 20 (1990) 389–395.

15 Winston, M. L.: The Biology of the Honey Bee. Harvard University Press, Cambridge Massachusetts, 1987.

16 Tautz, J. Maier, S. Groh, C. Rössler, W. Brockmann, A.: Behavioral performance in adult honey bees is influenced by the temperature experienced during their pupal development. *PNAS,* 100 (2003) 7343–7347.

17 Bell, J.: The heat production and oxygen consumption of pupae of *Galleria mellonella* at different constant temperatures. *Physiol. Zool.,* 13 (1940) 73–81.

18 Kuusik, A. Tartes, U. Harak, M. Hiiesar, K. Metspalu, L.: Developmental changes during metamorphosis in *Tenebrio molitor* (Coleoptera: Tenebrionidae) studied by calorimetric thermography, *Eur. J. Entomol.*, 91 (1994) 297–305.
19 Schmolz, E. Lamprecht, I.: Calorimetric investigations on activity states and development of holometabolous insects. *Thermochim. Acta,* 349 (2000) 61–68.
20 Prat, H.: Calorimetry of higher organisms. In: Brown, H. D.: Biochemical Microcalorimetry, Academic Press, New York and London 1969, pp. 181–198.
21 Schmolz, E., Kösece, F., Lamprecht I.: The energetics of honeybee development. In prep.
22 Schmolz, E.: Kalorimetrische Untersuchungen zu Wärmeproduktion und Thermoregulation der Hornisse *Vespa crabro*. Ph.D. – thesis, Free University of Berlin, 1997.
23 Harrison, J. F., Hall, H. G.: African-European honeybee hybrids have low nonintermediate metabolic capacities. *Nature,* 363 (1993) 258–260.
24 Dyer, F. C. Seeley, T. D.: Interspecific comparison of endothermy in honeybees (*Apis*): Deviations from the expected size-related pattern. *J. Exp. Biol.,* 127 (1987) 1–26.
25 Stavenga, D. G. Schwering, P. B. W. Tinbergen, J.: A three-compartment model describing temperature changes in tethered flying bowflies. *J. Exp. Biol.,* 185 (1993) 325–333.
26 Ludwig, H. G.: Der Sauerstoffverbrauch fliegender Coleopteren. *Verh. Dt. Zool. Ges.,* 1960 (1961) 96–99.
27 Gmeinbauer, R. Crailsheim, K.: Glucose utilitization during flight of the honeybee (*Apis mellifera*) workers, drones and queens. *J. Insect Physiol.,* 39 (1993) 959–967.
28 Nachtigall, W. Rothe, U. Feller, P. Jungmann, R.: Flight of the honeybee III. Flight metabolic power calculated from gas analysis, thermoregulation and fuel consumption. *J. Comp. Physiol.,* 158B (1989) 729–737.
29 Schmolz, E. Schricker, B. Lamprecht, I.:Direct carousel flight calorimeter for metabolic investigations of small insects. *J. Therm. Anal.,* 52 (1998) 33–44.
30 Ellington, C. Machin, K. E. Casey, T. M.: Oxygen consumption of bumbleebees in forward flight. *Nature,* 347 (1990) 472–473.
31 Nachtigall, W., Hanauer-Thieser, U., Mörz, M.: Flight of the honeybee VII. Metabolic power versus flight speed relation. *J. Comp. Physiol.*, 165B (1995) 484–489.
32 Schmolz, E. Brüders, N. Schricker, B. Lamprecht, I.: Direct calorimetric measurement of heat production rates in flying hornets (*Vespa crabro*; Hymenoptera). *Thermochim. Acta,* 328 (1999) 3–8.
33 Spiewok, S. Schmolz, E.: Changes in temperature and light alter the flight velocity of hornets. *Proc. R. Soc. Biol. Sci.,* submitted, 2003.
34 Harrison, J. F. Fewell, J. H. Roberts, S. P. Hall, H. G.: Achievement of thermal stability by varying metabolic heat production in flying honeybees. *Science,* 274 (1996) 88–90.
35 Schmolz, E. Geisenheyner, S. Schricker, B. Lamprecht, I.: Heat dissipation of flying wax moths (*Galleria mellonella*) measured by means of direct calorimetry. *J. Therm. Anal.,* 56 (1999) 1185–1190.
36 Tobler, I. Neuner-Jehle, M.: 24-h variation of vigilance in the cockroach *Blaberus giganteus. J. Sleep Res.,* 1 (1992) 231–239.
37 Shaw, P. J. Cirelli, C. Greenspan, R. J. Tononi, G.: Correlates of sleep and waking in *Drosophila melanogaster. Science,* 287 (2000) 1834–1837.
38 Kaiser W.: Busy bees need rest, too. Behavioural and electromyographical sleep signs in honey bees. *J. Comp. Physiol.,* A 163 (1988) 565–584.

39 Drucker-Colin, R.: The function of sleep is to regulate brain excitability in order to satisfy the requirements imposed by waking. *Behav. Brain Res.*, 69 (1995) 117–124.
40 Brown, R.: Muramyl peptides and the functions of sleep. *Behav. Brain Res.*, 69 (1995) 85–90.
41 Everson, C. A.: Functional consequences of sustained sleep deprivation in the rat. *Behav. Brain Res.*, 69 (1995) 43–54.
42 Smith, C.: Sleep states and memory processes. *Behav. Brain Res.*, 69 (1995) 137–145.
43 Berger, R. J. Phillips, N. H.: Energy conservation and sleep. *Behav. Brain Res.*, 69 (1995) 65–73.
44 Heinrich, B.: The Hot-blooded Insects. Springer, Berlin 1993.
45 Schmolz, E. Hoffmeister, D. Lamprecht, I.: Calorimetric investigations on metabolic rates and thermoregulation of sleeping honeybees (*Apis mellifera carnica*). *Thermochim. Acta,* 382 (2002) 221–227.
46 Gross, J. Schmolz, E. Hilker, M.: Thermal adaptations of the leaf beetle *Chrysomela lapponica* (Coleoptera: Chrysomelidae) to different climes of Central and Northern Europe. Environ. Entomol., submitted, 2003
47 Kittel, A.: Körpergröße, Körperzeiten und Energiebilanz II. Der Sauerstoffverbrauch der Insekten in Abhängigkeit von der Körpergröße. *Z. vergl. Physiol.* 28 (1941) 533–562.
48 Coelho, J. R. Moore, A. J.: Allometry of resting metabolic rate in cockroaches. *Comp. Biochem. Physiol.*, 94A (1998) 587–590.
49 Lehmann, F. O. Dickinson, M. H. Staunton, J.: The scaling of carbon dioxide release and respiratory water loss in flying fruit flies (*Drosophila* ssp.). *J. Exp. Biol.* 203 (2000) 1613–1624.
50 Gibo, D. L. Temporale, A. Lamarre T. P. Soutar, B. M. Dew, H. E.: Thermoregulation in colonies of *Vespula arenaria* and *Vespula maculata* (Hymenoptera: Vespidae) III. Heat production in queen nests. *Can. Ent.* 109 (1977) 615–620.
51 Makino, S. Yamane, S.: Heat production by the foundress of *Vespa simillima*, with description of its embryo nest. (Hymenoptera: Vespidae). *Insecta Matsumurana,* 19 (1980) 89–101.
52 Van Nerum, K. Buelens, H.: Hypoxia-controlled winter metabolism in honeybees (*Apis mellifera*). *Comp. Biochem. Physiol.*, 117A (1997) 445–455.
53 Lemke, M. Lamprecht, I.: A model for heat production and thermoregulation in winter clusters of honey bees using differential heat conduction equations. *J. theor. Biol.*, 142 (1990) 261–273.
54 Worswick, P. V. J. Comparative study of colony thermoregulation in the African honeybee, *Apis mellifera adansonii* Latreille, and the Cape honeybee, *Apis mellifera capensis* Escholtz. *Comp. Biochem. Physiol.*, 86A (1987) 95–102.
55 Newport, G.: On the temperature of insects, and its connexion with the functions of respiration and circulation in this class of invertebrated animals. *Phil. Trans. Roy. Soc. London* (1837) 259–338.
56 Gibo, D. L. Yarascavitch, R. H. Dew, H. E.: Thermoregulation in colonies of *Vespula arenaria* and *Vespula maculata* (Hymenoptera: Vespidae) under normal conditions and under cold stress. *Can. Entmol.*, 106 (1974) 503–507.
57 Martin S. J.: Nest thermoregulation in *Vespa simillima*, *V. tropica* and *V. analis*. *Ecol. Entomol.*, 15 (1990) 301–310.

58 Gibo, D. L. Dew, H. E. Hajduk, A. S.: Thermoregulation in colonies of *Vespula arenaria* and *Vespula maculata* (Hymenoptera: Vespidae). II. The relation between colony biomass and calorie production. *Can. Entomol.,* 106 (1974) 873–879.
59 Schmolz, E. Lamprecht, I. Schricker, B.: Calorimetric investigations on social thermogenesis in the hornet *Vespa crabro* L. (Hymenoptera: Vespinae). *Thermochim. Acta,* 229 (1993)173–180.
60 Schmolz, E. Lamprecht, I. Schricker, B.: A method for continuous direct calorimetric measurements of energy metabolism in intact hornet (*Vespa crabro*) and honeybee (*Apis mellifera*) colonies. *Thermochim. Acta,* 251 (1995) 293–301.
61 Wesolowski, T. Schaarschmidt, B. Lamprecht, I.: A poor man's calorimeter (PMC) for small animals. *J. Thermal. Anal.* 30 (1985) 1403–1413.
62 Schutze-Motel, P.: Heat loss and thermoregulation in a nest of the bumblebee *Bombus lapidarius* (Hymenoptera, Apidae). *Thermochim. Acta,* 193 (1991) 57–66.
63 Matsuura, M. Yamane, S.: Biology of the Vespine Wasps. Springer, Berlin 1990
64 Nagy, K. A.: Field metabolic rate and food requirement scaling in mammals and birds. *Ecol. Monogr.* 57 (1987) 111–128.
65 Janda, V. Kocián, V.: Über den Sauerstoffverbrauch der Puppen von *Tenebrio molitor* L. *Zool. Jb. 52, Abt. f. allg. Zool. u. Physiol.,*519–533, 1933.
66 Howe R. W.: Temperature effects on the embryonic development. *Ann. Rev. Entomol.,* 12 (1967) 15–42.
67 Bursell E.: Environmental Aspects - Temperature. In: Rockstein M. (ed.): The Physiology of Insecta Vol.II (2nd ed.), Academic Press, New York and London, pp 1–41, 1974
68 Bujok, B. Kleinhenz, M. Fuchs, S, Tautz, J.: Hot spots in the bee hive. *Naturwissenschaften* 89 (2002) 299–301.
69 Ishay J.: Thermoregulation by social wasps: behavior and pheromones. *Trans. N.Y. Acad. Sci.,* 35 (1973) 447–462.
70 Veith, H. J. Koeniger N.: Identifizierung von cis 9-Pentacosen als Auslöser für das Wärmen der Brut bei der Hornisse. *Naturwissenschaften* 65 (1978) 263.
71 MacLean, C. Schmolz, E.: Calorimetric investigations on the action of alarm pheromones in the hornet *Vespa crabro*. *Thermochim. Acta,* 414 (2004) 71–77.
72 Schmolz E. Brüders, N. Daum, R. Lamprecht, I.: Thermoanalytical investigations on paper covers of social wasps. *Thermochim. Acta,* 361 (2000) 121–129.
73 Coenen-Staß, D. Schaarschmidt, B. Lamprecht, I.: Temperature distribution and calorimetric determination of heat production in the nest of the wood ant, *Formica polyctena* (Hymenoptera, Formicidae). *Ecology,* 61 (1980) 238–244.
74 Horstmann K.: Zur Entstehung des Wärmezentrums in Waldameisennestern (*Formica polyctena* Förster; Hymenoptera, Formicidae). *Zool. Beiträge,* 33 (1990) 105–124.
75 Frouz J.: The effect of nest moisture on daily temperature regime in the nests of *Formica polyctena* wood ants. *Ins. Soc.* 47 (2000) 229–235.
76 Ruttner, F.: Biogeography and Taxonomy of Honeybees. Springer, Berlin 1988.
77 Fahrenholz, L. Lamprecht, I. Schricker, B.: Thermal investigations of a honey bee colony: Thermoregulation of the hive during summer and winter and heat production of members of different castes. *J. Comp. Physiol.,* B 159 (1989) 551–560.
78 Fahrenholz, L. Lamprecht, I. Schricker, B.: Calorimetric investigations of the different castes of honey bees, *Apis mellifera carnica*. *J. Comp. Physiol.,* B 162 (1992) 119–130.

79 Schmolz, E. Dewitz, R. Schricker, B. Lamprecht, I.: Microcalorimetric investigations of energy metabolism in European (*Apis mellifera carnica*) and Egyptian (*A.m.lamarckii*) honeybees. *J. Therm. Anal.*, 65 (2001) 131–140.
80 Schmidt, J. O.: Mass action in honey bees: Alarm, swarming and the role of releaser pheromones. In: Vander Meer, R. (Ed.): Pheromone Communication in Social Insects: Ants, Wasps, Bees, and Termites. Westview Press, Boulder Colorado, 1998.
81 Veith, H. J. Koeniger, N. Maschwitz, U.: 2-Methyl-3-butene-2-ol, a major component of the alarm pheromone of the hornet *Vespa crabro*. *Naturwissenschaften,* 71 (1984) 328–329.
82 Moritz, R. F. A. Bürgin, H.: Group response to alarm pheromones in social wasps and the honeybee. *Ethology,* 76 (1987) 15–26.
83 Schmolz, E. Scholz, T. Lamprecht, I.: Alarmpheromone bei sozialen Insekten. *Nachr. Chem. Techn. Lab.,* 47 (1999) 1095–1098.
84 Ono, M. Terabe, H. Hori, H. Sasaki, M.: Components of giant hornet alarm pheromone. *Nature,* 424 (2003) 637–638.
85 Ono, M. Igarashi, E. Ohno, E. Sasaki, M.: Unusual thermal defense by a honeybee against mass attack by hornets. *Nature,* 377 (1995) 334–336.
86 Boecking, O. Aumeier, P. Ritter, W. Wittmann, D.: Varroatosis-disease complex: is there any interrelation? *Apidologie,* 33 (2002) 486–487.
87 Schneider, P. Drescher, W.: Einfluß der Parasitierung durch die Milbe *Varroa jacobsoni* Oud. auf das Schlupfgewicht, die Gewichtsentwicklung, die Entwicklung der Hypopharynxdrüsen und die Lebensdauer von *Apis mellifera* L. *Apidologie,* 18 (1987) 101–110.
88 Garedew, A. Schmolz, E. Lamprecht, I. The energy and nutritional demand of the parasitic life of the mite *Varroa destructor*. *Apidologie,* in press.
89 Contzen, C. Garedew, A. Lamprecht, I. Schmolz, E.: Calorimetrical and biochemical investigations on the influence of the parasitic mite *Varroa destructor* on the development of honeybee brood, *Thermochim. Acta,* in press.
90 Garedew, A. Schmolz, E. Schricker, B. Polaczek, B. Lamprecht, I.: Energy metabolism of Varroa destructor mites and its implication on host vigour. *J. Apicult. Sci.* 46 (2002) 73–83.
91 Garedew, A. Schmolz, E. Schricker, B. Lamprecht, I.: The varroacidal action of propolis: a laboratory assay. *Apidologie,* 33 (2002) 41–50.
92 Garedew, A. Schmolz, E. Lamprecht, I.: Effect of bee glue (propolis) on the calorimetrically measured heat production rate and metamorphosis of the greater wax moth *Galleria mellonella*, *Thermochim. Acta*, 413 (2004) 63–72.
93 Garedew, A. Schmolz, E. Schricker, B. Lamprecht, I.: Microcalorimetric investigations of the action of propolis on *Varroa jacobsoni* mites. *Thermochim. Acta,* 382 (2002) 211–220.
94 Garedew, A. Schmolz, E. Lamprecht, I.: Microcalorimetric and respirometric investigation of the effect of temperature on the antivarroa action of Propolis. *Thermochim. Acta,* 399 (2003) 171–180.
95 Garedew, A. Schmolz, E. Lamprecht, I.: Microbiological and calorimetric studies on the antimicrobial actions of different extracts of propolis: an *in vitro* investigation. *Thermochim. Acta,* in press.
96 Cross, E. A. Mostafa, A. E.-S. Bauman, T. R. Lancaster, I. J.: Some aspects of energy transfer between the organ-pipe mud-dauber *Trypoxylon politum* and its araneid spider prey. *Environ. Entomol.* 7 (1978) 64–652.

Chapter 11

DSC examination of the musculosceletal system

P. Than[1], I. Domán[1] and D. Lőrinczy[2*]

[1]*Department of Orthopedics*
[2]*Department of Biophysics, University of Sciences of Pécs, Medical School, Pécs, Hungary*

DSC examination of the hyaline cartilage

THE HYALINE CARTILAGE

Characteristic features of the normal hyaline cartilage

Connective and supportive tissues of the human body are characterized by the large amount of basic substances filling the gap between cells and significantly influencing the mechanical properties of the tissue. In opposite to the connective tissues - where the ground substance is mainly formed by fibers - in case of supportive tissues, like for example cartilage, other organic and inorganic materials are typical for the basic substance.

Cartilage itself is a tissue where a relatively little number of bubble-like cells is bedded in an elastic but also strong basic substance. Because of this structure cartilage is very resistant against compression forces, but hardly withstands shearing stresses. The most rigid form is the hyaline type cartilage, covering the articulating surfaces of bones. This cartilage type has a homogenous basic substance, which surrounds groups of 2–4 cells having a very characteristic form. The plasma of these cells contains a large amount of glycogen and little fat bubbles. The ground substance has a fibrous frame built of II-type collagen. Within this frame complex proteins (proteoglycan) are placed, which contain a large amount of chondroitin sulfate, later being the most typical component of hyaline cartilage. The rest of the ground substance is composed of inorganic materials and in 70–80% of water.

The collagen-proteoglycan matrix serves as the mechanical frame of the cartilage and simultaneously determines its mechanical properties [1]. The collagen fibers having a triple helix form are responsible for tensile strength while proteoglycans are responsible for compressibility. Hyaline cartilage is a tissue having a complex and active metabolism from the biochemical point of view.

* denes.lorinczy@aok.pte.hu

D. Lőrinczy (ed.), The Nature of Biological Systems as Revealed by Thermal Methods, 285–305.

The cells produce both the basic substance and the collagen fibers. Since the cartilage does not have own blood supply, metabolism of the cells is possible only by diffusion from the peripheral layers.

The hyaline cartilage in osteoarthritis

Degenerative musulosceletal disorders are among the most severe and most frequent medical problems affecting large parts of the population. Due to these diseases, millions of people across the world are not able to carry out their work, spend their leisure time as they would like to. Recognizing this, the WHO has dedicated the first decade of the new century to the research of musculosceletal diseases, calling it the 'bone and joint decade'. The special meaning of the problem is underlined by the fact that in the United States more than half a million hip and knee joint endoprostheses are implanted every year due to degenerative joint disorders [2]. Besides weight bearing major joints the most affected part of the musculosceletal system is the spine, mainly in the lumbar part.

Osteoarthritis is probably the most important degenerative musculosceletal disorder. The etiology of its primary form is unknown, in these cases biomechanical abnormalities of the joint cannot be confirmed and no history of previous pathology (i.e. trauma, infection) exists. In osteoarthritis characteristic pathological changes occur in the tissue elements building up the hyaline cartilage. The derangement of the cartilage leads to the total destruction of the joint over several steps.

The basic histological and biochemical alterations in osteoarthritis have been clearly described before [3–6]. There is an increased cellular activity in the arthritic joint at the early stage combined with changes of the matrix and increased water uptake. The homeostasis shifts towards catabolic activity, cartilage is degraded, proteoglycan fragments are liberated which is followed by the fragmentation of collagen fibers and their structural change.

The structural changes occurring in the joint cartilage can be described as follows: the amount of chondroitine sulfate is decreasing, the surface shows cracks. There are also significant structural changes in the deeper layers, the number of cartilage cells gradually decreases in certain parts, in other parts, irregular cell proliferation can occur. The cell proliferation results in the increased release of proteolytic enzymes that cause a rapidly increased denaturation of proteins. In certain layers, calcium deposits are formed. As a result of all these changes, the integrity of the cartilage tissue gradually weakens. Due to the mechanical forces affecting the articular surface, parts of the cartilage are torn, ulcerations are developing. Because the cartilage is destroyed, the bony ends of the joints are left partially or completely cartilage-free [7–9]. (Fig. 1)

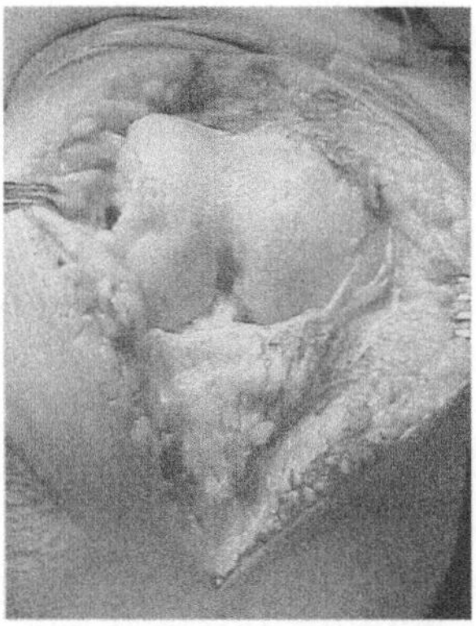

Fig. 1 Intraoperative view of severe osteoarthritis of the femoral condyles in the human knee joint *(See colour section, p. 350).*

CALORIMETRIC EXAMINATION OF THE HYALINE CARTILAGE

DSC of the hyaline cartilage in rabbits

The aim of our related studies was to experimentally create osteoarthritis in the knee joint by setting up an appropriate animal experiment model. The concept of the research program has been set up in cooperation between the Institutes of Biophysics, Pathology and the Department of Orthopedics of the Medical School of the University of Pécs [10–13].

The calorimetric examination of arthritis created this way was our target, which we hoped could give an answer to the following questions:

- Are we able to introduce DSC as a new method of cartilage research?
- Is it possible to use calorimetry in the experimental knee arthritis animal model and is it possible to reproduce results even if sample size is expected to be relatively small?
- Is it possible to show differences between samples regarding to the duration of osteoarthritis?

The experiment was performed under supervision and with permission of the local committee of the national animal experiment board. For the study we used 10 white, New Zealand type rabbits, each being 3–3.5 kg. Animals were approximately 5 months old at the time of first surgery. Anesthesia was performed using intramuscular injections containing 5 milliliters of ketamin hydrochlorate (Calypsol®) and 1 milliliter of diazepam (Seduxen®). After proper preparation of the surgical field (shaving, disinfection), further local anesthesia was performed by using 2 milliliters of lidocain hydrochlorate (Lidocain®) in the region of the skin incision.

Resection arthroplasty of the patella of one knee was carried out with an oscillating saw. Main purpose of this intervention was to remove the cartilage of the patella and to leave a rough surface articulating with the femoral cartilage. After repositioning the resected patellar surface into the femoral trochlea, reconstruction of the capsule and the surrounding tissues was carried out. The animals were granted free movement in a cage approximately 1×1 meters in size. Complications did not occur, all animals survived the intervention and were able for later examination.

The first animal was sacrificed 12 months after the surgery, by an overdose of ketamin hydrochlorate (Calypsol®), later on animals were sacrificed with 1 month intervals using the same protocol. As could be expected, on the femoral cartilage surface articulating with the resected patella macroscopically justifiable osteoarthritis developed. Formation of osteophytes, loss of cartilage thickness and remarkable fissuration could be observed. Macroscopic signs of osteoarthritis could be verified in all cases. We removed the whole distal femoral end in all cases for further preparation of the cartilage. In three animals, in order to have further verification of osteoarthritis, a part of the condyle underwent histological examination too. This took place after formaline fixation and decalcification with EDTA. Histological methods (hematoxylin-eosin and Giemsa) verified severe arthritis in all examined samples (Fig. 2). For control we explored the contralateral, unoperated knee as well, where macroscopically no cartilage degeneration was visible. The three animals, in which histological examination of the operated knee has been performed underwent the same in the contralateral knee as well. The histological slides showed slight signs of arthritis, the hyaline cartilage being intact, but thinner than normal.

The samples serving as a basis for research were derived from the femoral trochlear surface of the patellofemoral joint. We avoided removing subchondral bone, our aim was to harvest only pure hyaline cartilage. Samples were obtained

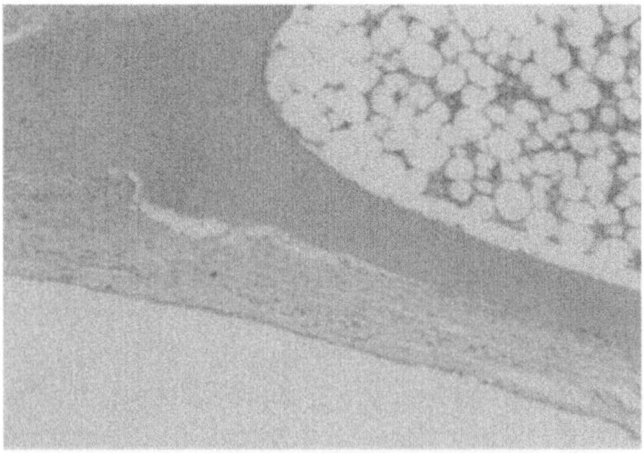

Fig. 2 Histological examination (hematoxylin-eosin) of osteoarthritis of the femoral hyaline cartilage in rabbits *(See colour section, p. 351)*

by a sharp scalpel, from the same anatomic region in all animals. The shape of cartilage pieces was flat with 1 mm of thickness and 5 mm of length. Most of the samples were identical by size. Samples were put into RPMI-1640 solution (SIGMA) containing 10 percent fetal bovine serum (HYCLONE laboratories), antibiotic, antimycotic solution (1U/ml penicilline, streptomycine, gentamycine and fungisone, GIBCO lab.), non-essential amino acids (GIBCO) and sodium carbonate. All the individual samples were stored separately at 4 degrees Celsius, no longer than 24 hours. Then samples were subjected to calorimetric measurement.

DSC was performed in cartilage samples of all operated knees and in 5 cases in the contra lateral, not operated knee joint as well. The calorimetric experiments were done as they were described earlier [14–17]. The thermal denaturation was monitored by a SETARAM Micro DSC-II calorimeter. All the experiments were performed between 0° and 100°C. The heating rate was 0.3°C/min. Conventional Hastelloy batch vessels were used during the denaturation experiments with 850 μl sample volume, on average. The pure RPMI-1640 solution served as reference. The sample and reference vessels were equilibrated with a precision of ±0.5 mg and there was no need to do any correction from the point of view of heat capacity between the sample and reference vessels. The data treatment after ASCII conversion was done by Origin 4.1.

DSC scans of arthritic samples clearly demonstrated that in every case a characteristic high temperature endothermic reaction could be traced at 62.5±1.2°C. This could be observed in every sample, the small size of sample did not affect the result of the examination (Fig. 3). Interestingly the curve detected in case of non-operated, thus seen as to be intact knees also showed an endothermic reaction

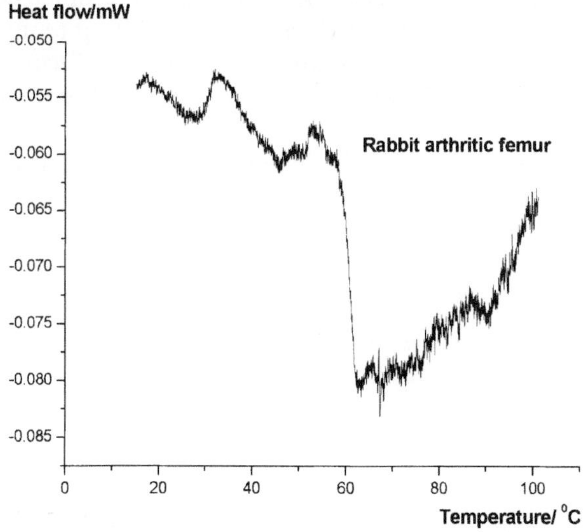

Fig. 3 Thermogram of osteoarthritic femoral sample sets in rabbits

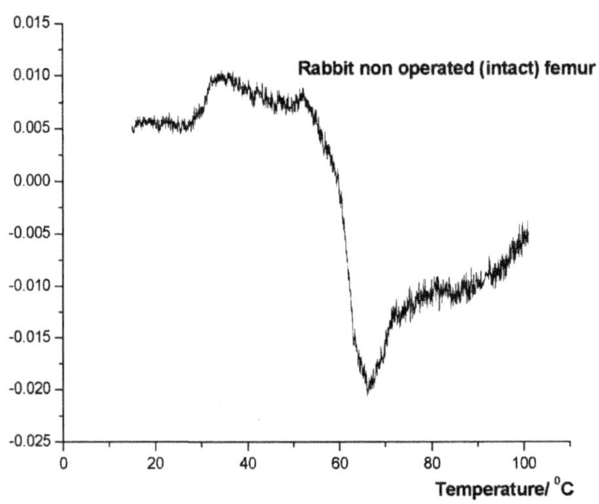

Fig. 4 Calorimetric scan of a femoral condyle without osteoarthritis in the rabbit knee

at 65°C (Fig. 4). One possible explanation for this finding of ours might be that the one-year-old animals might already have had degeneration in the hyaline cartilage of the unoperated knee. The fact of spontaneous progression of arthritis in the knee of elder rabbits is well known and could also be verified with the histological investigation of our samples.

According to the duration of the pathologic state no differences could be proven. Graphs showed similar characteristics in all animals, independently from the time period between the operation and the DSC investigation.

The pronounced heat capacity change in arthritic samples can be explained with the structural alterations. The values of main melting temperature fell into the range of stable biological macromolecules. This is a further verification of the possible cause being the denaturation of collagen and protein compounds of the cartilage.

With our investigations we could demonstrate that DSC is a well applicable method for the investigation of hyaline cartilage. The scans demonstrated that osteoarthritis has its characteristic calorimetric appearance.

DSC features of the human hyaline cartilage

DSC research of the cartilage is made exciting yet difficult by the circumstance that thermal stability of hyaline cartilage has never been examined by calorimetry, publications of such kind can not be found in literature. It was obvious that, since parameters of healthy hyaline cartilage were not at hand, examination of the arthritic condition could only be conducted after standards for healthy cartilage had been set.

Based on the animal studies, our hypothesis was that in human osteoarthritis there is a similar pathological abnormality in the tissue elements, which can also be detected by calorimetry. Besides examining healthy human cartilage with DSC, we planned to carry out preliminary investigations of cartilage destruction caused by osteoarthritis. Objectives of research were:

- Setting up of calorimetric standards of normal human hyaline cartilage.
- Applying calorimetric methods for the investigation of different samples from clinically proved degenerated articular surfaces.
- Presentation of the differences in the samples of normal and pathological conditions.

The healthy cartilage samples were of cadaver origin. This samples remain as waste materials when several preparates are dispensed for the bone bank of our orthopedic clinic, such as the joint cartilage surface remaining after the preparation of cancellous bone chips from cadaver femoral condyles. The donors taken into our study were all under the age of 40 at their death, we considered these persons to be free of degenerative changes in their joints. We took only samples where degeneration of the cartilage surface could not be verified macroscopically. Law paragraphs and valid permissions control the activity of the bone bank.

The pathologic human samples serving as a basis for research were derived from tissue fragments taken during operations and considered to be waste material. Such were the femoral condyle- and patella pieces removed during knee prosthesis implantations.

10 intact and 10 arthritic samples were examined, all were derived from different individuals. Gender ratios were approximately identical in both groups. In order to make measurements more objective, exclusively cartilage from the weight-bearing surface of the medial femoral condyle and from the medial facet of the patella were measured. From all patients and cadavers 2–3 samples were taken from the same place, on the one hand, to be able to repeat the measurement in case of any disturbance, on the other hand, to test the reproducibility of the measurements with different samples taken from the same patient.

The samples were obtained by devices especially designed for this task, from the same anatomic region (medial femoral condyle, patella) by standard methods both in the cadaver and patient samples. Although the form and size of the sample does not influence the examination, our target was to standardize these. The shape of samples was cylindrical with 3 mm of diameter and 15 mm of length. Most of the samples were identical by size and were stored and prepared in the same way, like described at the rabbit experiments. DSC measurements were also done as described earlier.

Before starting our human studies we have believed that the structural manifestations of osteoarthritis appear as a remarkable change of thermal stability of hyaline cartilage samples prepared from human femoral condyles and patellae. This kind of investigation is a perfectly new approach of this problem.

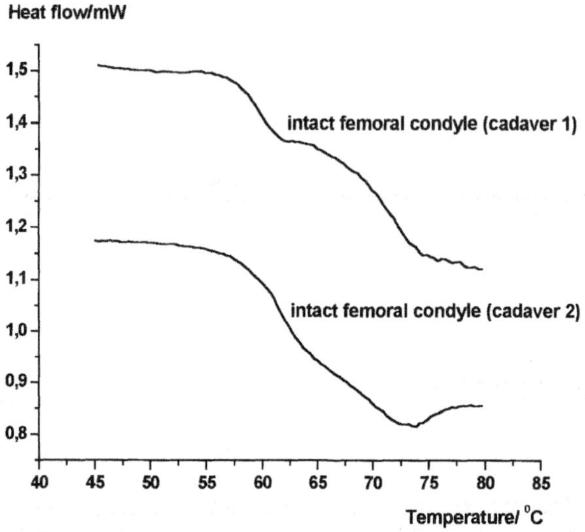

Fig. 5 Calorimetric curves of the intact human hyaline cartilage harvested from two different femoral condyles

DSC scans clearly demonstrated significant differences between the different types and conditions of cartilage samples. Studies of the intact hyaline cartilage stated that measurements could reliably be reproduced from a number of views. On the one hand, this was true to different samples taken from the same place, where measured graphs were almost completely identical. On the other hand, samples taken from different cadavers, considered to be intact, were identical in 9 out of 10 cases in terms of the medial femur condyle.

On Fig. 5 scans of two different cadavers can be seen, these identical curves are also representative for all the other measurements of the femoral condyle. This graph can be viewed as the DSC scan of the intact femoral hyaline cartilage, therefore considered as the healthy standard. Figure 6 represents the DSC scan of the intact patellar hyaline cartilage. The femoral condyle and the patella are parts of the same anatomical but not structural and functional unit, therefore the higher transition enthalpy for samples from the patella could be assigned to the different structure (i.e. cartilage thickness) of them.

Curves of the arthritic samples showed basic differences to the intact cartilage both in terms of the patella and the femur. Scans of the different samples deviated from each other in various respects, which is comprehensible since the patient sample group was heterogeneous. Independent of differences between samples, a characteristic endothermic reaction could be observed in the range of 60–70°C with every arthritic sample (Fig. 7). This thermodynamic effect occurred both in samples harvested from the femoral condyle and the patella.

The pronounced heat capacity change between intact and arthritic femur condyle samples can be explained with the structural alterations in osteoarthritis

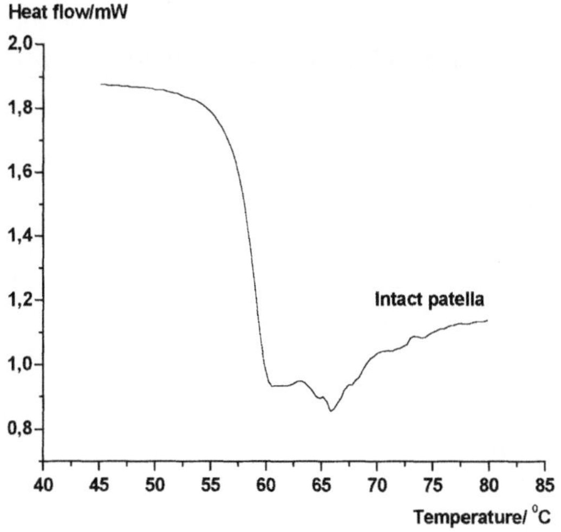

Fig. 6 Calorimetric curve of the intact hyaline cartilage of the human patella

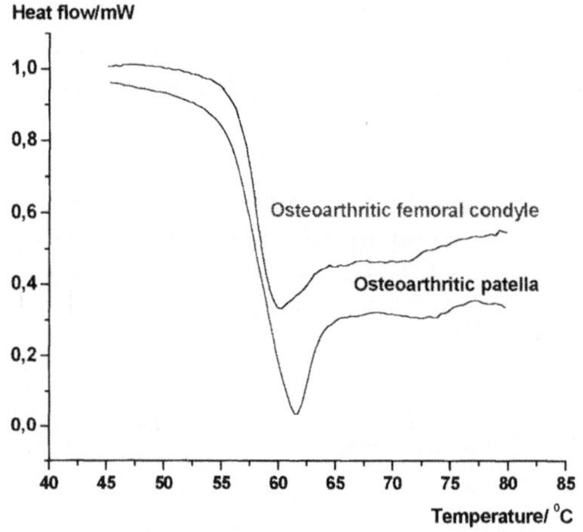

Fig. 7 Calorimetric feature of the osteoarthritic hyaline cartilage of the femoral condyle and the patella

caused by the biochemical processes. The values of main melting temperature fall into the range (~60°C) of more stable biological macromolecules therefore they could be assigned to the denaturation of collagen and protein compounds of the cartilage.

Since calorimetry has not been applied for this purpose, our measurements can not be compared to former publications, only own results can be discussed. The most important achievement, is in our opinion, the fact that it could be proven: calorimetry is applicable to the examination of human hyaline cartilage. It could be verified that intact cartilage has a characteristic and reproducible measurement result, which can serve as standard and reference value for further studies.

DSC characteristics of the arthritic human hyaline cartilage according to various stages

Since the first experiments were successful, the detailed and complex examination of osteoarthritis complemented the study. We especially focused on the calorimetric description of various stages of degenerative deformities, to maintain objectivity, additional histological examinations were carried out at the Institute of Pathology of the University. Sample harvesting, storage and calorimetric measurements were identical to that of the studies described before.

All samples were derived from cases operated on with knee osteoarthritis at the Department of Orthopedics. In all patients uni- or total condylar knee replacement was carried out. Examinations were performed exclusively on cartilage taken from the medial femoral condyle. 17 (6 male, 11 female) patients suffering from different stages of osteoarthritis underwent sample harvesting during operation, in 9 cases from the left, in 8 cases from the right side. Average age of patients was 64 years (46–79). The stage of osteoarthritis was classified during operation according to Outerbridge [18]. Stage I was diagnosed in 2, stage II in 9, stage III in 6 patients.

Measurements led to a very interesting result. Even in samples with slight osteoarthritis (Fig. 8) the endothermic reaction at around 60 degrees, experienced earlier, could be observed in all cases. Our basic observation was that curves of slight arthritis were more similar to the graphs defined as arthritic sample than to the graphs of intact cartilage of the previous study.

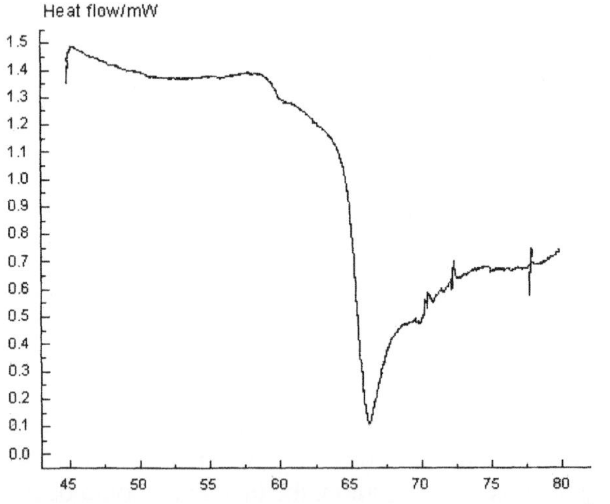

Fig. 8 Calorimetric scan of a human femoral condyle with slight osteoarthritis

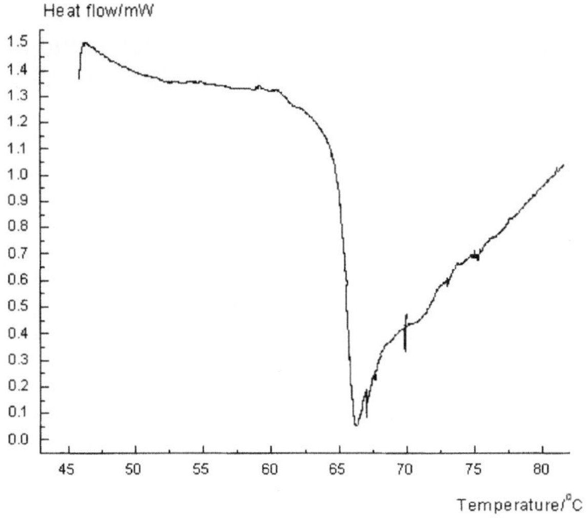

Fig. 9 Calorimetric curve of severe human knee joint osteoarthritis

The advanced cases (Fig. 9) also showed this thermodynamic effect between 60 and 70 degrees, but two important differences could be observed in all samples.

1) On the one hand with severe arthritis there was no significant difference between the thermal capacities of the starting and ending condition, to put it more simply, the graph 'returned' to the same heat capacity level where it started from. The underlying reason could be that the thermal capacity of any biologic system is basically dependent on the amount of water tied. In a tissue showing little water content at start (which is the case in advanced osteoarthritis) the heat capacity of starting and ending stage does not differ significantly. Contrarily, as can be seen in Fig. 8, the less affected cartilage showing larger water content will have differences in heat capacities, since during denaturation it will gradually loose its water content, the graph stabilizes at a lower value at the end of the heating process.

2) The other evident difference between Figs 8 and 9 is the fact that the endothermic reaction at 60 degrees shows a narrower 'peak' in case of the less affected cartilage. The reaction takes place quickly, while in advanced osteoarthritis the peak is wider, the thermal effect resulting presumably from collagen denaturation is significantly slower. This can theoretically be due to the reduced quantity of collagen, its abnormal structure or the deviations of the surrounding basic material.

The hyaline cartilage showing calorimetric graphs underwent histological examination as well. The assumptions stated above, regarding the differences of the calorimetric graphs have clearly been proven by histological examinations. Even by simple haematoxylin-eosin method, clear differences were evident. In slight osteoarthritis cases some irregularities could be observed in the basic material. Cartilage cells were typical, no cell destruction could be observed. In the arthritic sample massive destruction of the cartilage could be observed, more over, cartilage cells disappeared in large areas, in other places they formed groups as a sign of regeneration. There were also signs for the multiplication of inorganic material, the picture showed signs of advanced cartilage damage.

Our studies proved that the calorimetric scans of arthritic and healthy cartilage differ significantly, the thermal consequences of degeneration could be verified. Before discussing the possible causes of calorimetric differences we should state that the biochemical and histological processes developing during osteoarthritis are extraordinarily complex. Therefore it is almost impossible to explain the exact background of thermodynamic deviations observed with calorimetry when the cartilage is studied as a complex unit, only indirect conclusions can be made. The reaction observed at 60°C, falls into the range of stable macromolecules, the effect probably takes place because of the denaturation of the collagen or proteoglycan molecules of the cartilage. Since no deviation could be seen in healthy samples in this domain, we assume that collagen loses its thermodynamic stability in osteoarthritis, but it is too early to draw related conclusions.

Calorimetry itself is not suitable for structural studies, but together with histological examination the possible causes for the effects can be listed. The measurements carried out in our study verified that morphologic differences between various stages of osteoarthritis can be demonstrated in an indirect way, even if not based on classic clinical categories.

We are convinced that a number of questions have to be clarified which arise as a consequence of the studies carried out so far. The most recent problem is the question which component of the cartilage can be made responsible for the deviations measured with DSC, and if the method is sensitive enough to study these. To answer these questions the main components of the cartilage have to be separated and separately examined by calorimetry. The values gained this way can be compared to the values of the complete structure.

Another interesting point is the question what practical benefits can be achieved by the studies since we know that calorimetric data serve as basic reference values in several other (e.g. industrial) fields. The most probable option could be the use of the method in controlling the efficiency of certain medical treatments. Among these the demonstration of effectiveness of the widespread used oral chondroitin sulfate therapy, or the control of the effect of frequent intraarticular injections can be listed.

DSC examination of the lumbar intervertebral disc

THE IMPACT OF LOW BACK PAIN

Low back pain is one of the most common health problems affecting western society. At some point during their lifetime, 80% of the population will experience these symptoms, and as many as 60% may have suffered from it in the past year [19]. Beside the inconveniences, the cost of low back pain to the society is also estimated extremly high [20, 21].

Low back pain is most frequently due to the degeneration of the intervertebral discs [22, 23]. The degenerative processes begin early in adulthood and progress thereafter. It has been attributed to the accumulation of environmental effects, primarily mechanical insults and injuries, imposed on normal ageing changes [24–26]. Degeneration alters the morphology and the mechanical properties of the disc and leads to its gradual destruction. Certain features of degenerated discs are believed to play a role in the pathomechanism of pain production, leading severe disability and decrease in the quality of life [23]. In case of disc herniation, the irritation of neurologic structures can lead to back and/or leg pain. The degenerated disc itself can also be the source of non-specific back pain, or contribute to the loss of spinal canal dimensions by diffuse bulging. In the latter case, entrapment of the nerve roots is often associated with incapacitating pain in the back and lower extremities, difficulty ambulating, leg paresthesisas and weakness and, in severe cases, bowel or bladder disturbances [23].

ANATOMY AND BIOCHEMISTRY OF THE LUMBAR INTERVERTEBRAL DISC

The spine is a mechanical structure [25]. It consists of several vertebrae, which are connected to each other by ligaments, small joints (facet joints) and soft discs to form a strong yet mobile structure that is held in good balance by powerful muscles [27].

The intervertebral discs, which in aggregate, are the largest avascular structures in the human body, are discoid fibrocartilage tissues [26, 27]. They process viscoelastic properties allowing absorbance and dispersion of loads on the spinal column and providing for smooth movements of the spine. During these functions, they are subjected to a considerable variety of forces and moments [25, 28–30]. Since the spine transfers the weights and the resultant bending moments of the head, trunk, and any weights being lifted to the pelvis, the discs along with the facets joints are the major compression carrying component of the spine [25]. The discs are also subjected to other types of loads and stresses: tensile stresses are produced in certain portions of the disc during physiological motions of flexion, extension and lateral bending, while axial rotation causes torsional loads that results in shear stresses in the lumbar discs [25].

The intervertebral discs are generally considered to consist of a gel-like nucleus pulposus surrounded by sheets of interlacing lamellae of collagen forming the anulus fibrosus [26, 27, 31]. The discs are limited above and below by a

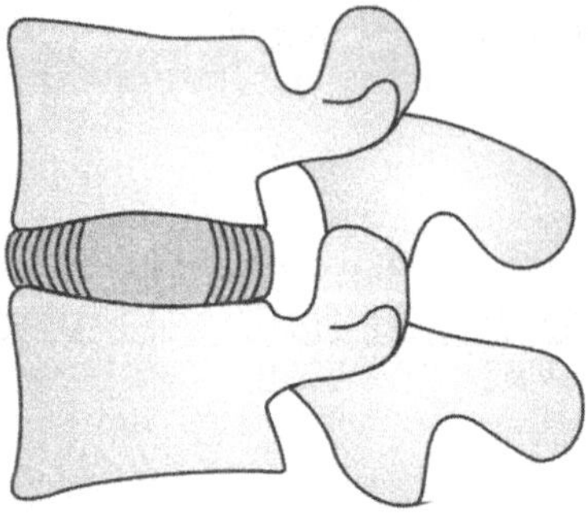

Fig. 10 The motion segment of the spine. It consists of two adjecent vertebrae and the connecting ligamentous tissues

sheet of hyaline cartilage, which constitutes the cartilage end-plate separating the nucleus pulposus from the adjacent vertebral body (Fig. 10). The nucleus pulposus consists of a gelatinous fluid containing a loose meshwork of randomly distributed collagen fibres (mainly type II) in a proteoglycan matrix and shows high affinity with water. It is highly hydrated, containing 80–90% of water. Approximately 65% of its dry weight is accounted for by proteoglycans (predominantly chondroitin-6-sulfate, chondroitin-4-sulfate, and keratin sulfate), 20% by collagens and the remainder by elastin and other minor components [26, 27]. The annulus fibrosus is a portion of the intervertebral disc that forms the outer boundary of the disc. This structure consists of interlacing lamellae of coarse collagen fibers (mainly type I) interconnecting the adjacent vertebral bodies, merging into the cartilaginous endplates and attaching to the vertebral bodies. The fibres interlace obliquely at a constant angle to form a three-dimensional collagen framework. Approximately 60% of its dry weight is collagen, and 20% is proteoglycans. Elastic fibres are also present as a minor component. The annulus fibrosus is less hydrated than the nucleus pulposus, with water content of 60–70%. Very few cells are present in the IVD, not more than 1–5% of the tissue volume [26, 27].

DSC FEATURES OF HEALTHY LUMBAR INTERVERTEBRAL DISCS

Because of its significant social and economic impact, the pathogenesis of intervertebral disc degeneration in humans has been the subject of ongoing research [32]. The major pathomorphological alterations in the intervertebral disc degeneration are well known [22–24, 26, 33–35]. Beside the well-established hystological, hystochemical and biochemical methods, differential scanning calorimetry represents a new approach in this field. By introducing this technique, our aim was to demonstrate that there is a definitive, reproducible difference in the structure of the healthy and pathological disc tissue. A calorimetric examination of this type has not been carried out in international level.

The material of our research were originated from cadavers. All samples were obtained during autposy within 24 h post mortem, with standard methods

Table 1. Description of the morphologic grades of intervertebral disc degeneration according to Thompson

Stage	Nucleus pulposus	Anulus fibrosus	End-plate	Vertebral body
I	Bulging gel	Discrete fibrous lamellas	Hyalin, uniformly thick	Margins rounded
II	White fibrous tissue periferially	Mucinous material between lamellas	Thickness irregular	Margins pointed
III	Consolidated fibrous tissue	Extensive mucinous infiltrations; loss of anular-nuclear demarcation	Focal defects in cartilage	Early chondrophytes or osteophytes at margins
IV	Horizontal clefts parallel to end-plate	Focal disruptions	Fibrocartilage extending from subchondral bone; irregularity and focal sclerosis in subchondral bone	Osteophystes less than 2 mm
V	Clefts extend through nucleus and anulus		Diffuse sclerosis	Osteophystes greater than 2 mm

and from the same anatomic regions (L4-L5 spinal segments). The applicability of the intervertebral discs taken from cadavers for IVD research is widely accepted. According to the visual evaluation of macroscopic changes affecting the motion segments we enrolled the discs into different stages described by Thompson [36]. This morphologic evaluating system is based upon the macroscopic appearance of the motion segment elements (i.e., AF, NP, endplate, vertebral body) on the transsection cut sagittally 5 mm lateral to the mediansagittal plane (Table 1). On the bassis of this evaluation system, stage I discs can be considered healthy, while stage V samples correspond to highly degenerated discs. The intermediate stages (stage II, III and IV) represent cummulative degeneration. Out of 40 samples, 6 were in stage I., 8 were in stage II., 8 were in stage III., 8 were in stage IV., and 10 were in stage V. according to Thompson classification. The mean age in the different groups was as follows: stage I. – 20 years (17–24), stage II. – 33 years (19–40), stage III. – 46 years (39–68), stage IV. – 57 years (43–84), stage V. – 78 years (70–88). According to medical reports all donors were free of clinical symptoms of connective tissue pathology in their life.

Following sample preparation all the individual samples were stored separately at 4°C, no longer than 24 h, then were subjected to calorimetric measurements. All the experiments were performed by a SETARAM Micro DSC-II calorimeter (SETARAM, France) between 0 and 100°C with a scanning rate of 0.3°C/min. Conventional Hastelloy batch vessels were used during the denaturation experiments with 850 μl sample volume in average. RPMI-1640 buffer was used as reference sample. The sample and reference vessels were equili-

Table 2. Calorimetric results of IVDs from different stages of Thompson evaluation system

Stage	No. of samlples	Average of age	Anulus fibrosus	Nucleus pulposus
I	6	20 years	T_m(°C): 60,5±0,3 ΔH (J/g): 0,87±0,04	T_m(°C): 60,7±0,4 ΔH (J/g): 0,45±0,07
II	8	33 years	T_m(°C): 60,6±0,4 ΔH (J/g): 0,80±0,1	T_m(°C): 60,4±0,2 ΔH (J/g): 0,43±0,07
III	8	46 years	T_m(°C): 61,1±0,4 ΔH (J/g): 0,62±0,07	T_m(°C): 59,5±0,2 ΔH (J/g): 0,37±0,09
IV	8	53 years	T_m(°C): 62,5±0,3 ΔH (J/g): 0,48±0,09	T_m(°C): 58,9±0,3 ΔH (J/g): 0,30±0,05
V	10	76 years	T_m(°C): 62,7±0,3 ΔH (J/g): 0,42±0,05	T_m(°C): 58,6±0,2 ΔH (J/g): 0,29±0,04

brated with a precision of ±0.1 mg. There was no need to do any correction from the point of view of heat capacity between the sample and reference vessels. The samples were irreversible denaturated during each cycle. In the evaluation, the repeated scan of denaturated sample was used as baseline reference, which was subtracted from the original DSC scan. Calorimetric enthalpy was calculated from the area under the heat absorption curves using two points setting SETARAM peak integration. For the statistical analysis Paired Student's *t*-test were used with a significance level of 0.05.

According to our results, no significant difference was found regarding the main transient temperature between the AF and NP in the healthy discs (stage I) (Table 2). This fact implies that the two tissues are both highly hydrated and re-

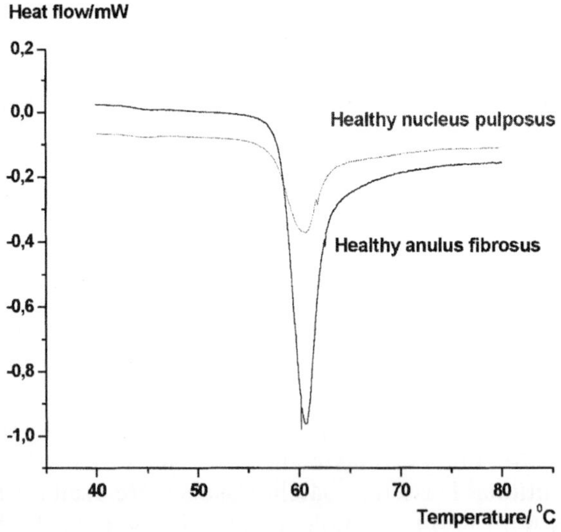

Fig. 11 Thermal stability of healthy anulus fibrosus and nucleus pulposus

tain a very integrated structure. It was suggested by the relatively narrow range of temperature for thermal denaturation and the almost symmetrical shape of the curves (Fig. 11). These facts hint at the strong cooperation between these components. The endothermic peak at about 60°C proposes the presence of stabile biological macromolecules and this phenomenon is probably due to the denaturation of the collagen and proteoglycan molecules of the disc [37]. The differences in the total calorimetric enthalpy is supposedly due to the different ratio of the two main components (collagen: AF:60%, NP:20%; proteoglycan: AF:20%, NP: 65%) and their structure (NP: gelatinous with high water content, AF: concentrically organized, more than 60 distinct collagen fibril layers running in alternating directions). The AF is more complex and in order to decompose its more compact structure significantly more energy is needed, thus it results in significantly higher enthalpy changes ($p<0.05$).

DSC CHARACTERISTICS OF DEGENERATED LUMBAR INTERVERTEBRAL DISCS

As a result of the degenerative process characteristic changes can be observed in the structure, composition and function of the IVDs (stage II-V). By the third decade of life the AF gradually becomes fibrotic, which is accompanied by the decline of the concentration of proteoglycans, water and other non-collagenous proteins and the increase of the ratio of collagen [23, 26, 33, 38]. The decrease of proteoglycans and their ability to aggregate, and as well as the loss of water is greater in the NP along the time [23, 39, 40]. The number of collagen fibrils (preferably type I collagen) and their size in diameter increases along the lifetime in every part of the IVD [41]. The increase of type I collagen within the nucleus results in increased stiffness of that structure. Repeating mechanical strain, probably via the increased production of proteolytic enzymes (especially matrix metallo proteinases), results in the degradation of the original structure of collagens and proteoglycans [23, 42, 43]. As a consequence, the IVD starts to disintegrate and several different morphological changes develop (Table 1) [23, 32, 34]. The degenerated NP, loosing its water content, becomes unable to complete its physiological function and to change its shape in a highly flexible manner [28] . The ability of the IVD to transmit compressive load becomes highly compromised. The compressive load is transferred from one vertebral end-plate to the other by way of the NP and the AF. In case of healthy discs (stage I), the NP has sufficient water content to act like a gelatinous mass. As the load is applied, a pressure develops within the NP, which pushes the surrounding structures in all directions away from the nucleus center [25]. In other words, the central portion of the two vertebral end-plates are pushed away from each other, and the AF is pushed radially outword (Fig. 12). The situation is quite different when the NP is dry and fibrotic. The load transferring mechanism is significantly altered because the NP is not capable of building sufficient fluid pressure. As a result, the end-plates are subjected to less pressure at the center, and the loads are

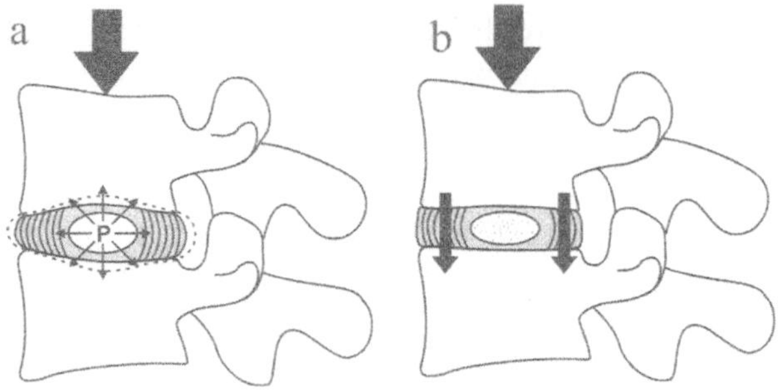

Fig. 12 Biomechanics of compression load transfer in healthy (a) and degenerated discs (b)

distribured more around the periphery. Therefore, considerable part of the compressive forces are transmitted to the neighboring vertebra through the AF [25, 29] (Fig. 12). The elevated main transition temperature of the AF is a consequence of the overload in the AF of stage II-V discs (Table 2). Presumably, due to the mechanical overload secondary bindings (intra- and intermolecular hydrogen bridges) develop in this enzymatically disintegrated structure. Thus, the entire structure becomes more tightly 'packed'. For the disintegration of this compact structure extra energy was necessary, thus the structural phase transformation began at a higher temperature. The decrease of the enthalpy of this structure along the degeneration (stage II-V) is attributed to the loss of bound water and thermal cooperation of the components (Table 2). It is also suggested by the widening of the thermal transition period and the asymmetry of the curves themselves (Fig. 13). The drop in the main transient temperature in the degenerated NPs is

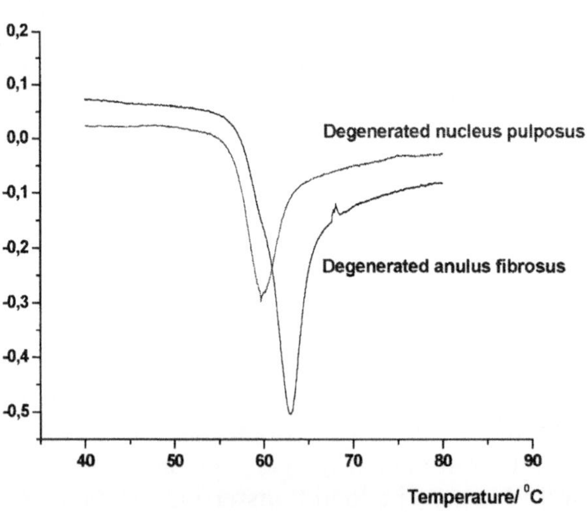

Fig. 13 Thermal denaturation of degenerated anulus fibrosus and nucleus pulposus

mostly due to the loss of the immensely hydrated proteoglycans (Table 2). The fragmentation of this structure results in the decrease of bound water clusters, and so consequently the decrease of the thermal capacity. In calorimetry, the significantly lower thermal capacity is an important sign of the loss of water clusters, resulting in a greater baseline shift when compared to the native stage (Figs 11 and 13). The consequence is the significantly smaller changes in the enthalpy and less thermical cooperation ($p<0.05$) in the degenerated NPs in comparison with the healthy (stage I) specimens (Table 2).

When comparing all the consecutive stages (stage I to V), highly significant differences were only found between stages I, III, and V in both the main transition temperature and the total calorimetric enthalpy changes ($p<0.05$). In other words, as opposed to the five stages defined morphologically, we were only establish three distinct stages when comparing the results from thermical analysis: the two marginal categories (stage I and V), and stage III in accordance to the Thompson evaluation system. This result can be explained with the fact, that calorimetry measures more complex changes affecting the biochemical and biophysical structure as a whole. IVD degeneration is a continuous process rather than a disease with distinct stages, and stages established by morphological means can not be distinguished when looked into in a completely different, peculiar thermical aspect [44].

References

1 Scott, J. E. *Biochem. J.,* 252 (1988) 313.
2 Booth, R. E. *Orthopedics,* 23 (2000) 903.
3 DeSimone, D. P. Parsons D. B. Johnson, K. E. Jakobs, R. P.: Type II collagen-induced arthritis. A morphologic and biochemical study of articular cartilage. *Arthritis Rheum.,* 26 (1983) 1245–58.
4 Gardner, D. L.: Problems and paradigms in joint pathology. *J. Anat.,* 184 (1994) 465–76.
5 Poole C. A.: Articular cartilage chondrons: form, function and failure. *J. Anat.,* 191 (1997) 1–13.
6 Sweet, M. B. E. Thonar, E. J.: Immelman AR, Solomon L. Biochemical changes in progressive osteoarthritis. *Ann. Rheum. Dis.,* 36 (1977) 387–98.
7 Meacock, S. C. Bodmer, J. L. Billingham, M. E.: Experimental osteoarthritis in guinea-pigs. *J. Exp. Pathol.* (Oxford), 71 (1990) 279–93.
8 Muehleman, C. Arsenis, C. H.: Articular cartilage. Part II. The osteoarthritic joint. *J. Am. Podiatr. Med. Assoc.,* 85 (1995) 282–86.
9 Schwartz, E. R. Oh, W. H.: Leveille CR. Experimentally induced osteoarthritis in guinea pigs: metabolic responses in articular cartilage to developing pathology. *Arthritis Rheum.,* 24 (1981)1345–55.
10 Than, P. Vermes, Cs. Schäffer, B. Lőrinczy, D.: Differential scanning calorimetric examination of the human hyaline cartilage. A preliminary study. *Thermochim. Acta,* 346 (2000) 147–51.

11 Than, P. Lőrinczy, D.: Differential scanning calorimetric examination of the osteoarthritic hyaline cartilage in rabbits. *Thermochim. Acta,* 404 (2003) 149–53.
12 Than, P. Domán, I. Lőrinczy, D.: DSC in the research of abnormalities of the organs of human locomotion. *Thermochim. Acta,* 2004 (in press)
13 Than, P. Kereskai, L.: A new option of examining the human hyaline cartilage: differential scanning calorimetry. *J.Biochem. Biophys. Methods,* 2004 (in press)
14 Lőrinczy, D. Belágyi, J.: Effect of nucleotid on sceletal muscle myosin unfolding in myofibrils by DSC. *Biochem. Biophys. Res. Com.,* 217 (1995) 592–8.
15 Lőrinczy, D. Belágyi, J.: Scanning calorimetric and EPR studies on thermal stability of actin. *Thermochim. Acta*; 259 (1995) 153–64.
16 Lőrinczy, D. Belágyi, J.: Comparative study of myosins in solutions and supramolecular complexes. Effect of nucleotides. *Thermochim. Acta,* 296 (1997) 161–8.
17 Lőrinczy, D. Könczöl, F. Gaszner, B. Belágyi J.: Structural stability of actin filaments as studied by DSC and EPR. *Thermochim. Acta,* 322 (1998) 95–100.
18 Outerbridge, R. E.: The etiology of chondromalacia patellae. *J. Bone Joint Surg.*; 43-B (1961) 752–57.
19 Vaddel, G.: Volvo Award in Clinical Sciences. A new clinical model for the treatment of low back pain. *Spine,* 12 (1987) 632–644.
20 Deyo, R. A. Tsui-Wu, Y. J.: Descriptive epidemiology of low back pain and its related medical care in the United States. *Spine,* 12 (1987) 264–268.
21 Frymoyer, J. W. Cats-Baril, W. L.: An overview of the incidences and costs of low back pain. *Orthop. Clin. North Am.,* 22 (1991) 263–71.
22 Adams, M. A. Dolan, P. Hutton, W. C.: The stages of disc degeneration as revealed by discograms. *J. Bone Joint Surg.,* 68-B (1986) 36–41.
23 Frymoyer, J. W. Moskowitz, R. W.: Spinal degeneration pathogenesis and medical management. In: Frymoyer JW, editor. The Adult Spine: principles and practice. New York, *Raven Press Ltd.,* (1991) 611–632.
24 Lyons, G. Eisenstein, S. M. Sweet, M. B.: Biochemical changes in intervertebral disc degeneration. Biochim. *Biophys. Acta,* 673 (1981) 443–453.
25 Panjabi, M. M. White, A. A.: Physical properties and functional biomechanics of the spine. In: White, A. A. Panjabi, M. M. editors. Clinical biomechanics of the spine. Philadelphia: JB. Lippicott Co, (1990) 3–83.
26 Pritzker, K. P. H.: Aging and degeneration in the lumbar intervertebral disc. *Orthop. Clin. North. Am.* 8 (1977) 65–77.
27 Larson, S. J. Maiman, D. J.: Lumbar anatomy. In: Surgery of the lumbar spine. New York, *Thieme.,* (1991) 1–12.
28 Nachemnson, A. L. Shultz, A. B. Berkson, M. H.: Mechanical properties of human lumbar spine motion segments. Influence of age, sex, disc level, and degeneration. *Spine,* 4 (1979) 1–8.
29 Shirazi-Adl, S. A. Sharivastava, Sc. Ahmed, A. M.: Stress Analysis of the lumbar disc-body unit in compression: A there-dimentional nonlinear finite element study. *Spine,* 9 (1984) 120–134.
30 Ziran, B. H. Pineda, S. Pokharna, H. Esteki, A. Mansour, J. M. Moskowitz, R. W.: Biomechanical, radiologic, and histopathologic correlations in the pathogenesis of experimental intervertebral disc disease. *Spine,* 19 (1994) 2159–2163.

31 Inoue, H.: Three-dimensional architecture of lumbar intervertebral disc. *Spine,* 6 (1981) 139–146.
32 Holbrook, T. L. Grazier, I. Kelsey, J. L. Stauffer, R. N.: The socioeconomic impact of selected musculoskeletal disorders. Am. Acad. Orthop. Surg. Chicago, IL, (1984).
33 Bernick, S. Walker, J. M. Paule, W. J.: Age changes to the anulus fibrosus in human intervertebral discs. *Spine,* 16 (1991) 520–524.
34 Fornasier, V. L. Garaffo, G. Denaro, L. Denaro, V.: Intervertebral disc degeneration - an autopsy study. *Eur. J. Orthop. Surg. Traumatol.,* 10 (2000) 159–165.
35 Prescher, A.: Anatomy and pathology of the aging spine. *Eur. J. Radiol.,* 27 (1998) 181–195.
36 Thompson, J. P. Schechter, M. T. Tsang, I. K. Y.: Preliminary evaluation of a scheme for grading the gross morphology of the human intervertebral disc. *Spine,* 5 (1990) 411–415.
37 Domán, I. Tóth, Gy. Lőrinczy, D. Illés, T.: Differential scanning calorimetric examination of the human intervertebral disc: a preliminary study. *Thermochim. Acta,* 376 (2001) 117–122.
38 Antoniou, J. Steffen, T. Nelson, F.: The human lumbar intervertebral disc: evidence for changes in the biosynthesis and denaturation of the extracellular matrix with growth, maturation, ageing, and degeneration. *J. Clin. Invest.,* 98 (1996) 996–1003.
39 Herbert, C. M. Lindberg, K. A. Jayson, M. I. V. Bailey A. J.: Proceedings: Intervertebral disc collagen in degenerative disc disease. *Ann. Rheum. Dis.,* 34 (1975) 467.
40 Herbert, C. M. Lindberg, K. A. Jayson, M. I. V. Bailey, A. J.: Changes in the collagen of the human intervertebral disc during ageing and degenerative disc disease. *J. Mol. Med.,* 1 (1975) 79–81.
41 Nerlich, A. G. Schleicher, E. D. Boos, N.: 1997 Volvo Award winner in basic science studies. Immunohistologic markers for age-related changes of human lumbar intervertebral discs. *Spine,* 22 (1997) 2781–2795.
42 Goupille, P. Jayson, M. I. V. Valat, G. P. Freemont, A.: Matrix metalloproteinases: The clue to intervertebral disc degeneration? *Spine,* 23 (1998) 1612–1626.
43 Fujita, K. Nakagawa, T. Hirabayashi, K. Nagai, Y.: Neutral proteinases in human intervertebral disc. Role in degeneration and probable origin. *Spine,* 18 (1993) 1766–1773.
44 Domán, I. Illés, T. Lőrinczy, D.: Differential scanning calorimetric examination of the human intervertebral disc: establishment of calorimetric standards of different stages of degeneration. *Thermocim. Acta,* 405 (2003) 293–299.

Chapter 12

Quantitative thermal analysis of carbohydrate–water systems

M. Pyda*

Department of Chemistry, The University of Tennessee, Knoxville, TN 37996-1600, and Chemical Sciences Division, Oak Ridge National Laboratory, Oak Ridge, TN 37831-6197 USA

Introduction

The thermodynamic functions such as heat capacities, C_p, enthalpies, H, free enthalpies, G, entropies, S, and temperatures and heats of transition of biomaterials are fundamental elements for the understanding and interpretation of their structure-property-processing behavior. Thermal analyses are often complicated by non-equilibrium effects, such as partial crystallinity, broadened transitions, including many aspects of glass transitions, irreversible changes during processing, and aging. The equilibrium functions of state and transition parameters need to be known as reference for this study. The thermodynamics of biomaterials is further complicated by their interaction with water [1–8].

In order to understand the thermodynamic properties of biomaterial-water systems such as carbohydrates, quantitative thermal analysis needs to be interpreted. In this chapter the calorimetry of carbohyrates without and with different low concentrations of water are presented and interpreted using the Advanced THermal Analysis System (ATHAS) [9]. A full understanding of carbohydrates requires both, knowledge of their microscopic structure, as obtained, for example, by X-ray diffraction, and its macroscopic energetics which can be gained by calorimetry and interpreted in terms of the microscopic molecular motion. The quantitative thermal analysis of biomaterial-water systems establishes the proper thermodynamic baselines of the solid and liquid heat capacity for the discussion of phase transitions, such as the glass transition as well as ordering and disordering transitions.

The heat capacity is the macroscopic, thermodynamic quantity that is based on molecule motion. The main contributions to the experimental heat capacity come from vibrations and large-amplitude molecular motions. The latter are mainly conformational, i.e., for flexible polymers they are caused by internal ro-

* pyda@novell.chem.utk.edu

D. Lörinczy (ed.), The Nature of Biological Systems as Revealed by Thermal Methods, 307–332.

tation. For small, rigid molecules, such as glucose or water, translational and rotational motions are the large-amplitude motions. At low-temperatures, usually below the glass or melting transitions, practically only vibrations contribute to the experimental heat capacity. The vibrations in solids are separated into an approximate group vibrational spectrum, which can usually be described by a normal mode calculation based on force constants derived from infrared and Raman spectra, and the skeletal vibrations, which can be approximated by the Tarasov equation that is based on Debye functions of different dimensionality [9, 10]. The thus determined vibrational heat capacity is then extended to high temperatures and serves as the baseline for quantitative thermal analysis of dry biopolymers and biopolymer-water systems. For dry biopolymers, glass transitions may occur in several stages and can be detected as a very broad, small deviation beyond vibrational heat capacity. These studies involve phase areas down to the nanometer scale and require first an understanding of the equilibrium thermodynamic properties. Quantitative calorimetry can provide such knowledge, but very little has appeared in the past in literature and no systematic, critically reviewed, data-bank is available [1–12].

The heat capacity of a liquid state is more complicated than that of a solid. Empirically it was found, however, that the heat capacity of liquid polymers is often a linear function of temperature. Furthermore, an empirical addition scheme based on contributions of the constituent chain segments could be derived within the framework of the Advanced THermal Analysis System, ATHAS [9]. For polymer-solvent systems, an empirical addition scheme was also developed by Prausnitz and coworkers [13].

For polymers and small molecules there is no adequate, simple microscopic theory of the heat capacity of liquids. Typical approaches are described by Flory [14–16], Sanchez [17–19], Simha [20–22], and Prigogine [23]. The motion in liquids involves, besides the vibrations discussed above, large-amplitude motions such as internal rotations (conformational motion), rotations, and translations. In the case of macromolecules, only vibrational and conformational motions give significant contributions to the liquid heat capacity, making them easier to assess than in small molecules which have appreciable contributions from rotational and translational degrees of freedom. A beginning has been made in the calculation of the liquid heat capacity for a few synthetic polymers and the partially liquid state of biopolymer-water systems [24–27]. This initial work on amorphous starch without and with different low concentrations of water are summarized and presented in this chapter. The experimental liquid heat capacity has been treated as a sum of vibrational, conformational, and anharmonic contributions. More details of this treatment will also be presented later. Knowing the solid and liquid heat capacities, the experimental, apparent heat capacity C_p can be analyzed. For many biomaterials, these quantitative thermal properties are not known [1–9, 28–31], except for the pioneering research from our laboratory where an initial study of the starch-water system was made [26, 27], and some solid proteins, and all solid poly(amino acid)s were characterized relative to their C_p [32–37].

In this chapter the quantitative thermal analyses of glucose, dry, amorphous starch and starch-water systems are presented. It is shown that the contribution to the experimental heat capacity of glucose, dry starch and starch-water originates from vibrational and large-amplitude molecular motion. At low-temperatures, usually below glass transition, only vibrational motion contributes to the experimental C_p. These are linked to an approximate group vibrational spectrum for α-D-glucose ($C_6H_{12}O_6$), repeating units of starch ($C_6H_{10}O_5$), and water (H_2O) in addition to the skeletal vibrations as described by the Tarasov equation [9]. The total vibrational heat capacity was then extended to high temperatures to serve as baseline to quantitative thermal analysis. The heat capacity of liquid α-D-glucose is known from experiment. Also this heat capacity of liquid of α-D-glucose was also used as a rough estimate of liquid heat capacity of starch and starch-water systems. The heat capacity above the observed partial glass transition, i.e., in the partially liquid state of dry and hydrated starch has been estimated from their vibrational, conformational, and external contributions. In the conformational contribution, the interaction of water molecules with chain of carbohydrate is interpreted as changes of stiffness, cooperativity, and degeneracy parameters caused by formation of monolayers or clusters about the starch molecule. The external contribution was computed from the known experimental thermal expansivity and compressibility as a function of temperature. The vibrational contribution was estimated the same as for solid state. The presented approach for carbohydrate-water systems should also be applicable for other biological materials.

Experimental heat capacity

Experimental data heat capacity should be the starting point to quantitative thermal analysis. Figure 1 shows the experimental heat capacity of α-D-glucose by adiabatic calorimetry and standard differential scanning calorimetry (DSC). The low-temperature experimental C_p of solid α-D-glucose, as collected from paper [38], was used to calculate the contribution of the group vibrations and the ATHAS Scheme allowed measurement of the contributions from the skeletal vibrations, as described below and the total vibrational heat capacity has been extended to high temperatures and serve as a baseline for the quantitative thermal analysis. Also Fig. 1 shows a comparison of the experimental heat capacity of amorphous with crystalline α-D-glucose by DSC for higher temperature regions [39]. Figure 2 shows the experimental specific heat capacity of dry amorphous starch and starch with different low concentration of water measured by adiabatic calorimetry and differential scanning calorimetry from 8 to 490 K. Details about all measured and recommended experimental C_p data can be found in the full paper [26].

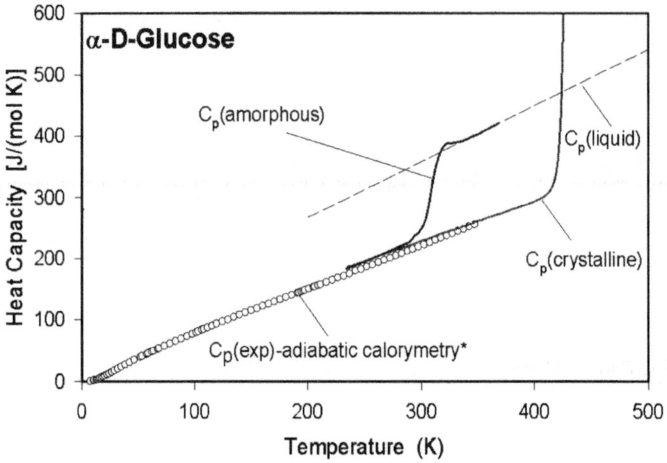

Fig. 1 Experimental heat capacities of α-D-glucose by adiabatic calorimetry (*data from Ref. [38]) and differential scanning calorimetry (DSC)

These experimental, macroscopic heat capacities presented in Fig. 1 and 2 were linked with molecular motion of α-D-glucose, starch and water in order to present quantitative thermal analysis of carbohydrate systems using two base-lines as references: the solid heat capacity and liquid heat capacity. The low temperature solid heat capacity was linked with the vibrational motions of α-D-glucose, starch and water. The interpretation of a partial liquid heat capacity of dry and hydrated starch will be in the terms of vibrational, conformational and external (anharmonic) contributions.

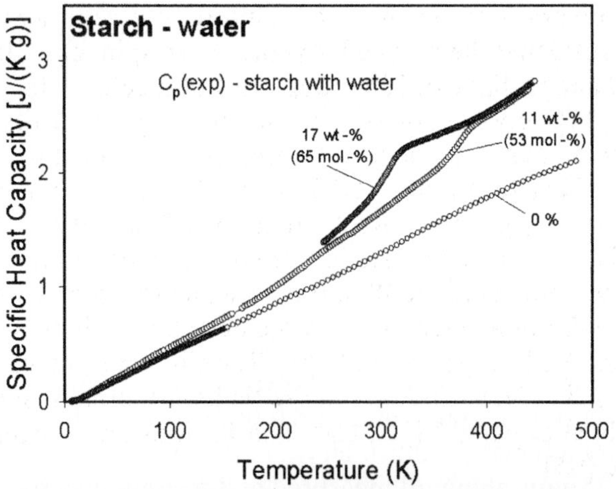

Fig. 2 Experimental specific heat capacities of dry starch and starch with 11 and 17 wt.-% of water by adiabatic calorimetry and differential scanning calorimetry (DSC)

SAMPLES

α-D-glucose was purchased from Aldrich Chem. Co. The amorphous starch was used as obtained from Blattmann AG, Switzerland (lot 51128), and after various pretreatments, as described in literature [40, 41]. In order to obtain dry starch, the sample was held in a vacuum oven at 353 K (80°C) for a minimum of 48 hours. To introduce known amounts of water either, the appropriate amounts were added to dry starch, or the dry starch was kept for 2–3 weeks in contact with the constant vapor pressure of different saturated inorganic salt solutions. Heat capacities of starch-water samples containing 53 mol-% (11 wt.-%) and 65 mol-% (17 wt.-%) of water were measured.

The concentrations are reported as weight fractions of water, W_W, and starch, W_S, or the corresponding mole fractions of water, X_W, and starch, X_S. Using the molar mass of water (M_W = 18.0152 g mol^{-1}) and the repeating unit of starch (M_S = 162.142 g mol^{-1}) the mole fractions of water and starch can easily be calculated from the weight fractions:

$$X_W = \frac{W_W / M_W}{W_W / M_W + W_S / M_S} \qquad X_S = \frac{W_S / M_S}{W_W / M_W + W_S / M_S} \tag{1}$$

The molar masses of the mixtures, M, were calculated as follows:

$$M = X_W M_W + X_S M_S \tag{2}$$

and the molar heat capacities of the mixtures are given by:

$$C_p^{\text{Starch-Water}} = c_p^{\text{Starch-Water}} M \tag{3}$$

where $c_p^{\text{Starch Water}}$ is specific heat capacity of the starch-water mixture in J K^{-1}g^{-1}. The partial molar heat capacities of water and starch can be estimated from:

$$C_p^{\text{Water}}(\text{partial}) = C_p^{\text{Starch-water}} + X_S \frac{\partial C_p^{\text{Starch-Water}}}{\partial X_W} \tag{4}$$

$$C_p^{\text{Starch}}(\text{partial}) = C_p^{\text{Starc-water}} - X_W \frac{\partial C_p^{\text{Starch-Water}}}{\partial X_W}$$

The samples were sealed in high-pressure stainless-steel pans (HPS) from the Perkin-Elmer Corp., Norwalk, CT, to prevent water loss during measurements by standard DSC.

Single-crystalline sapphire (Al_2O_3) was used for the calibration of the heat capacity at each temperature. Temperature calibrations were carried out at the phase transitions of indium (429.75 K), water (273.15 K), and tin (505.08 K).

INSTRUMENTATION AND MEASUREMENTS

For the measurement of the transition behavior and heat capacities two calorimeters were used: An adiabatic calorimeter for low-temperature measurements of heat capacity, and a heat-flux type MDSC 2920 of TA Instruments, Inc. for the DSC measurements at higher temperature. Heat capacities of starch and starch-water from 8 K to 330 K were measured with the adiabatic vacuum calorimeter which was completely automated and was fully described previously [42]. In short, the heat capacity of the sample was 60–70% of the total heat capacity of the calorimeter and the substance over the whole temperature range. The calorimetric ampule was a cylindrical vessel of platinum with a volume of *ca.* $15 \cdot 10^{-6}$ m^3. The heat capacity of an unloaded calorimetric ampule increased gradually from 0.0045 J K^{-1} to 1.440 J K^{-1} with increasing temperature from 5 K to 330 K. The temperature was measured with a platinum resistance thermometer. Liquid helium and nitrogen were used to obtain low temperature in the cooling system. Heat capacities were calibrated with benzoic acid standard before the measurements of starch and starch-water samples. The precision was estimated to be ±0.5% from 5 K to 330 K.

The measurement of heat capacity by standard DSC was carried out at a heating rate, q, of 10 K min^{-1}. Three runs were carried out, one of the empty-empty pans, one with empty-sapphire for calibration, and one with empty-sample. After steady state was attained, the heat capacity was determined from the following equation [43]:

$$mc_{\mathrm{p}} = K\frac{HF}{q} \tag{5}$$

where K is determined as a function of temperature from the sapphire calibration; HF is the heat-flow rate (proportional to the temperature difference between reference and sample, ΔT). The effects of calorimeter asymmetry and differences between empty reference and sample calorimeters are eliminated by using the empty pan run as baseline for the heat flow amplitudes. For measurements of heat capacity by standard DSC 10–30 mg of sample were employed. Three or more separate runs were made for each sample. The data of heat capacities were collected from second run. No water loss was detected after the measurements, as proven by the constant weight of the samples. The accuracy of the measurements is estimated to be ±3% or better.

Heat capacity of the solid state

The heat capacity of the solid biomaterials such as glucose, dry starch and starch-glassy water is to be computed from approximate vibrational spectra, as summarized earlier based on the experimental low-temperature heat capacity. This analysis represents the well-established ATHAS scheme [9, 43, 44] for synthetic polymers. There are $3N$ vibrational degrees of freedom in the solid state with N representing the total numbers of atoms in small biomolecules, or in the constitutional repeating unit of biopolymers, CRU. These are separated into group and skeletal vibrations ($3N = N_{gr} + N_{sk}$). The numbers and approximate types of group vibrations, N_{gr}, are derived from the chemical structure of the molecule or CRU as a series of single frequencies and box distributions over narrow frequency ranges. These frequencies are collected from normal-mode calculations on isolated molecules or CRUs. The skeletal vibrations, N_{sk}, are not well represented by normal-mode calculations, but they can be approximated by fitting the C_p(exp) at sufficiently low-temperatures to a Tarasov function which consists of an appropriate combination of Debye functions [45]. For this process, the C_p(exp) is converted to C_V(exp). The standard thermodynamic relationship between the two quantities involves the compressibility, β, and the expansivity, α is:

$$C_p\,(\exp) - C_V\,(\exp) = TV\alpha^2/\beta \tag{6}$$

where V represents the molar volume. All quantities must be known or estimated as a function of temperature. For most biopolymers the experimental data for the expansivity and compressibility are not available, especially for different crystallinities and in the presence of water. In this case, traditionally use is made of the Nernst-Lindemann approximation which was first developed for the description of C_V of metals and minerals. Later it was also fitted to macromolecules with T as the temperature, T_m, the equilibrium melting temperature, and R, the gas constant [46, 47]:

$$C_p\,(\exp) - C_V\,(\exp) = 3RA_\circ \frac{C_p^2\,(\exp)}{C_V\,(\exp)} T/T_m \tag{7}$$

Assuming that C_V(exp) contains only vibrational contributions at sufficiently low temperatures, it is separated into the heat capacities from group vibrations C_V(group) and skeletal vibrations C_V(skeletal):

$$C_V\,(\exp) = C_V\,(\text{group}) + C_V\,(\text{skeletal}) \tag{8}$$

The heat capacities from the group vibrations are represented by a sum of single Einstein frequencies or box-distributions [48–51]:

$$C_V\,(\text{group}) = C_V\,(\text{box}) + C_V\,(\text{Einstein}) \tag{9}$$

where C_V(box) is the part of the heat capacity linked to the spectrum represented by a box-like distribution functions, and C_V(Einstein) are the heat capacity contributions best approximated by separate Einstein modes. For the latter one can write:

$$C_V\,(\text{Einstein})/NR = \sum_i E(\Theta_{E_i}/T) = \sum_i \frac{(\Theta_{E_i}/T)\exp(\Theta_{E_i}/T)}{[\exp(\Theta_{E_i}/T)-1]^2} \tag{10}$$

with $\Theta_{Ei} = h\nu_i/k$, representing the given Einstein frequencies in kelvin, and h and k are Plancks and Boltzmanns constants, respectively. The C_V (box) is given by a sum over the identified areas of the vibrational spectrum. Each box is represented by a sum of one-dimensional Debye functions D_1 and corresponds to uniformly distributed vibrations within the frequencies from Θ_U to Θ_L [51, 52]:

$$C_V\,(\text{box})/NR = B(\Theta_U/T, \Theta_L/T) = \frac{\Theta_U}{\Theta_U - \Theta_L}[D_1(\Theta_U/T) - \tag{11}$$
$$-(\Theta_L/\Theta_U)D_1(\Theta_L/T)]$$

After subtracting the group vibration contributions from C_V (exp), the experimental C_V(skeletal) remains and is fitted at low temperatures to the general Tarasov function Ta[44, 45, 53–55]:

$$C_V\,(\text{skeletal})/NR = Ta(\Theta_1/T, \Theta_2/T, \Theta_3/T) = \tag{12}$$
$$D_1(\Theta_1/T) - (\Theta_2/\Theta_1)[D_1(\Theta/T) - D_2(\Theta_2/T)] - (\Theta_3^2/\Theta_1\Theta_2)$$
$$[D_2(\Theta_3/T) - D_3(\Theta_3/T)]$$

to obtain the three characteristic parameters Θ_1, Θ_2, and Θ_3 that represent the maximum frequencies of the corresponding distribution. The functions D_1, D_2 and D_3 are the one-, two-, and three-dimensional Debye functions, respectively [10, 43,44]:

$$C_V/NR = D_i(\Theta_i/T) = i(T/\Theta_i)^i \int_0^{(\Theta_i/T)} \frac{(\Theta/T)^{i+1}\exp(\Theta/T)}{[\exp(\Theta/T)-1]^2} d(\Theta/T) \tag{13}$$

where i is equal 1, 2 or 3. In the Debye functions $D_i(\Theta_i/T)$, N denotes the number of the vibrational modes for the frequency distribution, and R, the gas constant. The characteristic temperature Θ_3 describes skeletal contributions with a quadratic frequency distribution, as found in all solids at low frequency. For linear macromolecules the values of Θ_3 are usually less than 150 K. For planar molecular structures, an additional contribution with Θ_2 from 50 to 250 K is common. It yields in this frequency range a linearly increasing number of vibrational states with frequency [43, 53]. Linear macromolecules have commonly values of Θ_1 in the 200 to 900 K range with a constant density of states as a function of frequency (box distribution).

Knowing the values of Θ_1, Θ_2 and Θ_3 from a best fit of the experimental, skeletal heat capacities and with a list of group vibrations, one can calculate the heat capacity for the solid state at constant volume: C_V(total). Next C_V(total) is converted with Eq.(6) or Eq.(7) to the calculated heat capacity at constant pressure C_p(vib). This calculated heat capacity can, furthermore, be extended over a wider temperature range and serve as a baseline of the vibrational contribution to the heat capacity.

For glucose, dry starch and starch-glassy water of calculations of vibrational heat capacities are presented by Figs 3, 4 and 5 respectively [39, 26]. In order to evaluate the vibrational heat capacity of α-D-glucose ($C_6H_{12}O_6$), its 72 degrees of freedom resulting from its 24 atoms of the molecule, including from 6 degrees of freedom for rotation (3) and translation (3), were separated into 53 group vibrations (N_{gr} = 53) and 19 skeletal vibrations (N_{sk} = 19). With a fit of the experimental, skeletal heat capacity at low-temperatures to the Tarasov equation [Eq.(12)] the parameters Θ_1 = 664 K and $\Theta_2 = \Theta_3$= 119 K were calculated. With these parameters, the total, vibrational heat capacity for α-D-glucose was estimated, C_p^S(vib), and was extended from 0.1 to 1000 K. Figure 3 displays the evaluations of all contributions to vibrational C_p of α-D-glucose and shows that the major contribution in to the C_p^S(vib) is from skeletal heat capacity, C_p(skeletal) below 100 K which gets saturated around room temperature. The contribution from C_p(group) starts around 100 K and continue increasing in heigh temperature but groups vibrations are not excited even at 500 K. The sum of C_p(group) and C_p(skeletal) gives total heat capacity at constant volume, C_v(total) that after conversion to constant pressure gives a base line only from total vibrational motion, C_p^S(vib). Good agreement at low temperature is found between experimental and calculated C_p (error less than± 1%).

Figure 4 shows the comparison of experimental and calculated heat capacity of dry starch from 8 to 490 K. For evaluation of the vibrational heat capacity of dry starch, the 63 degrees of freedom resulting from the 21 atoms of the repeating

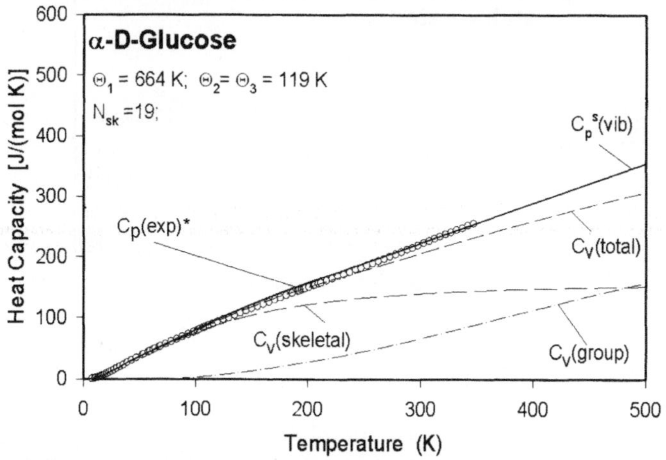

Fig. 3 Comparison of the experimental heat capacities, C_p(exp)* to the calculated vibrational heat capacities for α-D-glucose

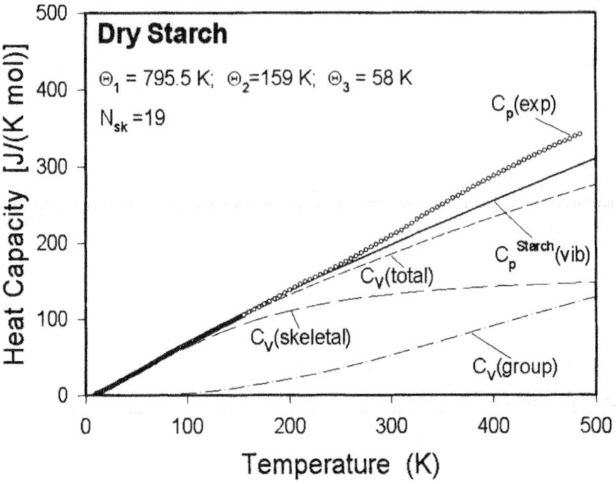

Fig. 4 Comparison of experimental and calculated, vibrational heat capacities for dry starch

unit ($C_6H_{10}O_5$) were separated into 44 group vibrations ($N_{gr} = 44$) and 19 skeletal vibrations ($N_{sk} = 19$). All approximate group vibrational frequencies that are relevant to the current study of starch were taken from normal-mode calculations of α-D-glucose [26]. The fit in the skeletal heat capacity shows a unique minimum in the chi-square statistical function at $\Theta_1 = 795.5$ K and $\Theta_2 = 159$ K and $\Theta_3 = 58$ K. With these parameters, the heat capacity for dry solid starch attributed to vibrations only, $C_p^{Starch}(vib)$, was collected from 0.1 to 1000 K in the ATHAS Data Bank [9]. Also Fig. 4 shows the full evaluation of group, skeletal, and total heat capacities contribution for dry amorphous starch. The calculated $C_p^{Starch}(vib)$ agrees with experimental $C_p(exp)$ to a precision of better than 3±% below 250 K.

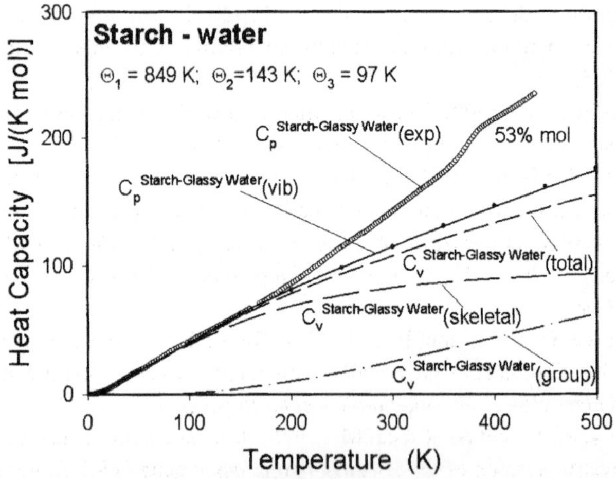

Fig. 5 Comparison experimental and calculated, vibrational heat capacities for starch with 53 mol-% of water

Figure 5 shows an example of the calculated vibrational heat capacity for starch-water systems, with 11wt-% (53 mol-%) of water. The heat capacities of the solid starch-water below the glass transition temperature was estimated by adding the vibrational heat capacity of dry starch $C_p^{\text{Starch}}(\text{vib})$ and the vibrational heat capacity of glassy water $C_p^{\text{Glassy Water}}(\text{vib})$ according to the equation:

$$C_p^{\text{Starch -Glassy Water}}(\text{vib}) = X_S C_p^{\text{Starch}}(\text{vib}) + X_W C_p^{\text{Glassy Water}}(\text{vib}) \qquad (14)$$

where X_S, and X_W are the molar fractions of starch and water, respectively, in the starch-water systems. The vibrational heat capacity of glassy water, $C_p^{\text{Glassy Water}}(\text{vib})$ was estimated based on the low-temperature experimental data of heat capacity glassy water used as evaluated by Suga [56]. For the calculation of the vibrational heat capacity of glassy water, the nine degrees of freedom resulting from the three atoms of the molecule H_2O were separated into the obvious three group vibrations and six skeletal vibrations. The group vibrational frequencies of glassy water were taken from normal-mode calculations [57] that are fitted to the experimental infrared and Raman frequencies, as shown in paper [26]. The Tarasov equation is used to estimate the heat capacity contribution due to skeletal vibrations ($\Theta_1 = 1105.5$ K and $\Theta_2 = 72.4$ K $\Theta_3 = 72.4$ K, $N_{\text{skeletal}} = 6$). Using these parameters for the skeletal contribution and combined with the contributions of the group vibrations, the total vibrational heat capacity for glassy water, $C_p^{\text{Glassy Water}}(\text{vib})$, was calculated from 0.1 to 1000 K and full results of this calculated heat capacity of glassy water is presented in the temperature range of interest, from 100 to 600 K [26], excluding the possible appearance of the glass transition for glassy water which was found between 120 and 134 K according to paper [56]. Figure 5 shows these data using Eq. (14) with $X_S = 0.47$, and $X_W = 0.53$, and solids heat capacities $C_p^{\text{Starch}}(\text{vib})$ and $C_p^{\text{GlassyWater}}(\text{vib})$ (•). Figure 5 displays also the calculated vibrational heat capacities using the best fit of experimental data to the skeletal vibrations $C_v^{\text{Starch- Glassy Water}}(\text{skeletal})$ (solid line). Both data are in good agreement.

The calculated vibrational heat capacity for starch with 53 mol-% of water, $C_p^{\text{starch-Glassy Water}}(\text{vib})$ can serve as the baseline to the thermal analysis as shown in Figs 7 and 11.

Calculations of the heat capacity in the liquid state of dry carbohydrates

Generally, the heat capacity of dry biopolymers such as carbohydrates in the liquid or rubbery states is calculated from the standard thermodynamic relationship [26, 27]:

$$C_p = C_v + TV\alpha^2/\beta \approx C_{\text{vib}}(\text{biopoly}) + C_{\text{conf}}(\text{biopoly}) + C_{\text{ext}}(\text{biopoly}) \qquad (15)$$

All quantities must be known as a function of temperature over the full range of calculation. The experimental heat capacity is, thus, separated into a vibrational heat capacity C_{vib}(biopoly), the conformational heat capacity C_{conf}(biopoly), and the so-called external (anharmonic) contribution C_{ext}(biopoly). For the heat capacities of solids, Eq. (15) is reduced to the two parts [C_{vib}(biopoly) and C_{ext}(biopoly)] since the chain conformation is largely fixed. The major part of the total heat capacity comes from vibrational motion for both the solid and the liquid. It is calculated as:

$$C_{vib}\,(\text{biopoly}) = C_v\,(\text{group}) + C_v\,(\text{skeletal})[\Theta_1, \Theta_2, \Theta_3] \quad (16)$$

where C_{vib}(biopoly) is the same as given in Eq. (7). The C_{ext}(biopoly) in Eq. (15), is calculated as for linear macromolecules either from Eq. (6) or (7).

The conformational contribution C_{conf}(biopoly) is calculated by making use of an earlier derived equation for synthetic polymers [25]. It is based on the one-dimensional Ising model [58] with the following simplifying assumption: The conformational states of the bonds or flexible segments of the biopolymer can occur in only two discrete states, a ground state and an excited state, with an energy-difference between the two states of B. The energy B is modified with the parameter A, describing the interaction of the nearest conformational neighbors. The parameters, A and B, have the meaning of stiffness and cooperativity. The conformations of the chain of a biopolymer with a total of N rotatable bonds and can be described by the one-dimensional Ising-type model with the total energy [25, 27, 58]:

$$E_I = A\sum_{j=1}^{N} m_j m_{j+1} + B\sum_{j=1}^{N} m_j \quad (17)$$

The energy A, depending on the state of the next neighbor, may be positive or negative. The conformation number $m_j = 0$ applies to the ground state with energy zero and degeneracy g_o. The conformation number $m_j = 1$ corresponds to the excited state with energy B and degeneracy g_1. In the present analysis, the ratio of the degeneracies of the conformational states, $\Gamma = g_1/g_o$, are determined by a fit to the experimental heat capacity.

Knowing E_I, one can calculate the partition function and the free energy per bond using the transfer-matrix method. The conformational heat capacity in closed form is then [27]:

$$C_{conf}\,(\text{biopoly}) = R\frac{(g_1/g_0)[B/(k_B T)]^2 e^{-B(k_B T)}}{\left[(g_1/g_0)e^{-B/(k_B T)} + 1\right]^2}[1 + \vartheta(A, B, \Gamma, T)] \quad (18)$$

where first part in Eq. (18) is identical to the rotational isomers model [59–62], the second expression, $\vartheta(A, B, \Gamma, T)$, is too extensive to be shown in detail, but gives contribution from the interaction of the nearest conformational neighbors. In Eq. (18) R is the gas constant and k_B, the Boltzmann constant. A full description of this calculation is given in [25, 27].

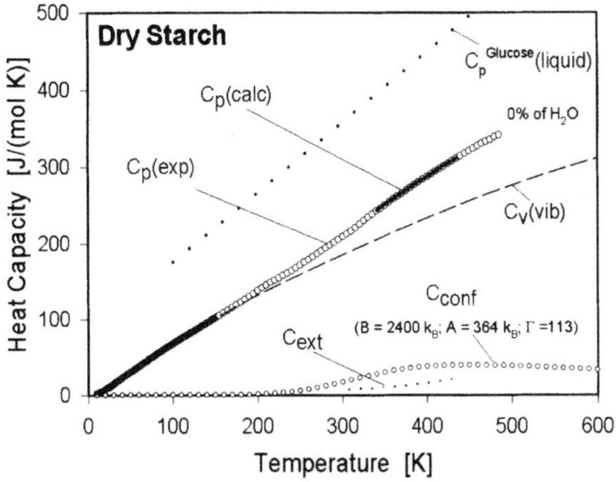

Fig. 6 Comparison of the experimental and calculated heat capacities for dry starch in the partially liquid state

The heat capacity of the amorphous, dry starch above the partial glass transition can be analyzed with Eqs (15)–(18), as is shown in Fig. 6 with a good agreement of experimental and calculated C_p. The experimental data in the temperature range from 320 to 436 K were separated as indicated by Eq. (15). The vibrational heat capacity was calculated as before, the same as for the solid state [26]. For the external heat capacity contribution, C_{ext}(biopoly), Eq. (6) was used with the expansivity ($\alpha = -\,\mathrm{d}\ln V/\mathrm{d}T$), and the compressibility ($\beta_c = \mathrm{d}\ln V/\mathrm{d}P$), derived from the experimental *P-V-T* diagram for liquid-like state, collected from literature [40, 41, 63]. The conformational contribution to total heat capacity of dry starch was then fitted to Eq. (18) after subtracting the vibrational and external portions of heat capacity to find the three characteristic parameters, B, A, and Γ which are listed in Fig. 6. Adding calculated conformational, vibrational, and external heat capacities, the total, calculated heat capacity, C_p(calc), was computed, as is illustrated in Fig. 6. Note that the parameters A and B are given in terms of Θ-temperatures in kelvins, so that the energy B of 2400 K is obtained in units of J mol^{-1} by multiplication with the gas constant $R = 8.314$ J K^{-1} mol^{-1} (=20 kJ mol^{-1}). The effective energy between the two conformational states $B + A$=23 kJ mol^{-1} is rather high which suggests that the segments or bonds in the glucose unit of starch are rather stiff, as expected from the abundant hydrogen bonds. The value of the degeneracy ratio Γ = 113 for dry starch refers to the number of conformational isomers of 4 bonds (3^4) from the total of 12 potentially mobile bonds in a residue of glucose. This is a similar value as derived from the experimental change of the heat capacity at T_g (32.7 kJ mol^{-1} K^{-1}) as is shown in

Fig. 10. This value corresponds to three mobile bonds in the glucose repeating unit when applying the empirical rule that each bond that becomes mobile at the glass transition contributes about 11 J K^{-1} mol^{-1} to ΔC_p. Figure 6 shows a comparison of the experimental and calculated heat-capacity contributions at partially liquid-like state of dry amorphous starch.

Heat capacity in the liquid state of the carbohydrate-water systems

The heat capacity of an amorphous starch with additional small molecules such as water is close to additive and can be represented in analogy to Eq. (15) by:

$$\begin{aligned} &C_p\,(\text{biopoly}-\text{sm}) = \\ &C_{vib}\,(\text{biopoly}-\text{sm}) + C_{conf}\,(\text{biopoly}-\text{sm}) + C_{ext}\,(\text{biopoly}-\text{sm}) \end{aligned} \tag{19}$$

with C_p(biopoly-sm) representing the liquid heat capacity at constant pressure, C_{vib}(biopoly-sm) is the vibrational heat capacity at constant volume, C_{ext}(biopoly-sm) is the external (anharmonic) heat capacity for the biopolymer-small molecules systems, and C_{conf}(biopoly-sm) is the conformational heat capacity of the biopolymer-small molecules complex. As before, the major part of the total heat capacity comes from the vibrational motion. Since the heat capacities of glasses and crystals are similar in value [25], it is assumed that the liquid has also largely the same vibrational contribution to C_p. Above 75 K the C_{vib}(biopoly-sm) can be approximated by addition of the vibrational contributions from the biopolymer C_{vib}(biopoly) and the small molecules, C_{vib}(sm), according to Eq.(14). The vibrational heat capacity of each component in Eq. (14) is calculated by the appropriate Eqs (6)–(13). The external contribution in Eq. (19) is calculated using the parameters for the proper system. The calculation of the conformational contributions is based on an earlier derived description [25, 27], as summarized below.

As a starting point for the conformational heat capacity of the biopolymer-small molecule system, one first considers the clusters of small molecules which interact with the flexible chain of the biomacromolecules to be in a single layer. One assumes the same discrete conformational states of the segments of polymers [24, 25, 27]. Each of the conformational states can be degenerate in multiple ways. The excluded volume effect is neglected [60]. To describe a conformation of the whole chain of the biopolymer, one uses again the one-dimensional Ising model [25, 27]. The small molecules appear as a monolayer on the polymer, having an energy $E < 0$. The second and all higher layers possess the energy K_1, where $K_1 \leq 0$ (K_1 is the energy of an intermediate state, and s is the maximum number of molecules in the cluster).

Each interaction energy of the small molecules with the segment of the polymer, E, can be modified by a parameter C, if the nearest segment of the chain is in an excited state. Then C describes the coupling between conformational states and the states of the small molecules. If the segment of the chain and its neighboring small molecule are considered as one center, the energy of the whole system can be represented as follows for monolayers ($s = 1$): and for cluster ($s > 1$):

$$E_{I_1}(\text{biopoly}-\text{sm}) = A\sum_{j=1}^{N} m_j m_{j+1} + B\sum_{j=1}^{N} m_j + C\sum_{j=1}^{N} m_j n_j + (E-\mu)\sum_{j=1}^{N} n_j \quad (20)$$

$$\begin{aligned} E_{I_s}(\text{biopoly}-\text{sm}) &= A\sum_{j=1}^{N} m_j m_{j+1} + B\sum_{j=1}^{N} m_j + C\sum_{j+1}^{N} m_j (1-\delta_{n_j 0}) + \\ &+(E-K_1)\sum_{j=1}^{N}\sum_{n'}^{s} \delta_{n_j n'_j} + (K_1-\mu)\sum_{j=1}^{N}\sum_{n'}^{s} n_j \delta_{n_j n'_j} \end{aligned} \quad (21)$$

In Eqs (20) and (21), $K_1 = K - \Delta K$ is the energy of the intermediate state, related to the condensation energy K of a small molecule, and μ is the chemical potential of the small molecules. Each site of the model is described by the pair of numbers, m_j,n_j, where conformational number $m_j = 0$ corresponds to the ground state with a degeneracy of g_0, and $m_j = 1$ corresponds to the excited states with degeneracy g_1 for the jth segment. In Eq. (20), the occupancies for small molecules n_j can be only zero, for empty centers, or one for occupied centers ($s = 1$ for monolayers). For cluster of small molecules in Eq. (21), the occupancy $n_j = 0,1,3, \ldots, s$ refer to empty centers (0) or ones occupied with one, two, three, or s small molecules with energies 0, E, $E + K_1, \ldots E + (s-1)K_1$, respectively. Each site of the system can be in $(g_1 + g_0)(s + 1)$ states. In Eq. (21), δ is the Kronecker delta. The free energy F per site of the system can be calculated exactly by using the well-known transfer matrix method [27, 58]:

$$F = -k_B T \ln \Xi = -k_B T \ln \left[Tr e^{-\frac{E_1(\text{biopoly}-\text{sm})}{k_B T}} \right] \quad (22)$$

where Ξ is the partition function and E_1(biopoly-sm) is given by Eq. (21). The result has the form:

$$F = -(\beta N)^{-1} \ln P_{\max}^{N} = -(\beta)^{-1} \ln \frac{1}{2}\left[U + (U^2 - 4V)^{0.5}\right] \quad (23)$$

whith:

$$U = 1 + ab' + k_1 k_2 x(1 + ab'c)\left[\frac{1-(k_2 x)^2}{1-k_2 x}\right] \quad (24)$$

and with:

$$V=(ab'-b')\left[1+k_1k_2x(1+ab'c)\frac{1-(k_2x)^2}{1-k_2x}\right]\left[1+ck_1k_2x(1+ab'c)\frac{1-(k_2x)^2}{1-k_2x}\right] \quad (25)$$

The parameters labeled k_1, k_2, a, b' and c are defined by the following relationships: $a = \exp(-\beta A)$; $b' = \Gamma\, b = \Gamma \exp(-\beta B)$; $c = \exp(-\beta C)$; $k_1 = \exp[-\beta(E - K_1)]$; $k_2 = \exp[-\beta(K - K_1)]$; with $\beta = (k_B T)^{-1}$ and $\Gamma = g_1/g_0$. The quantity $Tr \exp(-\beta E_I(\text{biopoly-sm}))$ in Eq. (22) denotes the trace of a matrix. The quantity $x = \exp[-\beta(E - \mu)]$ has the meaning of an activity of small molecules, which varies from zero to one. Knowing F from Eq. (23), it is possible to calculate all thermodynamic quantities of the system. The conformational heat capacity C_{conf}(biopoly-sm) is then given, with U_E as enthalpy, by:

Calculation of C_p is rather tedious because of the double differentiation of the partition function, but can be done by using Mathematica 3.0. The final results can be expressed in closed form as the conformational heat capacity, but is too involved to be shown here. The conformational heat capacity is a function of many parameters $C_{conf}(\text{biopoly-sm}) = f(T, B, A, C, E, \Gamma, K_1, K_2, s)$, with $B, A, C,$

$$C_{conf}(\text{biopoly}-\text{sm}) = \frac{dU_E}{dT}$$
$$U_E = -k_B T^2 \frac{d\left(\frac{F}{k_B T}\right)}{dT} = -k_B T^2 \frac{d(\ln \Xi)}{dT} \quad (26)$$

Γ describing the biopolymer and E, K_1, K_2, and s the small molecules.

The partial liquid or rubbery C_p of carbohydrae-water systems such as starch-water is next estimated similar to the dry starch as the sum of the vibrational, external, and conformational contributions. The experimental conformational contribution of the systems is estimated by subtraction of the vibrational and external contributions given in Eq. (19) from the partial liquid heat capacity in the glass transitions. The experimental conformational heat capacity is then fitted to the calculated C_{conf}(biopoly-sm) obtained from Eq. (26). An example of such thermal analysis is presented by Figs 7, and 8 for starch with low concentrations of water. The contributions to the experimental, apparent heat capacity of starch-water originate from vibrational and large-amplitude molecular motion and were determined over the range from 8 to 490 K.

The vibrational heat capacity at constant volume C_{vib}(biopoly-sm) [dashed lines in Figs 7 and 8] was estimated by adding the vibrational heat capacity from dry starch, C_V^{Starch}(vib), and the vibrational contribution from glassy water, $C_p^{\text{Glassy Water}}$(vib), as given in detail earlier in Eq.(14) [26]. For the calculation of the external heat capacity, C_{ext}(biopoly-sm), the second part of the center of Eq.

(19) was used, with V the total molar volume, α, the thermal expansivity, and βc, the compressibility for starch with 53 and 65 mol-% of water, respectively, as is available in the literature [26, 61]. The results of the external heat capacity are drawn in Figs 7 and 8 as dotted lines.

The conformational heat capacity of starch-water systems, C_{conf}(biopoly-sm) = f (*T, B, A, C, E,* Γ, K_1, *K, s*) was fitted to the experimental conformational contribution setting *B* and *A* to the values obtained for the dry starch. From the experimental heat capacities of starch-water the vibrational and external portions of the heat capacity were subtracted, and the remaining conformational part was fitted as

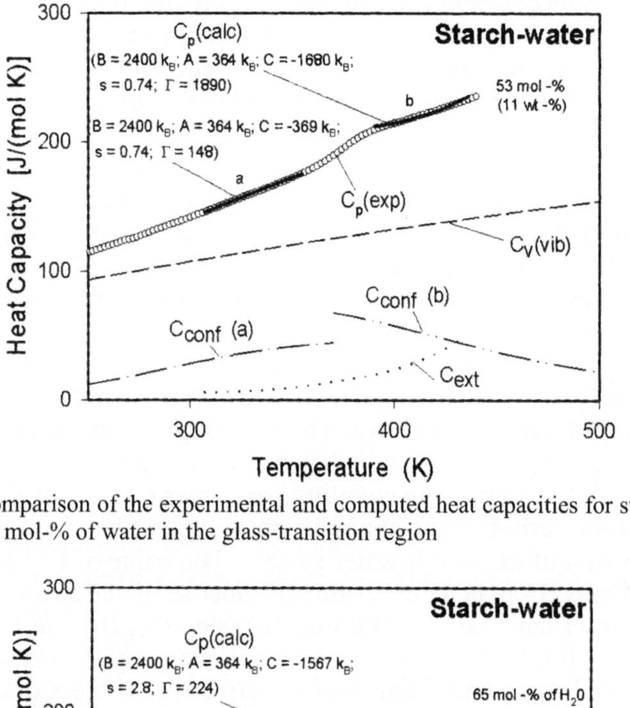

Fig. 7 Comparison of the experimental and computed heat capacities for starch with 53 mol-% of water in the glass-transition region

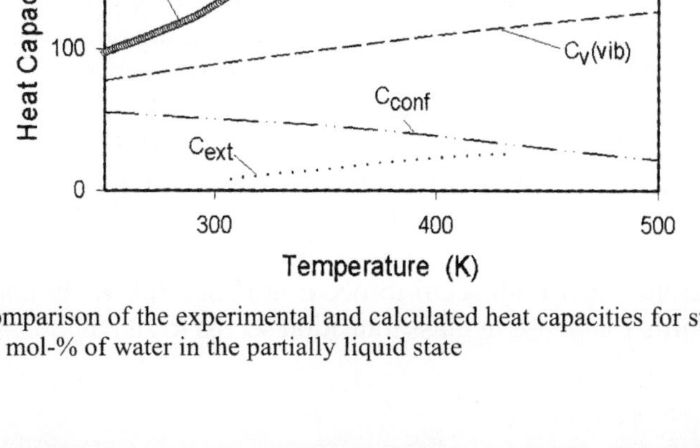

Fig. 8 Comparison of the experimental and calculated heat capacities for starch with 65 mol-% of water in the partially liquid state

outlined above. The parameters obtained from the fitting are shown in part in Figs 7 and 8. Finally, the vibrational, conformational, and external heat capacities were added and the total heat capacity was calculated and shown in Figs 7 and 8 as the heavy lines. The root-mean-square deviations of the experimental heat capacity from the calculated values are always less than ±2% and depend on the temperature ranges used.

In Fig. 7, from the best fit from 390 to 445 K (region *b*), the effective energy difference between the two conformational states $B + A + C = 1084$ K (9.01 kJ mol^{-1}) is reduced from the 23 kJ mol^{-1} for dry starch. A result one would expect, since the presence of water should increase the flexibility. Also, the best fit in region (*b*) gives a total number of 0.74 molecules of water per glucose repeating unit (*s*), a value not far from 1.13 expected for the experimental 53 mol-% of water since no crystallizable water was detected [64]. The value of the degeneracy ratio Γ is 1890 and corresponds to a number of conformational-isomers of almost 7 bonds (3^7) compared to the 4 in dry starch. The added mobility must be caused by the water. It corresponds also to the larger experimental change of the heat capacity in this section of the glass transition [$\Delta C_p^{AB} = 67.4$ J K^{-1} mol^{-1}, see Fig. 11. The values of other parameters, such as energies of E, and K1 correspond to water-carbohydrate and water-water interactions, respectively are listed in Table 1 in paper [27].

In addition, the experimental heat capacity of starch with 53 mol-% water was fitted below $T_g = 372$ K, from 305 K to 355 K (region a in Fig. 7), using the fixed values of *B*, *A* and *s* from above. The best fit gives an effective energy of 20 kJ mol^{-1} ($B + A + C = 2395$ K) which is higher than the value in region *b* due to the lower absolute value of the coupling parameter *C* (= –369 K). The effective stiffness of the carbohydrate chain becomes higher than in region *b* due to some frozen motion of the starch-water system. The value of Γ = 148 gives the suggestion that only 34 bonds per residue of glucose of starch are mobile. The difference of conformational contributions between C_{conf}(b) and C_{conf}(a) at the lower partial glass transition is, again, similar to the jump in heat capacity ΔC_p in the experimental data. The changes of the effective stiffness of chain carbohydrate below and above the glass transition can also be related to a change of the population of secondary bonds between the carbohydrate chain. The coupling parameter *C* can indirectly describe this process through the secondary subsystem which was fixed in our model. The Fig. 7 in the ranges of temperature 305 K to 445 K, demonstrates a good agreement between both, experimental and calculated heat capacity at constant pressure when assuming two steps of a glass transition which, however, is still incomplete at 445 K.

In Fig. 8 is presented of the comparison of the experimental and calculated heat capacities for amorphous starch with 65 mol-% of water in the partial liquid region. For this sample the water has moved the second glass transition step more than the first, so that now both seem to occur at about 300 K. In addition above about 400 K a further increase in glass transition seems to signal the last step of the glass transition, lowered by the water sufficiently to fall below the decomposition temperature. Again, a good agreement exists between calculated and experimental data between 330 and 380 K. For the calculation of the heat capacity for the partially liquid-like state was estimated in the same manner as before. The vibrational contribution according to Eq. (14) is given by the dashed line in Fig. 8, and the external contribution is plotted as the dotted line. All parameters are listed in Table 1in the paper [27]. As in Fig. 7, B and A parameters were fixed to the dry starch values. The effective energy difference between the two conformational states is 10 kJ mol^{-1} ($B + A + C = 1197$ K), similar as in region b of Fig. 7. This means that adding water beyond 53 mol-% does not increase the flexibility of the carbohydrate chain. From the best fit, the value of Γ was reduced to 224 compared to 1890 for starch with 53 mol-% of water so that it corresponds to a number of conformational-isomers of almost 5 bonds (3^5). This should be related to the increasing of numbers of water molecules per glucose unit in the starch-water system. From the best fit, $s =$ 2.8, maximum number of water molecules which can interact with one glucose unit is very close to the expected value: 3.0. This comparison to the experimental value of 1.9 molecules per glucose residue for 65 mol-% of water in starch shows that still is enough space to saturate the system.

Quantitative thermal analysis of carbohydrate and carbohydrate-water systems

Having established the two baselines as references: solid, vibrational heat capacity and liquid heat capacity, a final quantitative analysis can be presented.

Figure 9 shows the quantitative thermal analysis of amorphous α-D-glucose. The experimental data is analyzed in the frame of liquid heat capacity estimated as linear function of temperature and the vibrational heat capacity as resulting from skeletal and group vibrations of glucose. The glass transition temperature was found as the value of 310K corresponding to halfway between glassy and liquid states. Using solid and liquid C_p, the estimated changes of heat capacity of at the glass transition temperature $T_g = 310$ K was measured as $\Delta C_p =$ 141 J K1 mol^{-1} a value that corresponds to 12 bonds ('beads') becoming mobile in glucose at T_g.

Next, examples of such thermal analysis are presented by Figs 10, and 11 for starch without and with low concentrations of water. Figure 10 shows the comparison of experimental and calculated heat capacity of dry starch from 8 to 490 K. The contributions to the experimental, apparent heat capacity of dry starch originate from vibrational and large-amplitude molecular motion. At low temperature, C_p(exp) of amorphous dry starch agrees with the vibrational heat capacity C_p^{Starch}(vib). Deviations of C_p(exp) from C_p^{Starch}(vib) start at 250 K and continue into the high temperature region. Figure 10 suggests a halt in this broad increase in C_p(exp) at 32.7 J K^{-1} mol^{-1}, resulting in a partial glass transition at about $T_g = 320$ K, measured at the half-height of the change in C_p, from points A

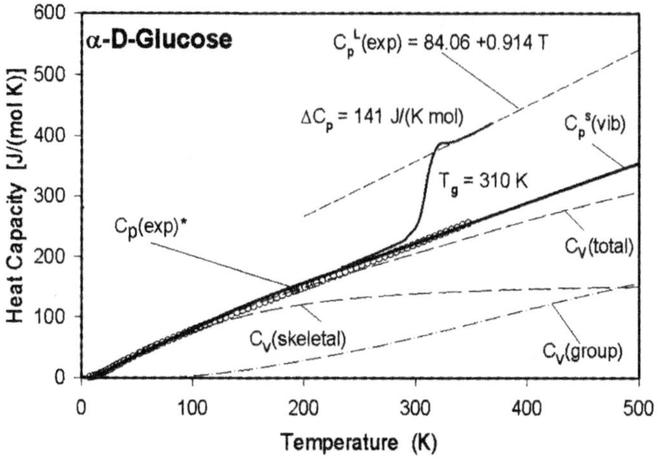

Fig. 9 Analysis of the experimental heat capacity of α-D-glucose in the frame of the solid, vibrational and liquid heat capacity

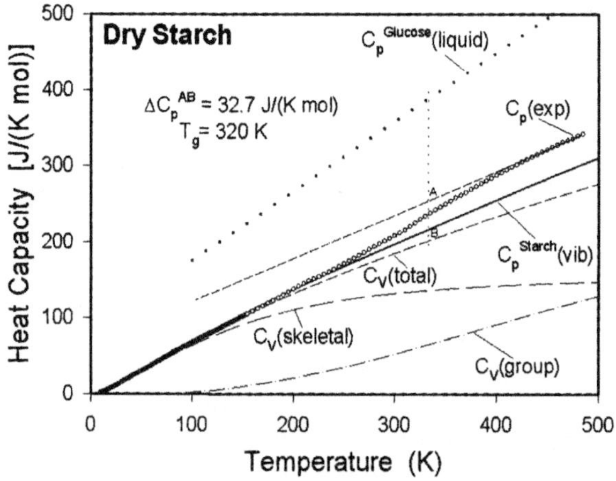

Fig. 10 Quantitative thermal analysis of the experimental heat capacity of dry starch in terms of vibrations to identify the partial glass transition

to B. This small value can be compared to the 141 J K^{-1} mol^{-1} at the glass transition temperature for amorphous glucose (T_g = 310 K) and gain mobility at the glass transition only 2–3 of such beads account for the heat-capacity increase. This change of heat capacity 32.7 J K^{-1} mol^{-1} must, thus, be related to some local conformational motion that is initiated within the repeating unit of starch. (The beads correspond to the 12 mobile bonds permitting hindered rotation at sufficiently high temperature in the glucose molecule: a total of five bonds of the type C–C, two bonds of the type C–O, four bonds of the type C–OH, and one bond of the type CH_2–OH. Each group needs a change of about

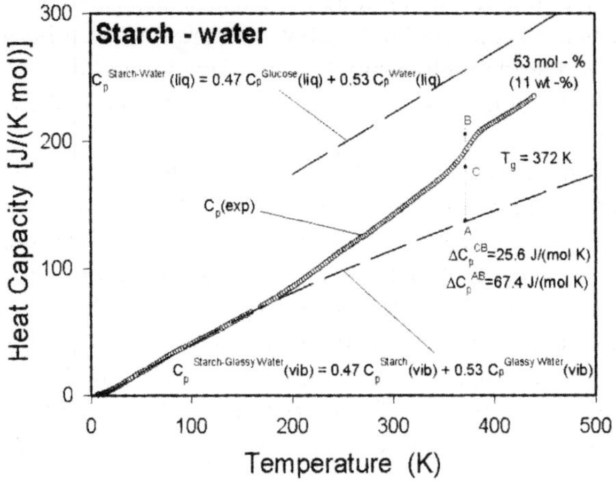

Fig. 11 Experimental heat capacity of starch with 53 mol-% of water in the frame of the vibrational and liquid heat capacity for the starch-water system

11.8 J K^{-1} mol^{-1} in heat capacity to become mobile, as is also common for most linear macromolecules [9].) In Fig. 10, the liquid heat capacity of glucose (dotted line) is used only as a reference to help in the discussion; it is not identified as the liquid heat capacity of starch. The surprising conclusion from this quantitative thermal analysis is the discovery that the glass transition of dry starch is rather broad and occurs partially before decomposition.

Similar quantitative analysis can be presented for the starch-53 mol-% of water systems. In Fig. 11, the experimental C_p of amorphous starch with 53 mol-% of water is analyzed in the frame of vibrational heat capacity of starch-glassy water, $C_p^{\text{Starch-Water}}$(vib) and liquid heat capacity starch-water, $C_p^{\text{Starch-Water}}$(liq). For $C_p^{\text{Starch-Water}}$(vib) good agreement exists at temperatures below 170 K, indicating that only vibrational motion of starch and glassy water contribute in this temperature region to C_p. The deviation of the experimental heat capacity from the vibrational baseline gradually increases until a clear jump occurs at a T_g, = 372 K. This is followed by a ΔCp parallel to that expected for the indicated liquid $C_p^{\text{Starch-Water}}$(liq), but at a lower than expected level. The measured ΔC_p at T_g = 372 K is only 25.6 J K^{-1} mol^{-1}, still not sufficient to account for the full glass transition. For a full step from *A* to *B*, in contrast, one expects 67.4 J K^{-1} mol^{-1}. The conclusion from this quantitative thermal analysis is that the glass transition of amorphous starch with 53 mol-% of water is in the two steps: gradual deviation from vibrational heat capacity and a jump in heat capacity due to conformational and anharmonic motions.

Figure 12 shows comparison of the heat capacities of starch mostly presented in this chapter with literature data on amylopectin [65] with different amounts of water. In the limited temperature range of overlap the two data sets are similar. The plasticizing of starch and amylopectin by water is obvious. An

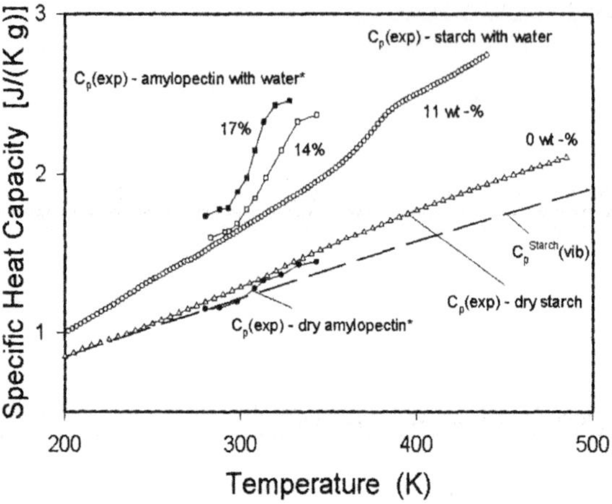

Fig. 12 Comparison of the experimental heat capacity of starch and starch-water with literature data for amylopectin (* data from Ref.[65])

increasing amount of water in starch and amylopectin lowers the main part of the glass transition. Even for dry amylopectin, a broad transition appears at a similar temperature region as for the dry, amorphous starch.

Conclusions

The quantitative thermal analysis and interpretation on a molecular level have shown that starch has a surprising multistage glass transition which stretches over 100 degrees. Even dry starch has such a very broad partial transition which starts at about 250K, far below the decomposition temperature where the glass transition is not yet completed. Adding water, as in food applications, moves this transition to lower temperatures. This is the plasticizing effect of water which allows the study of more of the transition, and is, naturally, also the base of the application of starches in the food industry. Three steps were observed in starch-water systems (11–17 wt-% water). The first is indicated by a gradual deviation from the vibrational heat capacity of the glassy starch. The second is a jump in C_p, followed by the third, another deviation from linear C_p. The latter two are due to conformational and anharmonic motions. This new research resolves the old problem of gel-starch aging with moisture (11wt%) during food storage. Amorphous starch at room temperature absorbs 10–12 wt-% water. This starts at the gradual beginning of the transition and is far from the jump in C_p at about 372 K and ultimately leads to a semicrystalline starch. A similar, low-temperature beginning seems to exist in the glass-transition behavior of other biomaterials, such as proteins and poly(amino acid)s.

Acknowledgments

This work was supported by the Division of Materials Research, National Science Foundation, Polymers Program, Grant # DMR-0312233. Some use of equipment and laboratory space was provided by the Division of Materials Sciences and Engineering, Office of Basic Energy Sciences, U.S. Department of Energy at Oak Ridge National Laboratory, managed and operated by UT-Battelle, LLC, for the U.S. Department of Energy, under contract number DOE-AC05-00OR22725.

References

1 Appelqvist, I. A. M. Cooke, D. Gidley, M. J. and Sally, J. L (1993) Thermal properties of polysaccharides at low moisture: 1-An endothermic melting process and water - carbohydrate
interactions, *Carbohydrate Polymers,* 20, 291–299.

2 Roos, Y. and Karel, M (1991) Water and molecular weight effects on glass transition in amorphous carbohydrates and carbohydrate solutions, *Journal of Food Science,* 566, 1676 –1681.

3 Makhatadze, G. I. Privalov, P. L (1993) Contribution of hydration to protein-folding thermodynamics. 1. The enthalpy of hydration, *J. Mol. Biology*, 232, 639.

4 Kalichevsky, M. T. Jaroszkiewicz, E. M. Ablett, S. Blanshard, J. M. V. and Lillford, P. J (1992) The glass transition of amylopectin measured by DSC, DMTA and NMR, *Carbohydrate Polymers,* 18, 77–88.

5 Lourdin, D. Coignard, L. Bizot, H. Colonna, P (1997) Influence of equilibrium relative humidity and plasticizer concentration on the water content and glass transition of starch materials, *Polymer*, 38, 5401–5406.

6 Thiewes, H. J. and Steeneken, P. A. M (1997) The glass transition and the sub-Tg endotherm of amorphous and native potato starch at low moisture content. *Carbohydrate Polymers,* 32, 123–130.

7 Borde, B. Bizot, H. Vigier, G. and Buléon, A (2002) Calorimetric analysis of the structural relaxation in partially hydrated amorphous polysaccharides. I. Glass transition and fragility, *Carbohydrate Polymers,* 48, 83–96.

8 Levine, H. Slade, L (1991) Beyond water activity - recent advances based on an alternative approach to the assessment of food quality and safety, *Crit. Rev. Food Sci. Nutrit.*, 30, 115–360.

9 The *ATHAS* data bank. For a general description, see: (a) Wunderlich, B (1995) The ATHAS database on heat capacities of polymers, *Pure Applied Chem.*, 67, 1919. For detailed information of the computed heat capacities see (b) Pyda, M (ed.) ATHAS Data Bank web-site at *http://web.utk.edu/~athas*. For experimental C_p data are critically evaluated, see (c) Gaur, U. Shu, H.-C Mehta, A. Lau, S.-F Wunderlich, B. B Varma-Nair, M. Wunderlich, B Heat-capacity and other thermodynamic properties of linear macromolecules, *J. Phys. Chem. Ref. Data* (1981), 10, 89, 119, 1001, 1051; (1982), 11, 313, 1065; (1983), 12, 29, 65, 91; and (1991), 20, 349. For tables of the addition scheme see (d) Gaur, U. Cao, M. Y. Pan, R. Wunderlich, B. (1986) An addition scheme of heat-capacities of linear macromolecules-carbon backone polymers, *J. Therm. Anal.,* 31 421; Pan, R. Cao, M.Y. Wunderlich, B. (1986)

An addition scheme of heat-capacities of linear macromolecules. 2. Backone-chains that contain other than C-bonds, *J. Therm. Anal.,* 31, 1319.
10 Debye, P.(1912) Zur Theorie der spezifischen Wärme, *Ann. Physik*, 39, 789.
11 Miyazaki, Y. Matsuo, T. and Suga, H (1993) Glass transition of myoglobin crystal, *Chem. Phys. Lett.*, 213, 303.
12 Hatakeyama, H. Hatakeyama, T (1998) Interaction between water and hydrophilic polymers, *Thermochimica Acta*, 308, 3–27.
13 Anderson, T. F. Prausnitz, J. M (1978)Application of the UNIQUAC equation to calculation of multicomponent phase equilibrums. 1. Vapor-liquid equilibriums *Ind. Eng. Chem. Process Des. Dev.,* 17, 552–561.
14 Flory, P. J (1965) Statistical thermodynamics of liquid mixtures, *J. Am. Chem. Sci.,* 87, 1833.
15 Flory, P. J. Orwal, R. A. Vrij, A (1964) Statistical thermodynamics of chain molecule liquids. I. An equation of state for normal paraffin hydrocarbons *J. Am. Chem. Sci.*, 86, 3507.
16 Eichinger, B. E. Flory, P. J (1968) Thermodynamics of polymer solutions. I. Natural rubber and benzene, *J. Trans. Faraday. Soc.*, 64, 2305.
17 Sanchez, I. C. Lacomb, R. H (1976) Elementary molecular theory of classical fluids- pure fluids, *J. Phys. Chem.,* 80, 2352.
18 Sanchez, I. C. Cho, J (1995) A universal equation of state for polymer liquids, *Polymer,* 36, 2929.
19 Cho, J. Sanchez, I. C (1998) An analytical free energy and the temperature-pressure superposition principle for pure polymeric liquids, *Macromolecules,* 31, 6650.
20 Simha, R. Somcynsky, T (1969) Statistical thermodynamics of spherical and chain molecule fluids, *Macromolecules,* 2, 342.
21 Simha, R (1977) Configurational thermodynamics of liquid and glassy polymeric states, *Macromolecules,* 10, 1025.
22 Simha, R (1976) Configurational statistical thermodynamics of polymer liquids, *Ann. NY Acad. Sci.,* 2, 279.
23 Prigogine, I. Trappeniers, N. Mathot,V (1953) Statistical thermodynamics of r-mers and r-mer solutions, *Discuss. Faraday. Soc.,* 15, 93.
24 Loufakis, K. Wunderlich, B (1988) Computation of heat capacity of liquid macromolecules based on a statistical mechanical approximation, *J. Phys. Chem.,* 92, 4205.
25 Pyda, M. Wunderlich, B (1999) Computation of heat capacities of liquid polymers, *Macromolecules,* 32, 2044–2050.
26 Pyda, M (2001) Conformational contribution to the heat capacity of the starch and water system, *J. Polymer Sci., Part B: Polymer Phys.*, 39, 3038–3054.
27 Pyda, M (2002) Conformational heat capacity of interacting systems of polymer and water, *Macromolecules*, 35, 4009–4016.
28 Angell, C. A (1995) Formation of glasses from liquids and biopolymers, *Science*, 267 1924.
29 Sochava, I. V. and Smirnova, O. I (1993) Heat - capacity of hydrated and dehydrated globular-proteins- denaturation increment of heat capacity, *Food Hyrocoloids,* 6, 513.
30 Nashed, G. Rutgers, P. P. G. Sopade, P. A (2003) The plasticisation effect of glycerol and water on the gelatinisation of wheat starch, *Starch-Starke,* 55, 131-137.
31 Mathew, A. P. Dufresne, A (2002) Plasticized waxy maize starch: Effect of polyols and relative humidity on material properties, *Biomolecules,* 3, 1101–1108.

32 Roles K. and Wunderlich, B (1991) Heat capacities of solid poly(amino acid)s, I. Polyglycine, poly-L-alanine, and poly-L-valine, *Biopolymers,* 31, 477.
33 Roles, K. Xenopoulos, A. and Wunderlich, B (1993) Heat capacities of solid poly(amino acid)s II., *Biopolymers,* 33, 753.
34 Zhang, G. Lebedev, B. V. Wunderlich, B. and Zhang, Jing-ye (1995) Heat-capacites of poly(amino acid)s and proteins, *J. Polym. Sci., B: Polym. Phys.*, 33, 2449.
35 Zhang, G. Gerdes, S. and Wunderlich, B (1996) Heat capacities of solid, globular proteins, *Macromolecular Chem. Phys.*, 197, 3791.
36 Di Lorenzo, M. L. Zhang, G. Pyda M. and Wunderlich, B (1999) Heat capacity of solid-state biopolymers by thermal analysis, *J. Polymer Sci., Part B: Polymer Phys.*, 37, 2093.
37 Bakk, A. (2002) Heat capacities of solid state proteins: implications for protein stability in solution, *Physica-A*, 313, 540.
38 Boerio-Goates, J (1991) Heat-capacity measurements and thermodynamic functions of crystalline α-D-glucose at temperatures from 10 K to 340 K *J. Chem. Thermodynamics,* 23, 403–409.
39 Pyda M. and Wunderlich, B (2001) Quantitative thermal analysis of biomaterial-water systems , *Proc. 29th NATAS Conf. in St. Louis, MO*, Sept. 24-26, K. J. Kociba and B. J. Kociba, edts., 29, 76–81.
40 Benczedi, D. Tomka, I. Escher, F. (1998) Thermodynamics of amorphous starch-water systems. 1. Volume fluctuations, *Macromolecules,* 31, 3055–3061.
41 Benczedi, D. Tomka, I. Escher, F. (1998) Thermodynamics of amorphous starch-water systems. 2. Concentration fluctuations, *Macromolecules,* 31, 3062–3074.
42 Lebedev, B. V. Kulagina, T. G. Smirnova, N. N (1988) Thermodynamics of 4-methylcyclo-hexene, glycollide, and 1,1,3,3,5,5-hexaethylcylotrisiloxane from 13.4 to 400 K, *J. Chem. Thermdyn.*, 20, 1383–1396.
43 Wunderlich, B (1990) *Thermal Analysis*, Academic Press, New York.
44 Pyda, M. Bartkowiak, M. and Wunderlich, B (1998) Computation of heat capacities of solids using a general Tarasov equation, *J. Thermal Analysis and Calorimetry*, 52, 631.
45 Tarasov, V. V. Yunitskii, G. A (1965) Theory of heat capacity of chain-layer structures, *Zh. Fiz. Khim.*, 39, 2077.
46 Nernst W. and Lindemann, F. A (1912) Specific heats and the theory of energy units, *Z. Electrochem.*, 17, 817.
47 Pan, R. Varma-Nair, M. and Wunderlich, B. (1989) On the C_p to C_v conversion for solid linear macromolecules II, *J. Thermal Analysis*, 35, 955.
48 Wunderlich, B. (2002) Heat capacity of polymers, in S. Z. D. Cheng, (ed), *Handbook of Thermal Analysis and Calorimetry, Vol.3, Applications to Polymers and Plastics*, Elsevier Science, Amsterdam, 1–47.
49 Grebowicz, J. Suzuki, H. and Wunderlich, B (1985) Heat capacities of polyethylene and linear aliphatic polyoxides, *Polymer*, 26, 561.
50 Cheng, S. Z. D. Lim, S. Judovits, L. H. and Wunderlich, B (1987) Heat capacities of high melting polymers containing phenylene groups, *Polymer*, 28, 10.
51 Lau S.-F. and Wunderlich, B (1983) Calculation of the heat capacity of linear macromolecules from Θ-temperatures and group vibrations, *J. Thermal Anal.,* 28, 59.

52 Pyda, M. Varma-Nair, M. Chen, W. Aldrich, H. S. Schlosberg, R. H. and Wunderlich, B (1996) Analysis of a symmetric neopentane ester. I. Measurement and calculation of heat capacity. *J. Therm. Anal.,* 46, 1093.
53 Wunderlich B. and Bauer, H (1970) Heat Capacities of Linear High Polymers, *Adv. Polym. Sci.*, 7, 151.
54 Cheban, Yu. V. Lau S. F. and Wunderlich, B (1982) Analysis of the contribution of skeletal vibrations to the heat capacity of linear macromolecules in the solid state, *Colloid Polymer Sci.*, 260, 9.
55 Zhang G. and Wunderlich, B (1996) A new method to fit approximate vibrational spectra to the heat capacity of solids with Tarasov functions, *J. Therm. Anal.*, 47, 899.
56 Sugisaki, M. Suga, H. Seki, G (1968) *Bull. Chem. Soc. Jpn.,* 41, 2591.
57 Diem, M (1993) *Interpretation and Processing of Vibrational Spectra*; John Wiley & Sons Ltd., New York.
58 Huang, K (1963*) Statistical Machanics*, ed., Wiley, New York.
59 O'Reilly, J. M (1977) Conformational specific-heat of polymers, *J. Appl. Phys.*, 48, 4043.
60 Flory, P. J (1953) *Principles of Polymers Chemistry*; Cornell Univ. Press, Ithaca, NY.
61 Birshtein, T. M. Ptitsyn, O. B (1966) *Conformations of Macromolecules*; Interscience, New York.
62 Volkenstein, M. V (1963) *Configurational Statistics of Polymer Chains*; Interscience, New York.
63 Benczedi, D (1995) Statistical thermodynamic inestigations on amorphous starch-water systems *Thesis*, ETH, Nr.11203 Zurich, Switzerland.
64 Androsch, R. Pyda, M (2001) unpublished data.
65 Noel, T. R. Ring, S. G (1992) A study of the heat capacity of starch-water mixtures. *Carbohydrate Polymers,* 277, 203–213.

Chapter 13

Statistical mechanical analysis of protein heat capacity accompanied with thermal transition

Shun-ichi Kidokoro[1-3*]

[1]*Department of Bioengineering, Nagaoka University of Technology, 1603-1 Kamitomioka, Nagaoka 940-2188, Japan*
[2]*Precision and Intelligence Laboratory, Tokyo Institute of Technology, 4259 Nagatsuta, Midori-Ku, Yokohama 226-8501, Japan*
[3]*Genome Science Center, Riken, 1-7-22 Suehiro-cho, Tsurumi-ku, Yokohama 230-0045, Japan*

Introduction

Legendre transformation in thermodynamics is known as useful tool to convert the thermodynamic variables. One of the most useful transformations is

$$G = H - TS \qquad (1)$$

where G is Gibbs energy, H is enthalpy and S is entropy.
It should be remarked that entropy is well defined in statistical thermodynamics as a function of internal energy and volume, while Gibbs energy is defined as a function of temperature and pressure, and enthalpy is defined as an average around the system under fixed temperature and pressure. These functions are approximately related in statistical thermodynamics. In thermodynamics, entropy is converted to a function of temperature and pressure using the Legendre transformation, Eq. (1).

In statistical thermodynamics, the relation of Legendre transformation such as Eq. (1) will be guaranteed only if the property of thermodynamic functions satisfies some conditions. Concerning to the transformation of Eq. (1), the necessary condition is that the probability function of enthalpy under fixed temperature and pressure has only one maximum.

In the thermal transition of proteins, the thermodynamic functions and partition functions accompanied with thermal transition have been precisely determined with calorimetry [1, 2]. These thermodynamic functions are useful to discuss the mechanism for protein folding and the stability. Using these functions, it has been shown that the probability function has several maximums around

* kidokoro@nagaokaut.ac.jp

D. Lörinczy (ed.), The Nature of Biological Systems as Revealed by Thermal Methods, 333–341.

the transition temperature [3]. This indicates apparently that the necessary condition for the Legendre transformation is not satisfied.

In this report, the microscopic enthalpy is introduced to clarify the two kinds of entropy functions. One entropy function is a function of macroscopic enthalpy and another is of microscopic enthalpy. When the total entropy of macroscopic enthalpy was evaluated without deconvolution method, it does not coincide to the total entropy of microscopic enthalpy apparently.

On the other hand, the methods to determine the thermodynamic functions of each thermodynamic state have been developed [1, 2]. They are know as 'deconvolution' method. From the heat capacity of the whole system, the thermodynamic functions of several thermodynamic states can be determined. In the present study, the entropy functions of each deconvoluted thermodynamic states were found to become good approximation of the entropy functions of microscopic enthalpy. This indicates that we should deconvolute the thermodynamic states where the Legendre transformation becomes a good approximation in order to discuss the system properly in statistical mechanics.

Methods

In this study, the heat capacity function of the thermal transition of protein was simulated using the experimental values of thermal transition of lysozyme [4, 5].

$$C_{\mathrm{N}}(T) = 2a_{\mathrm{N}}(T - T_0) + b_{\mathrm{N}} \tag{2}$$

$$C_{\mathrm{D}}(T) = 2a_{\mathrm{D}}(T - T_0) + b_{\mathrm{D}} \tag{3}$$

where the parameters for the two thermodynamic states, native (N) and denatured (D), a_{N}, b_{N}, a_{D}, b_{D} were assumed as 45 J K^{-2}mol^{-1}, 22 kJ K^{-1}mol^{-1}, 19 J K^{-2}mol^{-1} and 29 kJ K^{-1}mol^{-1}, respectively. T_0 was 320 K and the midpoint temperature of the transition was assumed to be equal to this temperature.

Using the thermodynamic Eqs (4) and (5), the enthalpy and Gibbs energy for these states were obtained as Eqs (6)–(9):

$$\frac{\partial}{\partial T} H(T, p) = C_{\mathrm{P}}(T, p) \tag{4}$$

$$\frac{\partial}{\partial T} \frac{G(T, p)}{T} = -\frac{H(T, p)}{T^2} \tag{5}$$

$$H_{\mathrm{N}}(T, p) = a_{\mathrm{N}}(T - T_0)^2 + b_{\mathrm{N}}(T - T_0) \tag{6}$$

$$H_{\mathrm{D}}(T, p) = a_{\mathrm{D}}(T - T_0)^2 + b_{\mathrm{D}}(T - T_0) + \Delta H_0 \tag{7}$$

$$G_{\mathrm{N}}(T, p) = -a_{\mathrm{N}}(T^2 - T_0^2) + b_{\mathrm{N}}(T - T_0) - (b_{\mathrm{N}} - 2a_{\mathrm{N}}T_0)T\ln(T / T_0) \tag{8}$$

$$G_{\mathrm{D}}(T,p) = -a_{\mathrm{D}}(T^2 - T_0^2) + b_{\mathrm{D}}(T - T_0) - \\ - (b_{\mathrm{D}} - 2a_{\mathrm{D}}T_0)T\ln(T/T_0) + \Delta H_0(1 - T/T_0) \tag{9}$$

where ΔH_0 was enthalpy difference between N and D states at T_0, which was assumed to 400 kJ mol^{-1} for this thermal transition of lysozyme [5]. As the absolute value of enthalpy and Gibbs energy can not be calculated from heat capacity as seen in Eqs (4) and (5), the enthalpy and Gibbs energy of N sate was set to be zero at $T = T_0$ in the above equations. By adopting these values, the entropy of N state becomes negative at the temperature below T_0, which formally violates the 3rd law of thermodynamics. However, the following results and discussion does not depend on the assumed value for the absolute enthalpy and Gibbs energy at the temperature, T_0, and these values are used in order to simplify the calculation and the illustration of the results.

Equation (4) is the definition of heat capacity and Eq. (5) can be easily derived in statistical thermodynamics as shown later. Therefore Eqs (4) and (5) hold without any approximation.

The Gibbs energy of the system that consists of N and D state is derived as

$$G(T,p) = -RT\ln\left[\exp\left(-\frac{G_{\mathrm{N}}(T,p)}{RT}\right) + \exp\left(-\frac{G_{\mathrm{D}}(T,p)}{RT}\right)\right] \tag{10}$$

Using Eq. (5), the enthalpy of the system is obtained from Eq. (10) as

$$H(T,p) = H_{\mathrm{N}}(T,p)f_{\mathrm{N}}(T,p) + H_{\mathrm{D}}(T,p)f_{\mathrm{D}}(T,p) \tag{11}$$

where the molar fractions of N and D state, $f_{\mathrm{N}}(T,p)$ and $f_{\mathrm{D}}(T,p)$, respectively, are defied as

$$f_{\mathrm{N}}(T,p) = \exp\left[-\frac{G_{\mathrm{N}}(T,p) - G(T,p)}{RT}\right] \tag{12}$$

$$f_{\mathrm{D}}(T,p) = \exp\left[-\frac{G_{\mathrm{D}}(T,p) - G(T,p)}{RT}\right] \tag{13}$$

Heat capacity is obtained using Eqs (4), (11) and (12) as

$$C_{\mathrm{p}}(T,p) = C_{\mathrm{N}}f_{\mathrm{N}} + C_{\mathrm{D}}f_{\mathrm{D}} + \frac{(H_{\mathrm{D}} - H_{\mathrm{N}})^2 f_{\mathrm{N}}f_{\mathrm{D}}}{RT^2} \tag{14}$$

Figure 1 shows this heat capacity function C_{p} with $C_{\mathrm{p,N}}$ and $C_{\mathrm{p,D}}$. Experimentally the heat capacity function C_{p} is obtained with calorimetry as a function of temperature, and the two heat capacity functions C_{N} and C_{D} are assumed for the heat capacity of N and D states [1]. Figure 2 shows the enthalpy and Gibbs energy of the system and those of N and D states.

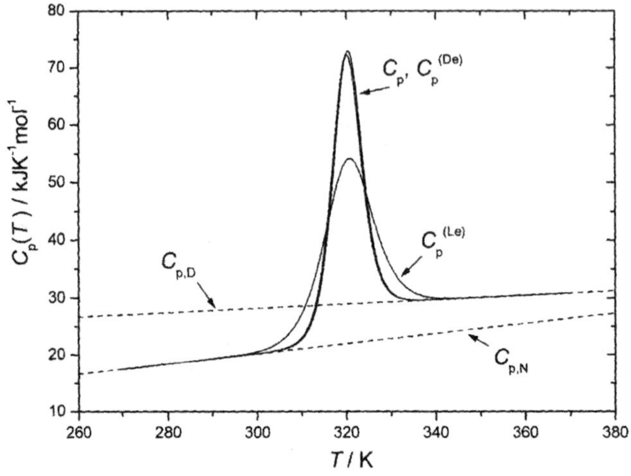

Fig. 1 The heat capacity function that simulates the 2-state transition of lysozyme, $C_p(T)$. The heat capacity functions of native (N) and denatured (D) sate, $C_N(T)$ and $C_D(T)$ are also displayed. The heat capacity, $C_p^{(Le)}(T)$ and $C_p^{(De)}(T)$, which are calculated from statistical thermodynamics based on $S^{(Le)}(H,p)$ and $S^{(De)}(H,p)$ are shown. $C_p(T)$ agrees very well with $C_p^{(De)}(T)$

The entropy using Legendre transformation, $S^{(Le)}(T,p)$ is obtained as a function of temperature using Eq. (1) from $G(T,p)$ and $H(T,p$ *as*

$$S^{(\mathrm{Le})}(T,p) = \frac{H(T,p) - G(T,p)}{T} \tag{15}$$

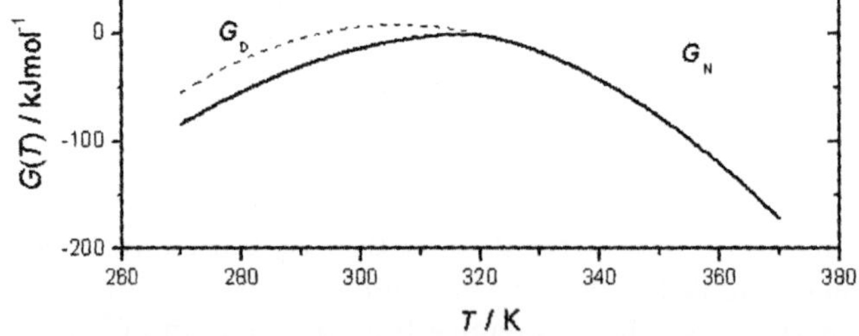

Fig. 2 Enthalpy (upper plane) $H(T)$ and Gibbs energy $G(T)$ (lower plane) that are corresponding to the heat capacity function $C_p(T)$ in Fig.1 are displayed in thick lines. Those of N and D state that are corresponding to the heat capacity function of each state are shown in broken lines

As the enthalpy H is monotonously increasing function of temperature, the entropy can be represented as a function of enthalpy instead of temperature, $S^{(\mathrm{Le})}(H,p)$.

In this simulation, $S^{(\mathrm{Le})}(H,p)$ was obtained as follows. The enthalpy value was fixed first and the temperature was searched with Simplex method where the enthalpy of Eq. (11) agrees with the fixed value of enthalpy. The entropy at that temperature was calculated with Eq. (15). The calculation range of enthalpy of the fixed enthalpy was scanned from –3 MJ $\mathrm{mol^{-1}}$ to 3 MJ $\mathrm{mol^{-1}}$ that is widely enough to cover the enthalpy from 270 K to 370 K as seen in Fig. 2. The number of enthalpy values that divided the whole enthalpy range was tried from 10 to 10000. The 100 points were found to be enough as the results of the calculation did not change above 100 points. In this paper, 4000 points were used for the simulation.

Here it is worthwhile to derive the statistical thermodynamic relations necessary for this paper. Under the fixed temperature, T, and pressure, p, the probability of a microstate whose energy e and volume v: $P(e,v)$ is assumed to obey T–p distribution in statistical thermodynamics:

$$P(e,v) = \frac{\exp\left(-\dfrac{e+pv}{RT}\right)}{Y(T,p)} \tag{16}$$

where $Y(T,p)$ is T–p partition function:

$$Y(T,p) = \sum_e \sum_v \exp\left(-\frac{e+pv}{RT}\right)\Omega(e,v) \tag{17}$$

where $\Omega(e,v)$ is the number of microstates that have energy, e, and volume, v. The double summation of the right hand side of Eq. (17) can be simplified when the new function of number of microstates, $\Omega(h,p)$ is introduced as

$$\Omega(h,p) = \sum_{h=e+pv} \Omega(e,v) \tag{18}$$

where the microscopic enthalpy, h, can be introduced using the microscopic energy and volume as $h=e+pv$. As seen in Eq. (17), the statistical weight of the microscopic state is determined by the microscopic enthalpy. Therefore it is convenient to introduce the new microscopic variable. Using this new function, $Y(T,p)$ is represented as

$$Y(T,p) = \sum_h \exp\left(-\frac{h}{RT}\right)\Omega(h,p) \tag{19}$$

The Gibbs energy, $G(T,p)$ and entropy $S(h,p)$ are defined statistical thermodynamically as

$$G(T, p) = -RT \ln Y(T, p) \tag{20}$$

$$S(h, p) = R \ln \Omega(h, p) \tag{21}$$

The probability function $f(h,p;T)$ is defined as

$$f(h, p; T) = \frac{\exp\left(-\frac{h}{RT}\right)\Omega(h, p)}{Y(T, p)} \tag{22}$$

The enthalpy function is defined as the average of the enthalpy of microstates as

$$H(T, p) = \sum_h hf(h, p; T) \tag{23}$$

It is easy to show Eq. (5) from Eqs (20) and (22). From now, the average value using the probability function of the system is represented as $\langle\ \rangle(T,p)$. For example, the Eq. (23) can be represented as $H(T,p) = \langle h\rangle(T,p)$.

By partial derivative to both sides of Eq. (23) with temperature, the following well-known equation can be obtained:

$$C_{\mathrm{p}}(T, p) = \frac{\langle h^2\rangle - \langle h\rangle^2}{RT^2} \tag{24}$$

The Eqs (21–24) show that the enthalpy and heat capacity can be calculated by statistical thermodynamics when the entropy is obtained as a function of microscopic enthalpy.

It should be noted that the entropy previously obtained using Eq. (15) as a function of macroscopic enthalpy, $S(H,p)$, is different from the entropy as a function of the microscopic enthalpy, $S(h,p)$. If $S^{(\mathrm{Le})}(H,p)$ is a good approximation for $S(h,p)$, the heat capacity $C_{\mathrm{p}}^{(\mathrm{Le})}(T)$ with Legendre transformation can be calculated from the entropy of Eq. (15). In Fig. 1, the calculated heat capacity is shown. It is largely deviated from the initial heat capacity function, $C_{\mathrm{p}}(T)$, indicating that $S^{(\mathrm{Le})}(H,p)$ is not a good approximation for $S(h,p)$.

On the other hand, the total entropy as the function of microscopic enthalpy can be approximated from the entropy of each thermodynamic state. The entropy functions of N and D sate, $S_{\mathrm{N}}(T,p)$ and $S_{\mathrm{D}}(T,p)$ are derived from Eqs (6–9). The Legendre transformation is applied to each thermodynamic state such as:

$$S_{\mathrm{N}}(T, p) = \frac{H_{\mathrm{N}}(T, p) - G_{\mathrm{N}}(T, p)}{T} \tag{25}$$

Using the entropy and the enthalpy function of temperature, the entropy can be obtained as a function of the macroscopic enthalpy in the same way as above discussed in the case of $S^{(\mathrm{Le})}(H,p)$. Therefore $S_{\mathrm{N}}(H,p)$ and $S_{\mathrm{D}}(H,p)$ can be obtained. If these entropy functions of the macroscopic enthalpy can approximate those of the microscopic enthalpy, these entropy functions are related to the number of states as

$$S_{\mathrm{N}}(h, p) = R \ln \Omega_{\mathrm{N}}(h, p) \tag{26}$$

$$S_{\mathrm{D}}(h, p) = R \ln \Omega_{\mathrm{D}}(h, p) \tag{27}$$

As the number of states for the total system, $\Omega(h,p)$ satisfied the relation:

$$\Omega(h, p) = \Omega_{\mathrm{N}}(h, p) + \Omega_{\mathrm{D}}(h, p) \tag{28}$$

the entropy of the system $S^{(\mathrm{De})}(h,p)$ is calculated as

$$S^{(\mathrm{De})}(H, p) = R \ln\left[\exp\frac{S_{\mathrm{N}}(h, p)}{R} + \exp\frac{S_{\mathrm{D}}(h, p)}{R}\right] \tag{29}$$

where the suffix (De) means that this entropy is derived from 'deconvoluted' entropy functions.

Using $S^{(\mathrm{De})}$, the heat capacity was calculated as $C_{\mathrm{p}}^{(\mathrm{De})}$ in the same way as discussed above. $C_{\mathrm{p}}^{(\mathrm{De})}(T)$ in Fig.1 agreed well with the original heat capacity $C_{\mathrm{p}}(T)$. The heat capacity functions of N and D state, $C_{\mathrm{N}}^{(\mathrm{De})}(T)$ and $C_{\mathrm{D}}^{(\mathrm{De})}(T)$, were calculated from $S_{\mathrm{N}}(h,p)$ and $S_{\mathrm{D}}(h,p)$, respectively. They completely agreed with the original heat capacity functions, $C_{\mathrm{N}}(\mathrm{T})$ and $C_{\mathrm{D}}(\mathrm{T})$, respectively (data not shown).

Results and discussion

From a heat capacity function, $C_{\mathrm{p}}(T)$, a wrong heat capacity function $C_{\mathrm{p}}^{(\mathrm{Le})}(T)$ was obtained with the total entropy function of the macroscopic enthalpy. It clearly shows that the total entropy function of the macroscopic enthalpy cannot approximate the entropy of the microscopic enthalpy that was necessary to calculate the heat capacity function with Eq. (24).

Figure 3 shows the entropy as a function of enthalpy. All the entropy functions are represented as difference from N state. In this figure, the entropy $S^{(\mathrm{De})}$ is plotted as the function of the microscopic enthalpy, and the entropy $S^{(\mathrm{Le})}$ is shown as the function of the macroscopic enthalpy. The difference between $S^{(\mathrm{Le})}$ and $S^{(\mathrm{De})}$ at maximum in this case is 8 $\mathrm{JK^{-1}mol^{-1}}$, which is comparable to the gas constant and causes the large difference in heat capacity as seen in Fig.1.

In Fig.4, the probability function is calculated from both entropy $S^{(\mathrm{Le})}$ and $S^{(\mathrm{De})}$. These functions were already displayed with a different method by A. Cooper which is more complicated than this paper [3]. The probability function at the midpoint temperature has two maximums and resemble to that of A. Cooper, while the incorrect probability function, $f^{(\mathrm{Le})}$, shows only one maximum. Mathematically it is obvious that $f^{(\mathrm{Le})}$ has only one maximum because the following equation can be derived by partial derivative of both sides of Eq. (22)

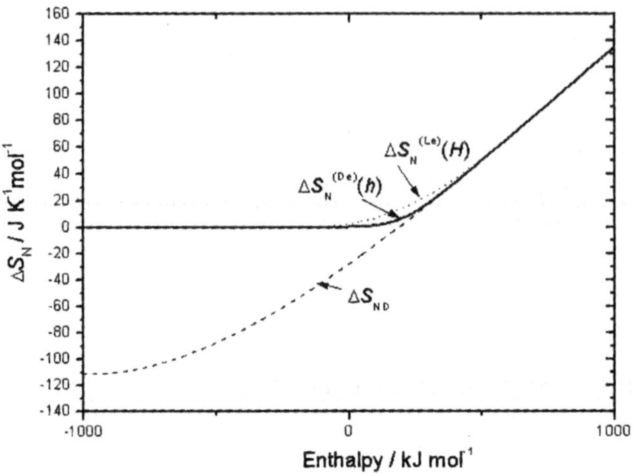

Fig. 3 Entropy functions are shown as functions of enthalpy with reference to N state. $\Delta S_N^{(Le)}=S^{(Le)}-S_N$, $\Delta S_N^{(De)}=S^{(De)}-S_N$ and $\Delta S_{ND}=S_D-S_N$. $S^{(Le)}$ is the entropy function of the macroscopic enthalpy directly calculated from the total Gibbs energy and total entropy by Legendre transformation. $S^{(De)}$ is the entropy function of the microscopic enthalpy composed from the entropy of N and D state (see text in detail)

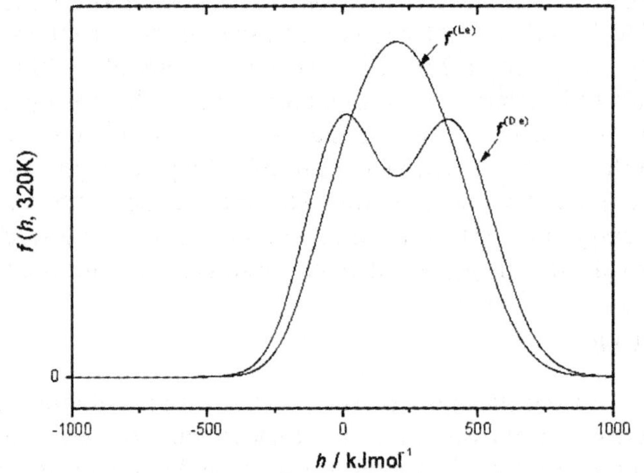

Fig. 4 Probability function of enthalpy at the midpoint of thermal transition (320K). $f^{(Le)}$ is calculated from $S^{(Le)}$ in Figure 3 and $f^{(De)}$ is calculated from $S^{(De)}$ in the figure

$$\frac{\partial}{\partial h} f(h, p; T) = -R\left(\frac{1}{T} - \frac{1}{T_H}\right) f(h, p; T) \tag{30}$$

where T_H is the temperature where $\langle h\rangle(T_H)=h$ satisfies. Because the $\langle h\rangle$ is a monotonously increasing function of T, the right hand side of Eq.(30) is positive when h is smaller than $\langle h\rangle(T)$, while it is negative when h is larger than $\langle h\rangle(T)$.

Therefore the function has one maximum where $h = \langle h \rangle(T)$. It shows that the entropy will cause only one maximum of probability function of enthalpy when the transformation is applied.

As shown above, the entropy function of the macroscopic enthalpy does not approximate that of the microscopic enthalpy around the thermal transition of proteins while it does very well for each thermodynamic state. It indicates that the thermal transition of proteins cannot be treated as one phase but treated as a phase transition in spite of the continuity of the thermodynamic functions.

Usually the first order phase transition requires the discontinuity of the first derivatives of Gibbs energy. Strictly speaking, however, the complete discontinuity will be achieved only for the infinite system. When the system becomes small to the size of proteins, the discontinuity cannot be observed. However the discrepancy between the two entropy functions may be observed in this system. It indicates that this discrepancy may become a good index for phase transition for such a system.

The deconvolution method was proposed for the thermal transition of biopolymers [1, 2]. This report clarifies the statistical mechanical meaning for the deconvolution method. If one thermodynamic state includes the thermal transition in it, the discrepancy of these two entropy functions can be obtained. Then the system can be deconvoluted to each thermodynamic state where the Legendre transformation becomes a good approximation to get the entropy function of the microscopic enthalpy.

References

1 Kidokoro, S. Wada, A. *Biopolymers,* 26 (1987) 213–229.
2 Kidokoro, S. Uedaira, H. Wada, A. *Biopolymers,* 27 (1988) 271–297.
3 Cooper, A. *Prog. Biophys. Mol. Biol.,* 44 (1984) 181–214.
4 Privalov, P. L. Makhatadze, G. I. *J. Mol. Biol.,* 213 (1990) 385–391.
5 Makhatadze, G. I. Privalov, P. L. *J. Mol. Biol.,* 232 (1993) 639–659.

1. [illegible]
2. Kitahara, [illegible]
3. [illegible]

Subject index

Colour Section

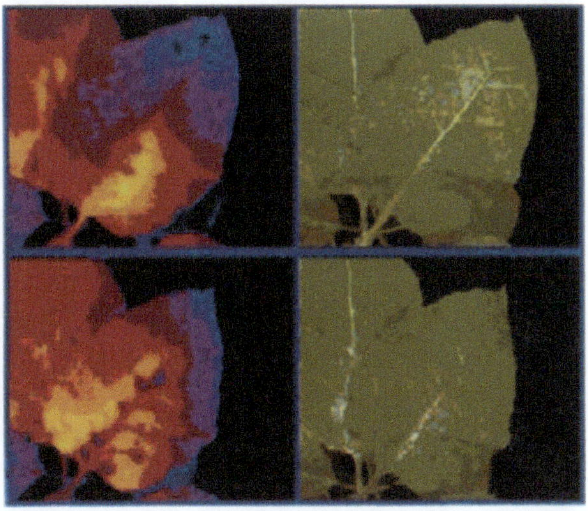

Fig. 11 Thermographic and visual imaging of cell death (yellow parts) in bacterio-opsin tobacco 32 h (upper two pictures) and 40h (lower two) after first detection of a thermal effect. The maximum temperature difference amounts to 0.6 K. With permission from [85] p. 205.

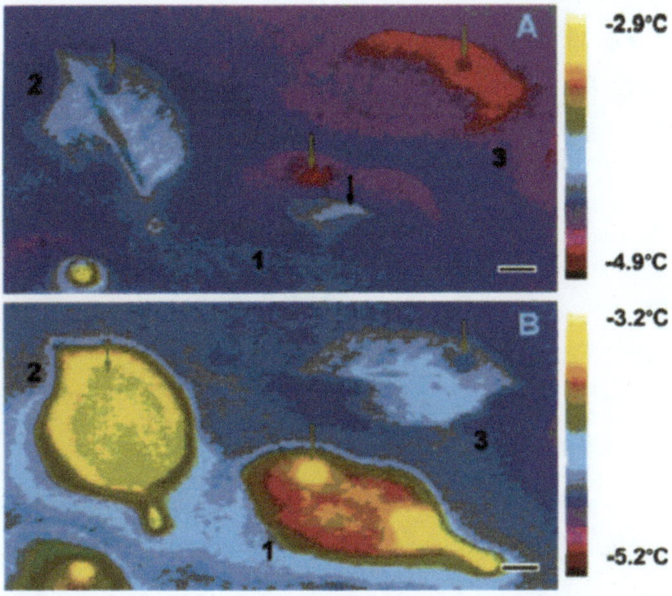

Fig. 12 Holly leaves (*Ilex* sp.) during freezing shown in false colours of a 2 K temperature range. Picture B was taken about 3 min after A. The pale blue to whitish areas (A,B) indicate an initial exothermic effect of low intensity, the yellish colours (B) a second stronger exothermic effect. Green arrows point to water droplets put on the leaves before cooling started. With permission from [90] p. 205

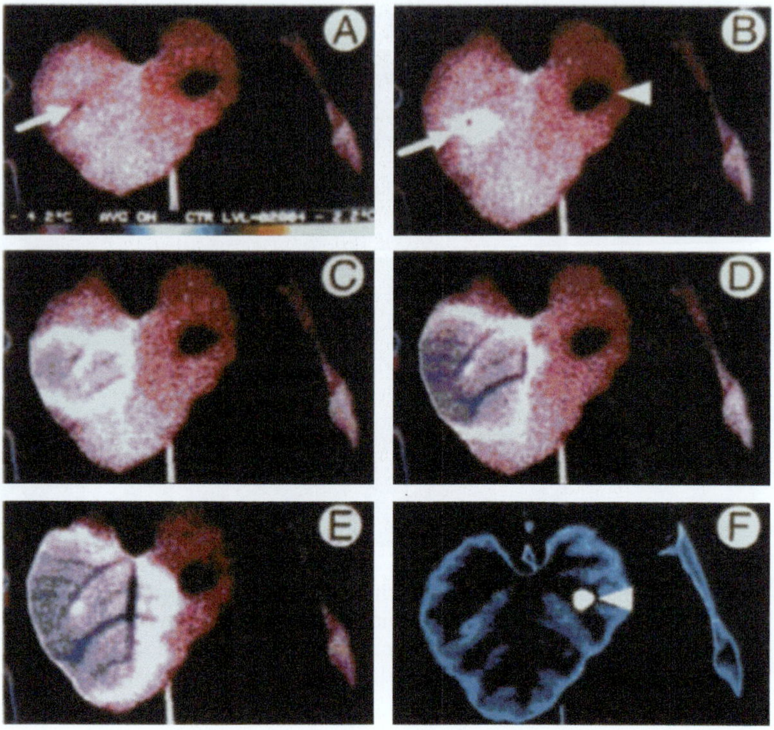

Fig. 13 Ice nucleation and propagation in a bean leaf shown by false-colour thermography. The temperature range was chosen 2 K. Black and white parts are out of range at the lower and the upper end, resp. For further explanations see text. With permission from [91] p. 207.

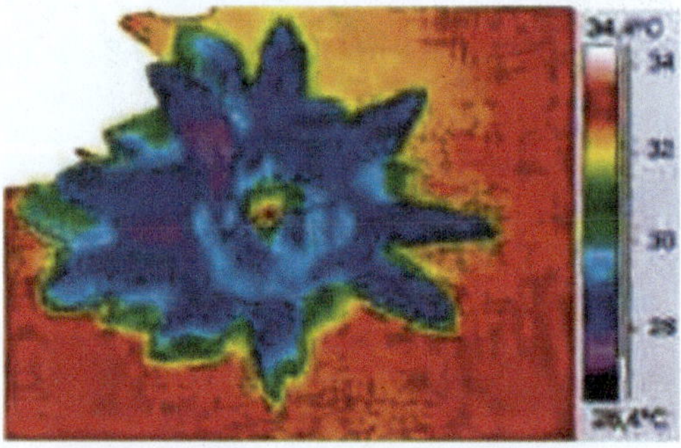

Fig. 14 Thermogenic active evening flower of the giant water lily *V. cruziana* in false colour. At air and water temperatures of 24.0 and 31.0°C, resp., the centre of the blossom shows a temperature from 30.9 to 33.5, significantly above the air temperature. The white area in the left upper corner represents the arm of the investigator. With permission from [23] p. 207.

Fig. 8 IR thermography of a flying hornet (*Vespa crabro*) worker in a wind channel experiment. The inserted photo shows a hornet in the same flying position for comparison. Unpublished data from the authors p. 261.

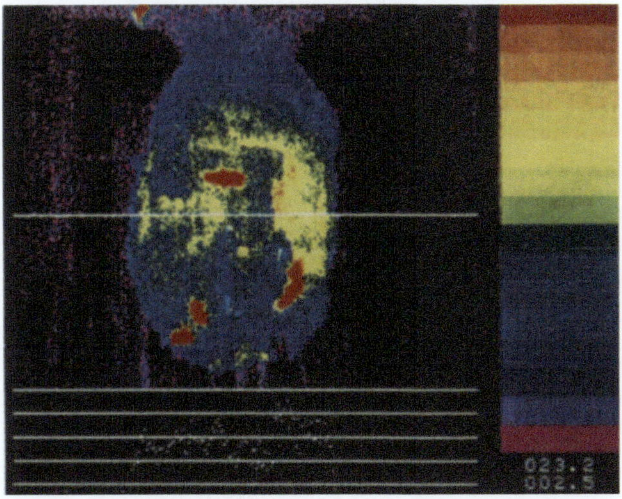

Fig. 14 IR thermography of a small hornet nest. The site of the combs inside the nest is clearly indicated by the yellowish area in the middle (about 22.4°C). The warm red spots are hornets walking on the nest envelope. Taken from [72] p. 270.

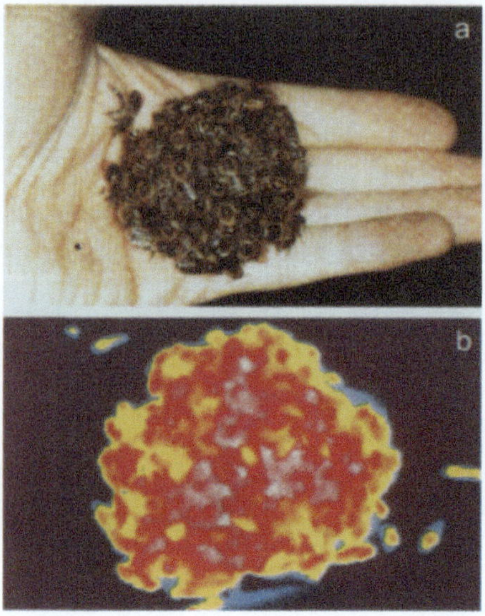

Fig. 16 a) IR thermograph showing a hot defensive ball of the Eastern Honeybee *Apis cerana*. b) Defensive ball with about 400 bees which engulf a predatory hornet. (Courtesy of Masato Ono) p. 275.

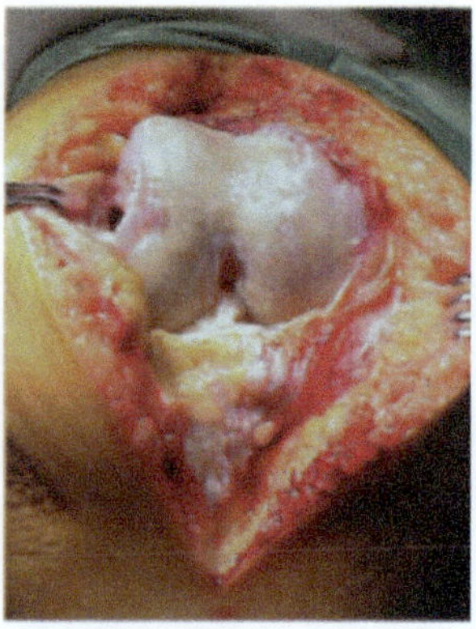

Fig. 1 Intraoperative view of severe osteoarthritis of the femoral condyles in the human knee joint p. 287.

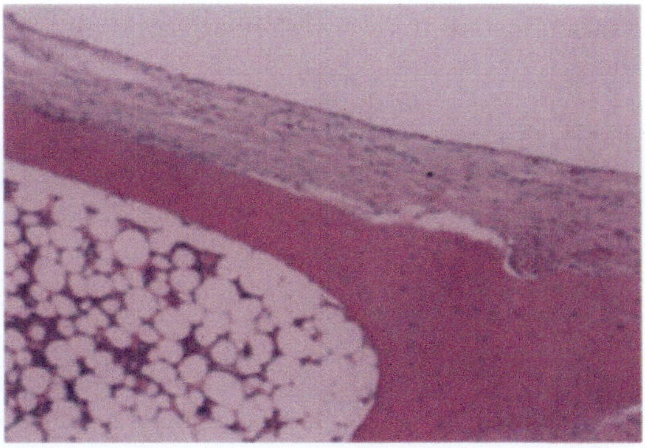

Fig. 2 Histological examination (hematoxylin-eosin) of osteoarthritis of the femoral hyaline cartilage in rabbits p. 288.

Hot Topics in Thermal Analysis and Calorimetry

Series Editor:
Judit Simon, *Budapest University of Technology and Economics, Hungary*

1. Michael E. Brown: *Introduction to Thermal Analysis, Second Edition*. 2001
ISBN 1-4020-0211-4; Pb 1-4020-0472-9
2. W. Zielenkiewicz and E. Margas: *Theory of Calorimetry*. 2002
ISBN 1-4020-0797-3
3. O. Toft Sørensen and J. Rouquerol (eds.): *Sample Controlled Thermal Analysis: Origin, Goals, Multiple Forms, Applications and Future*. 2003
ISBN 1-4020-1563-1
4. T. Hatakeyama and H. Hatakeyama: *Thermal Properties of Green Polymers and Biocomposites*. 2004
ISBN 1-4020-1907-6
5. D. Lörinczy (ed.): *The Nature of Biological Systems as Revealed by Thermal Methods*. 2004
ISBN 1-4020-2218-2

KLUWER ACADEMIC PUBLISHERS – DORDRECHT / BOSTON / LONDON

GPSR Compliance

The European Union's (EU) General Product Safety Regulation (GPSR) is a set of rules that requires consumer products to be safe and our obligations to ensure this.

If you have any concerns about our products, you can contact us on ProductSafety@springernature.com

In case Publisher is established outside the EU, the EU authorized representative is:

Springer Nature Customer Service Center GmbH
Europaplatz 3
69115 Heidelberg, Germany

Batch number: 10584688

Printed by Printforce, the Netherlands